ISBN 978-0-266-99085-7
PIBN 10920045

A

NEW AND EASY

INTRODUCTION TO THE MATHEMATICS:

CONTAINING,

I. A SYSTEM OF THEORETICAL AND PRACTICAL

ARITHMETIC,

In which the Rules are simplified, the Theory of them is explained, and the methods of operation are amply illustrated by many examples and familiar explanations, calculated to facilitate the improvement of the Learner and diminish the labor of the Teacher.

II. RULES FOR THE MENSURATION OF SUPERFICES AND SOLIDS, with their application to various practical purposes—illustrated by appropriate examples.

III. RULES FOR SOLVING A NUMBER OF USEFUL AND INTERESTING MATHEMATICAL, PHILOSOPHICAL, AND CHRONOLOGICAL PROBLEMS.

IV. A COLLECTION OF INTERESTING MATHEMATICAL QUESTIONS FOR EXERCISE.

V. USEFUL TABLES, &c.

DESIGNED FOR THE USE OF

Schools, Academies, and Private Learners.

BY IRA WANZER.

" Make the rough path of Science plain to find,
And through its mazes lead the pleasur'd mind."

DANBURY, CONN.
PRINTED AND PUBLISHED BY W. & M. YALE,
SOLD AT THEIR BOOKSTORE, AND BY THE PRINCIPAL BOOKSELLERS IN THE UNITED STATES.

1831.

PREFACE.

About two years since, the author of the following work, undertook the arduous task of composing an Introduction to the Mathematical Sciences, (which he intended should comprise Treatises on Arithmetic and Mensuration, and an Introduction to Algebra,) for the use of the citizens of the United States. Soon after he had determined on the plan of the work, he wrote the following Treatise on Arithmetic, and some other parts of the proposed work, but he was prevented, by ill health and other circumstances, from completing it according to his original design. He has since been advised by several respectable mathematicians to publish the System of Arithmetic, without delay, for the use of schools; and having lately revised the Treatise, and added a copious Appendix, containing much other useful matter, he now presents the work to the public.

The work now offered to the public, contains a complete System of common Arithmetic, (including many recent improvements in the science, some of which have never before been published,) and also, some other parts of the Mathematics which are of general utility.

In composing this system of Arithmetic, great pains have been taken to express the rules in plain, familiar language, and to adapt the illustrations of them to the capacities of learners. The most important rules are demonstrated, and their practical applications are amply illustrated. Many wrought examples are given, to elucidate the different rules; and these examples are, in general, accompanied by explanations, which will make the numerical operations easy to be understood by learners of the most ordinary capacities. Practical illustrations are also annexed to many of the questions for exercise, where difficulties occur.

It is believed that some of the rules in this Treatise are new, and preferable to any other rules of their kind, and that some others are reduced to more commodious forms than they have heretofore been in books on Arithmetic. The author has given a general method of computing simple interest, and a method of extracting the cube roots of numbers,* which he believes are more concise and convenient than any others now in use.

Questions on the nature of the principal rules of Arithmetic, are inserted in the work, the answers to which are contained in the rules to which the questions relate. These questions are calculated to lead the learner to examine those rules attentively, in order to find out the proper answers; and it will, perhaps, be well for teachers, in the general way, to see that their pupils are able to answer, correctly, the questions on each particular rule, before they allow them to proceed to the next.†

The following work contains, besides the System of Arithmetic, Rules for the Mensuration of the different kinds of Superfices and Solids, with directions for applying them to various practical purposes; and also, Rules for solving a number of useful and interesting Mathematical, Philosophical, and Chronological Problems—all of which are illustrated by suitable examples: Likewise, a collection of curious and instructive miscellaneous questions for exercise, and several useful Tables. The great practical utility of the rules of Mensuration, and of the subsequent Problems, &c. will, it is presumed, be considered a sufficient reason for inserting them in the work, at the end of the treatise on Arithmetic.

In composing this work, the writer has taken the liberty to make use of some short extracts from other mathematical works; and he acknowledges himself indebted to other writers for some of the illustrations of the principles of Arithmetic, and for a number of the questions for exercise. Some

---

* The method of extracting the cube root, given in this work, and also an accurate general method of finding the roots of compound algebraical equations, were discovered by the author a number of years ago; and subsequently, he and some of his friends made those methods known to several mathematicians, who resided in different parts of the U. States Those rules, (or very similar ones,) have since been published in this country, by several persons with whom the author of this work has had no acquaintance; and he has not yet been able satisfactorily to ascertain whether the rules were, or were not, known to any mathematician prior to the discovery of them by himself.

† Teachers who think it advisable for their pupils to acquire a tolerable knowledge of *Practical Arithmetic* before they undertake to learn the *theory* of the rules, can direct them to omit the *Theoretical Questions*, and the *demonstrations* of the rules, until they have made considerable progress in the study of Arithmetic, and then to attend to them as a review.

of those extracts are contained in so many different works that the original writers of them could not easily be ascertained; and the rest have, for the most part, been so disposed of, as to make it very inconvenient to refer the reader to the works from which they were taken; for which reasons such references are not generally inserted. The author has not aimed at novelty where he saw no prospect of making improvements; and he hopes the use which he has made of the labors of his predecessors will be considered pardonable by those who know how difficult it would be for any person, at the present day, to write a Treatise on Arithmetic without repeating much which has already been written by others. The practice of copying has indeed been general with almost all writers on Arithmetic, though few of them have candidly acknowledged it.

The author is aware that there is a number of works of this nature already in use, in different parts of this country. It does not appear, however, that any one of them is *now*, or is likely *very soon* to be, in general use in schools throughout the United-States; and hence it may be infered that none of them have met the approbation of the public, generally. The author is not disposed to expatiate on the defects of those systems, nor on the merits of his own; being persuaded that enlightened persons will *examine* and *judge* for themselves, and that *his* work, as well as others, must stand or fall, according to its real merit or demerit.

In submitting the work to the public, the author cannot forbear expressing his gratitude for the liberal patronage already bestowed upon this edition. To those persons who have encouraged the publication of the work, the author tenders his sincere thanks; and he would request of them, and others, that indulgence towards its imperfections, which the nature of the work must necessarily require. Should the present edition prove in some respects imperfect, it is hoped that such improvements will hereafter be made in the work, as will render the subsequent editions more deserving of public patronage.

IRA WANZER.

New-Fairfield, Conn., *Oct.* 1, 1831.

A

NEW

INTRODUCTION TO THE MATHEMATICS.

INTRODUCTORY OBSERVATIONS

ON THE

MATHEMATICS IN GENERAL.

MATHEMATICS is the science which treats of all kinds of *quantity* whatever, that can be numbered or measured. By *quantity*, is meant any thing that will admit of increase or decrease; or that is capable of any sort of calculation or mensuration: such as *numbers, lines, space, time, motion, weight.*

Note.—In the Mathematics, a quantity is said to be *simple*, when its value or magnitude is expressed by one term or denomination only; and *compound*, when its value is expressed by two or more terms.

Those parts of the Mathematics on which all the others are founded, are *Arithmetic*, *Algebra*, and *Geometry*.

ARITHMETIC is the science of *numbers*. Its aid is required to complete and apply the calculations in almost every other department of the mathematics.

ALGEBRA is a general method of computing by *letters* and other symbols. It is of extensive use in mathematical investigations.

GEOMETRY is that part of the mathematics which treats of *magnitude*. By *magnitude*, in the appropriate sense of the term, is meant that species of quantity which is *extended*; that is, which has one or more of the three dimensions, *length*, *breadth*, and *thickness*.

Mathematics are either *pure*, or *mixed*. In *pure* or *abstract* mathematics, quantities are considered independently of any substances actually existing. But, in *mixed* mathematics, the relations of quantities are investigated, in connection with some of the properties of matter, or with reference to the common transactions of business. Thus, in *Surveying*, mathematical principles are applied to the measuring of land; and in *Astronomy*, to the motions of the heavenly bodies, &c.

Mathematics are also distinguished into *Theoretical*, or *Speculative*, and *Practical*, viz. *Theoretical*, when concerned in investigating and demonstrating the various properties and relations of quantities; and *Practical*, when applied to practice and real use concerning physical objects.*

EXPLANATION

OF CERTAIN MATHEMATICAL CHARACTERS, OR SIGNS.

There are various characters or marks which are used in Arithmetic and Algebra, to denote the operations of Addition, Subtraction, &c.; the chief of which are the following:

CHARACTERS.	EXPLANATIONS.
+	*Plus*,† or *more*; the sign of Addition, or the positive sign, signifying that the number to which it is prefixed,§ is to be added to some other number. Thus, 2+5, denotes that 5 is to be added to 2; and the expression is read thus, 2 *plus* 5, or, *the sum of* 2 *and* 5.
—	*Minus*,† or *less*; the sign of Subtraction, or the negative sign, signifying that the number to which it is prefixed is to be subtracted from some other number. Thus, 8—5, signifies that 5 is to be subtracted from 8, and is read thus, 8 *minus* 5, *or the difference of* 8 *and* 5.
×	*Into*, or *with*; the sign of Multiplication, signifying that the numbers between which it is placed, are to be multiplied together. Thus, 8×2, signifies that 8 is to be multiplied by 2, and is read, 8 *into* 2, or *the product of* 8 *and* 2.

* The preceding "Introductory Observations" have been mostly extracted from Hutton's and Day's Courses of Mathematics.

† *Plus* and *minus* are Latin words; the former signifying *more*, and the latter *less*.

§ To *prefix*, is to place before: to *annex*, is to place after.

The product of numbers which are represented by letters, is usually denoted by placing the letters close together like a word, without any mark between them. Thus, suppose the letters a, b, and c, to represent any three numbers whatever; then abc will denote the continued product of those numbers, the same as $a\times b\times c$.

÷ *By*, or *divided by*; the sign of Division, signifying that the former of the two quantities between which it is placed, is to be divided by the latter. Thus, $8\div 2$, denotes that 8 is to be divided by 2, and is read, 8 *by* 2, or, 8 *divided by* 2.

Division is also frequently denoted by placing the dividend above, and the divisor below a horizontal line. Thus, $\frac{8}{2}$, denotes the same as $8\div 2$, viz., that 8 is to be divided by 2; and $\frac{6+8}{4-2}$ denotes that the sum of 6 and 8 is to be divided by the difference of 4 and 2.

: :: : The sign of Proportion. Thus, 2 : 4 :: 7 : 14, denotes that the ratio of 2 to 4 is the same as the ratio of 7 to 14: read thus, *as* 2 *is to* 4, *so is* 7 *to* 14.

4^2 Denotes the second power or square of 4; and 4^3 denotes the third power or cube of 4, and so on for higher powers. Also, a^2, or aa, denotes the second power of a; and a^3, or aaa, denotes the third power of a, &c.

√ The *radical sign*, signifying that the quantity before which it is placed is to have some root of it extracted. Thus, $\sqrt{4}$ denotes the second or square root of 4; $\sqrt[3]{8}$ denotes the third or cube root of 8; and $\sqrt[4]{8}$ denotes the 4th root of 8, &c.

= *Equal to*; the sign of equality, signifying that the quantities, or sets of quantities, between which it is placed, are equal to each other. Thus, 100 cents = 1 dollar, signifies that 100 cents are equal to 1 dollar: Also, $2+5=7$, denotes that the sum of 2 and 5 is equal to 7, and is read thus, 2 *plus* 5 *equal to* 7.

— A *horizontal line* drawn or placed over two or more quantities, signifies that all the quantities under it are to be considered jointly as *one* quantity; and the line is called a *vinculum*. Thus, $\overline{2+5}\times6$, denotes that the sum of 2 and 5 is to be multiplied by 6. The same thing is also frequently denoted by including in a parenthesis the several quantities which are to be considered as one quantity. Thus, $\sqrt{}(7+9)$ signifies the same as $\sqrt{7+9}$, viz. the square root of the sum of 7 and 9.

The *reciprocal* of any quantity, is that quantity inverted, or unity divided by it. Thus, the reciprocal of $\frac{4}{5}$ is $\frac{5}{4}$, and the reciprocal of a or $\frac{a}{1}$ is $\frac{1}{a}$.

.. Two points, standing beside each other, are used in this Work to separate the different denominations of compound numbers.

. A single point is used to separate integers, or whole numbers, from decimal parts.

The signs + and —, when annexed to the answers to mathematical questions, denote that such answers are not exact; the former sign denoting that the answer to which it is annexed should be a small fraction greater, and the latter, that the answer should be a little less.*

Note.—I have thought proper to explain the meaning and use of the foregoing characters in this part of the Book, though most of them are explained in the following Treatise, in the places where they are first used. It may be sufficient for the student, at first, to learn the use of only the signs + and — and the sign of equality, =; and afterwards to make himself acquainted with the use of the other characters as soon as he shall have occasion for them.

---

* It is probable that some of the answers in this Work which have the sign + annexed to them are too great. In some cases it was difficult to determine whether the omission or addition of small fractions, in performing the numerical operations, made the answers or results too small, or too great; and all such answers, as well as those which were known to be too great, have the sign + annexed to them.

ARITHMETIC.

ARITHMETIC is the *Science of Numbers*, and the *Art of using them*.

Arithmetic consists of two parts, *Theoretical*, and *Practical*. The *Theoretical*, considers the nature and quality of numbers, and demonstrates the reason of practical operations. The *Practical*, merely shows the method of working by numbers, so as to be most useful and expeditious for business. Theoretical Arithmetic is properly a *Science*, and Practical Arithmetic is an *Art*.

NUMBER is that which is used to express the *relations of quantity*.

UNITY, a UNIT, or ONE, is the beginning of number, and signifies a single, or an individual thing, of any kind. One and one more, taken collectively, make a number called *two:* two and one more, make the number *three:* this increased by one, composes the number *four:* and thus, by the continual addition of unity or one, we may obtain the higher numbers, *five, six, seven, eight, nine, ten, &c.*

An *Integer*, or a *Whole Number*, is some precise quantity of units; as *one, two, three, &c.* *Whole numbers* are so called as distinguished from *Fractions*, which are broken numbers, or parts of numbers; as *one-half, two-thirds, three-fourths, &c.*

NOTATION AND NUMERATION

OF WHOLE NUMBERS.

Notation teaches how to write down in characters any number proposed in words.

Numeration teaches how to read, in proper words, any number expressed by characters.

The characters now generally used to denote numbers, are the ten Arabic numeral characters, commonly called *Figures*. These characters, and their names, which are the numbers they represent, are as follows, viz. 1 *one*, 2 *two*, 3 *three*, 4 *four*, 5 *five*, 6 *six*, 7 *seven*, 8 *eight*, 9 *nine*, 0 *nought*, *cipher*, or *zero*.* The first nine of these characters are called *significant figures*, or *digits*, to distinguish them from the cipher, 0, which of itself, is quite *insignificant*, or does not express any number. By these ten figures any numbers may be expressed.

The value of any figure when alone, is called its *simple* value, and is invariable. Thus, the figure 1, when alone, always denotes *one*; the figure 2, *two*; &c. Figures have also a *local* value; that is, a value which depends upon the place they stand in when two or more of them are combined or joined together, as in the following table.

NUMERATION TABLE.

Hundreds of millions.	Tens of millions.	Millions.	Hundreds of thousands.	Tens of thousands.	Thousands.	Hundreds.	Tens.	Units.
9	8	7	6	5	4	3	2	1
	9	8	7	6	5	4	3	2
		9	8	7	6	5	4	3
			9	8	7	6	5	4
				9	8	7	6	5
					9	8	7	6
						9	8	7
							9	8
								9

Note.—The words at the head of this table, (viz. *Units*, *Tens*, *Hundreds*, &c.) show the *local* values of those figures over which they stand, and must be committed perfectly to memory.

ILLUSTRATION.

Here, each figure in the *first place*, at the right hand, denotes only its own simple value; but each figure in the *second place*, (counting from right to left,) denotes ten times its simple value; each figure in the *third place* denotes a hundred times its simple value, and so on;—the local value of a figure in any particular place being ten times its value in the next place to the right. Thus, the figure 9, in the *first place*, or column, at the right hand, signifies *nine units*, or simply *nine*; but in the *second place* it denotes *nine tens*, or *ninety*; in the *third place* it denotes

* These ten characters were formerly all called by the general name of *Ciphers*; whence it came to pass that the art of Arithmetic was then often called *Ciphering* The invention of these characters is usually ascribed to the Arabians: and it is said that they were first brought into Europe by the Moors, in the ninth century of the Christian era. The Roman method of Notation, by letters, had previously been in general use in Europe.

nine hundreds, and so on. The local values of all the other significant figures vary in the same manner, according to their distance from the place of units. Thus, in the number 9876, the 6, standing in the first or units' place, denotes *six units*, or *six;* the 7, in the second place, denotes *seven tens*, or *seventy;* the 8, in the third place, denotes *eight hundreds;* and the 9, in the fourth place, denotes *nine thousands:* and hence the expression 9876 is read thus, *nine thousand, eight hundred and seventy-six.*

In all expressions of whole numbers, the local values of the places of figures increase from right to left in the same tenfold ratio.*

Although the cipher, 0, does not of itself denote any number, yet, every cipher annexed to significant figures increases the local values of the latter in a tenfold ratio, by throwing them into higher places. Thus, 4 denotes only *four;* but 40 denotes *forty;* 400 denotes *four hundreds;* and 4000 denotes *four thousands*, &c.

The process of Numeration is more amply illustrated by the following table.

TABLE 2.†

&c.	Hund. of quintill.	Tens of quintill.	Quintillions.	Hund. of quadrill.	Tens of quadrill.	Quadrillions.	Hund. of trill.	Tens of trill.	Trillions.	Hund. of bill.	Tens of bill.	Billions.	Hund. of mill.	Tens of mill.	Millions.	Hundreds of thou.	Tens thousands.	Thousands.	Hundreds.	Tens.	Units.
	6	5	1,	2	3	7,	4	2	8,	7	1	4,	9	7	0,	3	0	5,	0	8	2
	Period of Quintillions.			Period of Quadrillions.			Period of Trillions.			Period of Billions.			Period of Millions.			Period of Thousands.			Period of Units.		

* *Whole numbers* are counted by tens and combinations of tens; and hence the reason why we make use of ten different characters to denote numbers. It may be well to observe that there is no reason in the nature of numbers, that they should be made to increase in a tenfold ratio: they might have been made to increase in 2, 3, 4, &c. fold, or any

Three places of figures, beginning on the right hand, are called a *period*, and each successive three places, another period; the first period, on the right hand, being called the *period of units;* the second, the *period of thousands*, and so on, as in the Table. There is an obvious reason for this division into periods; for at the beginning of each period, there is a new denomination of units, of which the tens and hundreds are numerated, as in the first period. The names of the periods are derived from the Latin numerals, and they may be continued without end. They are as fol-

---

other ratio. The method of counting by tens, which is now in general use, originated from the practice of counting on the fingers, which are ten in number.

† This method of enumerating figures to the left hand of the place of hundreds of millions, differs from that which has hitherto been in general use in this country. This method of dividing lines of figures into periods, and of naming those periods, is used by the French and Italians; and it has lately been adopted by a number of respectable British and American Authors on Arithmetic. The method is strongly recommended by its simplicity and elegance, and it is probable that it will soon be much more universally used.

The usual method is to divide the numbers into periods of six figures each; which periods have the same names as those in Table 2d, except *thousands*, for which there is not a distinct period. The common method of enumerating figures is exemplified in the following

TABLE.

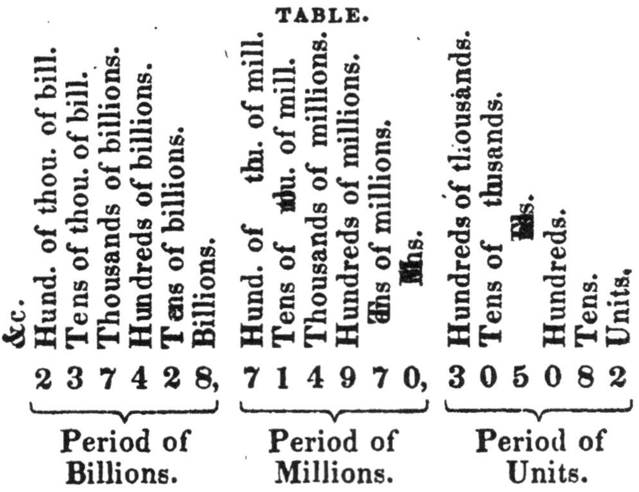

It will be seen that the two methods of enumerating lines of figures, agree as far as hundreds of millions; and, as it is very rarely necessary to name larger numbers, it is evident that the proposed change from the old to the new method, cannot be attended with much inconvenience.

lows, for twenty-two periods, viz. Units, Thousands, Millions, Billions, Trillions, Quadrillions, Quintillions, Sextillions, Septillions, Octillions, Nonillions, Decillions, Undecillions, Duodecillions, Tredecillions, Quatuordecillions, Quindecillions, Sexdecillions, Septendecillions, Octodecillions, Novemdecillions, Vigintillions.

For the convenience of reading large numbers, the periods of figures which express them are often separated by commas, as in Table 2d.

To enumerate any parcel of figures.

RULE.

1. In order to ascertain the local values of the given figures, divide them into periods of three figures each, beginning at the right hand, and proceeding towards the left, till not more than three figures remain. Then the first period towards the right hand will contain units or ones, the second thousands, the third millions, &c. as in Table 2d.

2. Then, in reading the number, begin at the left hand, and to the value expressed by the figures of each period, join the name of the period, excepting the *name* of the period of units, which need not be mentioned.

Ex. The number in Table 2d, viz. 651,237,428,714,970,-305,082, is read thus, *six hundred fifty-one quintillions, two hundred thirty-seven quadrillions, four hundred twenty-eight trillions, seven hundred fourteen billions, nine hundred seventy millions, three hundred five thousand, and eighty-two.*

Note.—By practice, the pupil will soon be able to enumerate numbers, or lines of figures, which are not very large, without dividing them into periods; for the local values of the figures may be readily ascertained by merely enumerating them from right to left, thus, *tens*, *hundreds*, *thousands*, &c. as in the Numeration Tables.

More Examples in Numeration.

10=Ten.	20=Twenty.	70=Seventy.
11=Eleven.	21=Twenty-one.	76=Seventy-six.
12=Twelve.	30=Thirty.	80=Eighty.
13=Thirteen.	32=Thirty-two.	87=Eighty-seven.
14=Fourteen.	40=Forty.	90=Ninety.
15=Fifteen.	43=Forty-three.	98=Ninety-eight.
16=Sixteen.	50=Fifty.	100=One hundred.
17=Seventeen.	54=Fifty-four.	200=Two hundred.
18=Eighteen.	60=Sixty.	1000=One thousand.
19=Nineteen.	65=Sixty-five.	2000=Two thousand.

148, read, one hundred and forty-eight.
407 - - - four hundred and seven.
950 - - - nine hundred and fifty.
4,549 - - - four thousand, five hundred and 49.
90,170 - - - ninety thousand, one hundred and 70.
800,000 - - - eight hundred thousand.
7,154,918 - - - 7 millions, 154 thousand, 918.
987,654,321 - - - 987 millions, 654 thousand, 321.
5,000,217,080 - - - 5 billions, 217 thousand, and 80.

Read, or write in proper words, the following numbers.

48	670304	123456789
576	8500000	5032600878
1072	10025460	87200280260
45005	856023408	706054000064

To set down in figures any number proposed in words, observe the following

RULE.

Set down the figures instead of the words, or names belonging to them, taking care to supply with ciphers all the places below the highest whose names are not mentioned in words.

Exercises in Notation.

Write down, in proper figures, the following numbers.

Four hundred and forty-seven. - - - - - - - 447.
Four thousand four hundred and eight. - - - - 4,408.
Ninety-seven thousand six hundred and fourteen. - - - - - - - - - - - - - - -
Sixty thousand and twenty one. - - - - - - 60,021.

Four hundred thousand. - - - - - - - - - -
Ninety-eight millions. - - - - - - - - 98,000,000.
Nine hundred and six millions, five hundred thirty thousand, two hundred and one. - -
Nine hundred fifty billions, twenty-four millions, six hundred ten thousand and ninety. - - - - - - - - - - - 950,024,610,090.

ROMAN METHOD OF NOTATION.

The Romans and several other ancient nations, made use of certain letters of the alphabet to express numbers. The Romans used only seven numeral letters, being the seven following capitals, viz. I for *one*, V for *five*, X for *ten*, L for *fifty*, C for an *hundred*, D for *five hundred*, and M for a *thousand*. The other numbers they expressed by various repetitions and combinations of these letters, after the following manner:—

1=I	40=XL
2=II	50=L
3=III	60=LX
4=IV	70=LXX
5=V	80=LXXX
6=VI	90=XC
7=VII	100=C
8=VIII	200=CC
9=IX	300=CCC
10=X	400=CCCC
11=XI	500=D
12=XII	600=DC
13=XIII	700=DCC
14=XIV	800=DCCC
15=XV	900=DCCCC
16=XVI	1000=M
17=XVII	2000=MM
18=XVIII	3000=MMM
19=XIX	4000=$\overline{\text{IV}}$
20=XX	5000=$\overline{\text{V}}$
30=XXX	&c.

Note.—As often as any letter is repeated, so many times is its value repeated: thus, II=2, III=3, XX=20, &c.

A less literal number placed before a greater, diminishes the value of the greater: thus, V denotes 5, but IV only 4; and X=10, but IX=9, &c.

A less literal number placed after a greater, increases its value: thus, VI=6, and XI=11, &c.

A bar or line over any literal number, increases it a thousand fold: thus, $\overline{\text{V}}$=5000, and $\overline{\text{XX}}$=20000, &c.

THEORETICAL QUESTIONS.

1. What is Arithmetic? 2. Into how many parts is it divided? 3. What does Theoretical Arithmetic explain? 4.

What does Practical Arithmetic teach? 5. What is a number? 6. What is unity, or a unit? 7. What is an integer, or a whole number? 8. What are fractions? 9. What is meant by *quantity?* 10. What is a simple quantity? 11. What is a compound quantity? 12. What is Notation? 13. What is Numeration? 14. What is meant by the simple value of any figure? 15. What by its local value? 16. What does the right hand figure of any whole number denote? and what the next figure to the left? and so on? 17. Does the cipher denote any number? 18. What effect does it have when annexed to other figures? 19. What is the rule for enumerating any parcel of figures? 20. What is the rule for expressing by figures any number proposed in words?

FUNDAMENTAL RULES OF ARITHMETIC.

There are four rules which are called the *Fundamental Rules*, because all operations in Arithmetic are performed by the use of them. They are *Addition*, *Subtraction*, *Multiplication*, and *Division*.

ADDITION,

Is putting together two or more numbers, so as to find their amount, or total value, which is called their *sum*. It is called *Simple Addition*, or, *Addition of Whole Numbers*, when the numbers to be put together are all simple, or whole numbers, of the same denomination; and *Compound Addition*, when the numbers are compound, or of different denominations.

SIMPLE ADDITION,

Is putting together two or more *whole* numbers, or quantities, of the same denomination; as 3 dollars and 2 dollars, added together, make a sum of 5 dollars: Or, it is simply adding together two or more whole numbers, without regard to their signification; as 5 added to 4, makes 9.

To perform the operation of addition, it is necessary that the learner should be able to assign the sum of any two of the small numbers less than 10; and for this purpose he should be exercised in the following table.

ADDITION TABLE.

2+0= 2	4+0= 4	6+0= 6	8+0= 8
2+1= 3	4+1= 5	6+1= 7	8+1= 9
2+2= 4	4+2= 6	6+2= 8	8+2=10
2+3= 5	4+3= 7	6+3= 9	8+3=11
2+4= 6	4+4= 8	6+4=10	8+4=12
2+5= 7	4+5= 9	6+5=11	8+5=13
2+6= 8	4+6=10	6+6=12	8+6=14
2+7= 9	4+7=11	6+7=13	8+7=15
2+8=10	4+8=12	6+8=14	8+8=16
2+9=11	4+9=13	6+9=15	8+9=17
3+0= 3	5+0= 5	7+0= 7	9+0= 9
3+1= 4	5+1= 6	7+1= 8	9+1=10
3+2= 5	5+2= 7	7+2= 9	9+2=11
3+3= 6	5+3= 8	7+3=10	9+3=12
3+4= 7	5+4= 9	7+4=11	9+4=13
3+5= 8	5+5=10	7+5=12	9+5=14
3+6= 9	5+6=11	7+6=13	9+6=15
3+7=10	5+7=12	7+7=14	9+7=16
3+8=11	5+8=13	7+8=15	9+8=17
3+9=12	5+9=14	7+9=16	9+9=18

The foregoing table may be read thus, 2 and 0 are 2; 2 and 1 are 3; 2 and 2 are 4; &c.

To enable the learner to acquire accuracy and dispatch in addition, it is proper to train him to add in the following manner, till he can do it with facility.

Since 4 and 5 are 9, 14 and 5 are 19; 24 and 5 are 29; 34 and 5 are 39; &c.—the right hand figure of each amount being 9.

Since 5 and 5 are 10, 15 and 5 are 20; 25 and 5 are 30; 35 and 5 are 40; &c.

Since 7 and 6 are 13, 17 and 6 are 23; 27 and 6 are 33; 37 and 6 are 43; &c.

In this way pupils may soon learn to put numbers together readily, so as not to hesitate, or stop to count, whilst adding up a column of figures.

When only a few *very small* numbers are to be put together, the addition may be readily performed in the *mind*,*

* Very young pupils may be taught the art of putting numbers together, by the following method: Let the pupil take one or more cents, (or

without writing down the numbers. The addition of large numbers may be performed by the following

RULE,

For the Addition of Simple or Whole Numbers.

1. Write the numbers which are to be added together, one under another, so that units shall stand under units, tens under tens, &c. and draw a line under the lowest number, to separate the numbers from their sum, when it shall be found.

2. Begin at the bottom of the right hand (or units) column of figures, and add together all the figures in the column: If the amount does not exceed 9, (or one figure,) set it down under that column, and then proceed to the next column; but if the amount exceeds 9, set down only the right hand (or units) figure of it; and the number expressed by the other figure, or figures, (which will be the number of tens in the amount,†) carry to the next column, or column of tens.

3. Then, add up the figures in the second column, (or column of tens,) together with the number, if any, carried from the first column; and, if the amount is only one figure, set it down under the second column; but if it is more than one figure, set down only the right hand figure, (which will denote tens,) and carry the rest, (which will denote hundreds,) to the third column, or column of hundreds.

4. Proceed in the same manner through all the remaining

---

beans, &c.) into each hand, and tell how many cents *each* hand contains; then request him to tell the number of cents in *both* hands, &c.

A similar method may be practised in Subtraction; viz. request the learner to tell how many *more* cents one hand contains than the other, &c.

Familiar methods of instruction, similar to the above, may sometimes be practised to advantage, in various other parts of Arithmetic.

† In any expression of a whole number, consisting of several figures, the figures exclusive of the one at the right hand, express the number of tens, and the right hand figure the remaining units, or ones, contained in the given number: the former figures being (as will hereafter be seen) the quotient, and the other figure, the remainder, which would be found by dividing the given number by 10. Therefore, if in adding together several numbers, we set down the right hand figure of the sum of each column of figures, and carry to the next column the number expressed by the other figure or figures of the sum, we carry from each column the number of tens contained in the sum, and set down the ones which remain over.

columns of figures, and set down the whole amount of the last column.*

Note.—In writing down the numbers, it will be convenient to begin on the right hand, and proceed towards the left, setting down the units, or right hand figure of each number, first, the tens next, and so on.

To prove Simple Addition.

Begin at the top of the right hand column of figures, and perform the addition of this, and of all the other columns downwards, proceeding in other respects according to the foregoing Rule; and, if the work be right, the total sum, thus found, will be the same as that found by adding the columns upwards.—There are several other methods of proving addition, but this is the most convenient one.

EXAMPLES.

1. Find the sum of the numbers 4262, 7683, and 91.

Operation.

	Thousands.	Hundreds.	Tens.	Units.
	4	2	6	2
	7	6	8	3
			9	1
Sum, 1	2	0	3	6
Proof, 1	2	0	3	6

Explanation.—First, I write down the given numbers, one under another, so that units stand under units, tens under tens, &c.; then I draw a line under them, and add as follows:

I begin at the bottom of the right hand column, or column of units, and say 1 and 3 are 4, and 2 are 6. So, the amount of the first column is 6, which being only one figure, I set it down under that column, and proceed to the next. Then, proceeding to the bottom of the second column, or column of tens, I add thus, 9 and 8 are 17, and 6 are 23. As this amount exceeds 9, I set down only the right hand figure, 3, and carry the other figure, 2, to the next column. Then, proceeding to the third column,

* This Rule is founded on the known axiom, "the whole of any quantity is equal to the sum of all its parts." The method of placing the given numbers, and carrying for the tens contained in the sum of the figures in each place, is evident from the nature of notation; for any other disposition of the given numbers would alter their value; and carrying one for every ten, from an inferior to the next superior column, is evidently

or column of hundreds, I say, 2 carried to 6 makes 8, and 2 are 10: I set down 0 under the third column, and carry 1 to the next. Then, 1 carried to 7 makes 8, and 4 are 12; which being the amount of the last column, I set down the whole of it, and the work is done. So, the total sum, or answer, is 12036.

To prove the work, I add all the columns downwards, thus: I begin at the top of the right hand column, and say, 2 and 3 are 5, and 1 makes 6. I set down this amount below the first column, and then I proceed to the top of the second column, and add thus; 6 and 8 are 14, and 9 are 23. I set down 3 below this column, and carry 2 to the next. Then, 2 carried to 2, makes 4, and 6 are 10: I set down 0, and carry 1 to the next column. Then, 1 carried to 4 makes 5, and 7 are 12, which I set down. So the total sum is 12036, the same as before; and therefore I conclude the work is right.

Note.—As young beginners are apt to make mistakes in carrying from one column to another, as well as in adding up the columns, it will be well for them to set down the sum of each column, placing the several sums one under another, in regular order, as in the next following example. These sums the learner should retain on his slate until he has finished the work; and then if it should contain any errors, he may the more easily correct them when he comes to prove the work.

2. What is the sum of the numbers 3106, 812, 48, 500, and 2019?

	3106	
	812	25 Amount of the units, or first column.
	48	8 Do. of the tens, or 2d do.
	500	14 Do. of the hundreds, or 3d do.
	2019	
Answer,	6485	
Proof,	6485	

right; because one unit in the latter case is equal to ten units in the former; that is, ten in the column of units are equal to one in the column of tens, and ten in the column of tens, to one in the column of hundreds, &c.

	(3)	(4)	(5)	(6)
	58	510	990	31
	32	400	7240	1058
	4	28	500	2000
	5	702	40	1082
	160	31	700	98030
Sum,	259	1671	9470	102201
Proof,	259	1671	9470	102201

(7)	(8)	(9)
4820	3874	48300825040
2527	56401	65800325080
65	84230	2300802070
702	2709	32700328000
432	984328	2800752000

(10)
287542
983421
887983
28984
57692
875427

(11)
8027056
7320829
4878543
3270528
2643175

(12)	(13)
59876	83487
4398	94876
3475	83259
8297	98798
5467	84398
19693	97698
78939	83989
54986	76546
43978	68548
98769	98589
87349	87674
4329	73465
96898	64389
89879	87698
6758	87387

14. Add together the numbers 8, 43, 201, 5278, 35, and 71507. Ans. 77072.

15. Add together the numbers 26, 60, 782, 2548, 92812, 913, and 83. Ans. 97224.

16. What is the sum total of 287, 34, 8210, 304, 510258, 83000, and 7? Ans. 602100.

17. What is the sum total of 90032, 1320, 2870, 4855, 875, and 48? Ans. One hundred thousand.

18. What is the sum total of 362, 88735, 9542, 521, and 100850? Ans. Two hundred thousand, and ten.

19. What is the sum total of the following numbers, viz.
Five hundred and twenty-eight, - - - - - -
Three thousand two hundred, - - - - - -
Seven thousand nine hundred and fifty, - -
Forty-two thousand, - - - - - - - - - -
Three hundred twelve thousand, four hundred and thirty-one? - - - - - - - - - - -

Ans. 366109.

20. Required the sum of the following numbers, viz.
Five hundred and sixty-eight, - - - - -
Eight thousand eight hundred and five, -
Seventy-nine thousand six hundred, - -
Nine hundred and eleven thousand, - -
Nine hundred ninety-nine millions and twenty-six. - - - - - - - - - -

Ans. 999999999.

PRACTICAL QUESTIONS.

1. A merchant, on settling his accounts, finds that he owes A 415 dollars, B 38 dollars, C 1248 dollars, and D 52 dollars: What is the amount of all these debts?

Answer, 1753 dollars.

2. The contents of five boards are as follows: The first board contains 24 square feet, the second 19, the third 22, the fourth 13, and the fifth 11. How many square feet do they all contain? Ans. 89.

3. A man built a dwelling house, a barn, a grist mill,

and a shed. The dwelling house cost 1520 dollars, the barn 218 dollars, the mill 1432 dollars, and the shed 75 dollars. What did all those buildings cost?

Ans. 3245 dollars.

4. A man borrowed of his neighbor thirty dollars, at one time; three hundred and five at another; and four thousand and twenty at another. What do all these sums amount to? Ans. 4355 dollars.

5. In the year 1821, the number of inhabitants in England was 11260555, in Wales 717108, in Scotland 2092014, and in Ireland 6847000. How many inhabitants were there in all of those countries? Ans. 20916677.

6. From the creation of the world to the general deluge, was 1650 years; from this time to the call of Abraham was 427 years; from that to the departure of the Israelites out of Egypt, 430; from that to the building of the temple by Solomon, 479; from that to the founding of Rome, 266; from that to the birth of our Saviour Jesus Christ, 752; and it is now 1831 years since the birth of Christ. How many years since the creation? Ans. 5835.

7. The four largest cities in Europe, are London in England, Paris in France, Constantinople in Turkey, and Naples in Italy. London contains about one million two hundred and twenty-five thousand inhabitants; Paris seven hundred and fifteen thousand; Constantinople five hundred thousand; and Naples three hundred and thirty thousand. What is the whole number of inhabitants in these four great cities? Ans. 2770000.

8. The population of the World is estimated as follows: Europe, one hundred and eighty millions; Asia, three hundred and eighty millions; Africa, fifty millions; America, thirty-eight millions; Austral Asia and Polynesia, two millions: How many in all?

Ans. Six hundred and fifty millions.

9. The Hudson and Erie canal, in the State of New-York, extends from the Hudson river at Albany, to Buffalo on Lake Erie. The distances on this canal are as follows, viz. From Albany to Schenectady 30 miles; from thence to Utica 80 miles; from thence to Syracuse 61; from thence to Rochester 99; and from thence to Buffalo 93. There are, in this canal, 26 locks between Albany and Schenectady, 25 between Schenectady and

Utica, 25 between Utica and Rochester, and 6 between Rochester and Buffalo. How long is the canal, and how many locks does it contain?

Ans. The canal is 363 miles long, and it contains 82 locks.

QUESTIONS ON THE FOREGOING.

1. What are the fundamental rules of Arithmetic? 2. Why are they so called? 3. What is addition? 4. What is *simple* addition? 5. When you would add together several numbers, how do you write them down? 6. Which column of figures do you add first? 7. Do you begin at the bottom, or at the top of the column? 8. When you have added up the column of units, what do you do with the amount? 9. How then do you proceed? 10. When you have added up the last column, what do you do with the amount? 11. How is addition proved?

SUBTRACTION,

Is taking a less number from a greater, so as to find their difference.

The greater number is called the *Minuend;* the less number, the *Subtrahend;* and their difference, or what is left after the subtraction is performed, is called the *Remainder.*

Subtraction is either *simple* or *compound.*

SIMPLE SUBTRACTION,

Is taking a less *whole* number from a greater of the same denomination; as 2 dollars taken from a sum of 5 dollars, will leave a remainder of 3 dollars: Or, it is simply subtracting a less whole number from a greater, without regard to their signification; as 4 taken from 7, leaves 3.

SUBTRACTION TABLE.

2—2=0	6—3=3	———	———
3—2=1	7—3=4	5—5=0	7—7=0
4—2=2	8—3=5	6—5=1	8—7=1
5—2=3	9—3=6	7—5=2	9—7=2
6—2=4	10—3=7	8—5=3	10—7=3
7—2=5	———	9—5=4	———
8—2=6	4—4=0	10—5=5	8—8=0
9—2=7	5—4=1	———	9—8=1
10—2=8	6—4=2	6—6=0	10—8=2
———	7—4=3	7—6=1	———
3—3=0	8—4=4	8—6=2	9—9=0
4—3=1	9—4=5	9—6=3	10—9=1
5—3=2	10—4=6	10—6=4	———

This Table may read thus, 2 from 2 leaves 0; 2 from 3 leaves 1; 2 from 4 leaves 2, &c.

When the numbers are small, as in the foregoing table, the subtraction of the less number from the greater may be readily done in the mind, without writing down the numbers. When the given numbers are large, their difference may be found by the following

RULE.

1. Set the subtrahend, or less number, under the minuend, or greater number,* so that units shall stand under units, tens under tens, &c. as in Addition, and draw a line underneath.

2. Begin with the right hand figures, and take, if possible, the lower figure from the upper, and set down the remainder underneath, or if nothing remains, set down a cipher; and in like manner proceed with all the other figures when the figures in the lower number are all less than the corresponding figures in the upper number.

3. But, if any figure in the lower number be greater than the one above it, borrow 10, which add to the upper figure,

* Although it is usual, and generally most convenient, to write the subtrahend below the minuend, it is by no means essential; and it may be well for the student, in solving some of the following questions in Subtraction, to write the subtrahend above the minuend, and so perform the subtractions *downwards*, in order that he may be able to perform subtraction in this manner, when it may happen, in complex operations, that the numbers are so arranged.

and from the amount subtract the lower figure, and set down the remainder below. Then carry 1, (as an equivalent to the 10 borrowed,) and add it to the next figure to the left, in the lower number, and proceed as before.*

To prove Subtraction.

Add the remainder, or answer, to the subtrahend; and, if the work be right, the amount will be equal to the minuend.

EXAMPLES.

1. Subtract 315 from 457.

Operation.	
Minuend,	457
Subtrahend,	315
Remainder,	142
Proof,	457

Explanation.—I first write down the numbers as directed in the 1st article of the Rule. Then I subtract as follows: I begin with the right hand figures, and say, 5 from 7 leaves 2; which I set down underneath: then, 1 from 5 leaves 4, which I set down: then, 3 from 4 leaves 1, which I set down, and the work is done. So, the whole remainder, or answer is, 142.—Then, in order to prove the work, I add the remainder to the subtrahend; and the sum being equal to the minuend, I conclude the work is right.

---

* *Demonstration.*—When all the figures in the subtrahend are less than their corresponding figures in the minuend, the difference of the figures in the several like places, must, all taken together, make the true difference sought; because, as the sum of the parts is equal to the whole, so must the sum of the differences, of all the similar parts, be equal to the difference of the whole.

2 The reason of the method of proceeding when any figure in the subtrahend is greater than the corresponding figure in the minuend, may be shown as follows: If any two given numbers be increased equally, by the addition of any one number to each of them, their difference, when thus increased, will still be the same as before. Thus, if to each of the numbers, 6 and 4, we add 10, or any other number, their difference, after being thus increased, will evidently be the same as before, viz. 2. Now, in performing subtraction, according to the above Rule, when any figure in the subtrahend is greater than the corresponding figure in the minuend, we, in effect, add 10 to each of those figures; for instead of adding 10 to the figure in the subtrahend, we add 1 to the next place to the left, which, by the nature of notation, is equal to ten units in the former place: Therefore, we increase the minuend and subtrahend equally, by adding the same number to both, and hence their difference remains the same

The reason of the method of proof is evident; for if the difference of two numbers be added to the less number, it must manifestly make up a sum equal to the greater number

Ex. 2.

From	6053
take	5846
Rem.	207
Proof,	6053

Here, as I cannot subtract 6 from 3, I borrow 10 and add to the 3, which makes 13; then I say, 6 from 13 leaves 7, which I set down. Then I carry 1 to the 4, (on account of the 10 which I borrowed,) and say, 1 to 4 is 5, and 5 from 5 leaves 0, which I set down. Then, 8 from 0 I cannot, but,(I borrow 10, and then say,) 8 from 10 leaves 2, which I set down. Then, 1 carried to 5 is 6, and 6 from 6 leaves 0; and, as there are no more figures to the left, I do not set down the 0, because ciphers at the left hand of whole numbers are useless.

Note.—In Simple Subtraction, when it is necessary to borrow, instead of adding the 10 borrowed to the upper figure, and then subtracting the lower figure from the sum, as directed in the foregoing Rule, you may, if you choose, subtract the lower figure from 10, and add the remainder to the upper figure, and set down the sum below: then carry 1 to the next figure in the lower number; and so proceed.

	(3)	(4)	(5)	(6)
From	7602	47026	8050790	860500
take	431	30858	32827	734102
Rem.	7171	16168		

	(7)	(8)	(9)
From	632006	43760078	7020005601
take	489008	5284090	2452708
Rem.			

	(10)	(11)
From	1000000000	1000010001110
take	987654321	10009999
Rem.		

12. From seventy-four thousand and twenty, take thirty-three thousand one hundred and ten. Ans. 40910.

13. From eighty thousand, subtract five hundred and eight. Ans. 79492.

14. From two hundred and twenty-eight thousand, take two hundred and twenty-eight. Ans. 227772.

15. Subtract one from a million, and show the remainder. Ans. 999999.

PRACTICAL QUESTIONS.

1. If a man's yearly income amounts to 2950 dollars, and his expenses to 2400 dollars, how much is added to his estate yearly? Ans. 550 dollars.

2. A owes B one thousand and five dollars: If eight hundred and seventy dollars of this debt should be paid, how much would then remain due? Ans. 135 dollars.

3. A vinter bought 20 casks of wine, containing in all 1260 gallons; and afterwards sold 14 casks, containing 948 gallons: How many casks, and how many gallons of wine had he left? Ans. 6 casks, and 312 gallons.

4. A owes B 756 dollars, and C owes B 1273 dollars; how much does C's debt exceed A's? Ans. 517 dollars.

5. The city of Philadelphia was founded by William Penn, in the year 1683: how many years from that time to the year 1831? Ans. 148 years.

6. Homer, the famous Greek epic poet, died about 900 years before the birth of Christ, and Virgil, the prince of the Latin poets, died 19 years before Christ: How many years from the death of Homer to the death of Virgil? Ans. 881 years.

7. Gunpowder was invented by Roger Bacon, in the year 1280; how long was this before the invention of printing by Laurentius of Harlæm, which was in the year 1430? Ans. 150 years.

8. In the year 1820, the State of Connecticut contained 275248 inhabitants; Rhode-Island 83059; New-Hampshire 244161; Maine 298335; and Vermont 235764. The city of London, in the same year, contained a population of about 1225000 souls: How many more inhabitants were there in London than in the five States above mentioned? Ans. 88433.

9. The number of inhabitants in the whole world is supposed to be about six hundred and fifty millions: About

two hundred and seventeen millions of them are professors of Christianity; one hundred millions are Mahometans; three millions are Jews; and the rest Heathens: What is the number of Heathens, and how many more Heathens are there than Christians?

Ans. The number of Heathens is 330 millions, and there are 113 millions more Heathens than Christians.

QUESTIONS ON THE FOREGOING.

1. What is subtraction? 2. What is the greater number called? 3. What is the less number called? 4. What is their difference called? 5. What is *simple* subtraction? 6. When you would subtract one number from another, how do you set them down? 7. Where do you begin to subtract; and how do you proceed? 8. What is to be done when any figure in the lower number is greater than that above it? 9. How is subtraction proved?

MULTIPLICATION,

Is a compendious method of performing Addition; teaching how to find the amount of any given number when repeated a certain number of times; as 3 times 4, which is 12; that is, 4 repeated 3 times, (viz. 4+4+4,) makes 12.

The number to be multiplied, or repeated, is called the *Multiplicand;* the number to multiply by, or the number of repetitions, is called the *Multiplier;* and the number produced by multiplying the former by the latter, being the total amount of all the repetitions, is called the *Product.* Also, the multiplicand and multiplier are both called *Factors.*

SIMPLE MULTIPLICATION,

Is the multiplying of any two or more numbers together, without having regard to their signification; as 4 times 5 is 20.

Before proceeding to any operations in multiplication, it is necessary to learn very perfectly the following table of all the products of the first 12 numbers, commonly called the *multiplication table.*

MULTIPLICATION TABLE.*

1	2	3	4	5	6	7	8	9	10	11	12
2	4	6	8	10	12	14	16	18	20	22	24
3	6	9	12	15	18	21	24	27	30	33	36
4	8	12	16	20	24	28	32	36	40	44	48
5	10	15	20	25	30	35	40	45	50	55	60
6	12	18	24	30	36	42	48	54	60	66	72
7	14	21	28	35	42	49	56	63	70	77	84
8	16	24	32	40	48	56	64	72	80	88	96
9	18	27	36	45	54	63	72	81	90	99	108
10	20	30	40	50	60	70	80	90	100	110	120
11	22	33	44	55	66	77	88	99	110	121	132
12	24	36	48	60	72	84	96	108	120	132	144

To commit this table to memory, begin with the 2 in the top line of figures, and multiply each figure in the left hand column by it, and you will find each product in the 2d column right against the figure you multiply; thus, twice 1 is 2, twice 2 is 4, twice 3 is 6, and so on. Then take the 3 in the top line for a multiplier, and multiply each figure in the left hand column by it, as before, and you will find the several products in the 3d column, against the figures you multiply; thus, 3 times 1 is 3, 3 times 2 is 6, &c. In like manner proceed through all the columns in the table, and each product will be found in the same column with the multiplier, right against the multiplicand.

CASE I.

To multiply any number by a single figure, or by any number not greater than 12.

RULE.

1. Set down the multiplier under the right hand figure, or figures, of the multiplicand, and draw a line underneath.

* The invention of this table is ascribed to Pythagoras, a Grecian philosopher, who died about five centuries before the birth of Christ. The table is formed by addition, in the following manner: The numbers 1, 2, 3, 4, 5, 6, 7, 8, 9, 10, 11, 12, are written in the first column; then each of these numbers is added to itself, and the several sums are written in

2. Multiply the right hand figure of the multiplicand by the multiplier; if the product is but one figure, set it down in the place of units below the multiplier; but if it consists of more figures than one, set down only the right hand figure of it, and carry the rest to be added to the product found by multiplying the next figure of the multiplicand.

3. Multiply the next figure of the multiplicand by the multiplier; to the product add the number, (if any,) carried from the former product; set down the right hand figure of the sum, and carry the rest, (if any,) to be added to the product in the next place, as before; and so proceed in multiplying all the figures of the multiplicand by the multiplier, and set down the whole amount found in the last place at the left hand.†

EXAMPLES.

1. Multiply 30256 by 4.

Operation.

Multiplicand,	30256	} The Factors.
Multiplier,	4	
Product,	121024	

Explanation.—I first place the multiplier under the right hand figure of the multiplicand, and draw a line underneath. Then I begin at the right hand to multiply, and say 4 times 6 is 24; I set down 4 and carry 2: then, 4 times 5 is 20, and the 2 I carried makes 22; I set down 2 and carry 2: then, 4 times 2 is 8, and the 2 I carried makes 10; I set down 0 and carry 1: then, 4 times

the 2d column, which contains each number in the first column doubled. Each number in the first column is then added to the number standing against it in the 2d column; and the several sums are written in the 3d column, which contains each number in the first tripled; and in this manner the table is formed throughout.

† The reason of this rule is the same as for the process in Addition, in which 1 is carried for every 10, to the next place, gradually as the several products are produced one after another, instead of setting them all down one below another, as in the annexed example.

$$\begin{array}{r} 8742 \\ 6 \\ \hline 12 = 2\times6 \\ 240 = 40\times6 \\ 4200 = 700\times6 \\ 48000 = 8000\times6 \\ \hline 52452 = 8742\times6 \end{array}$$

0 is 0,* but the 1 I carried makes 1; I set down 1: then, 4 times 3 is 12, the whole of which I set down, because there are no more figures to multiply. So, the work is done, and the whole product is 121024.

The learner will readily perceive that the rule for carrying the tens from the amount in each place to that in the next higher place, is the same in Multiplication as in Addition. I would advise the learner, at first, in multiplying, to set down the amount found in each place, as in Addition; and after having found the total product, to look over the whole work carefully again, in order to correct the errors in it, if there should happen to be any. See the next example.

Several methods of proving Multiplication will be given in Case II.

(2)

Multiply	8742	12 Product	of the units.
by	6	25 do.	of the tens.
		44 do.	of the hundreds.
Product,	52452		

	(3)	(4)	(5)
Mult.	54296	8705080	62050300
by	2	3	4
Prod.	108592	26115240	248201200

(6)	(7)	(8)	(9)
800924	9980077	750289	657884
5	6	7	8

(10)	(11)	(12)
123456789	7654321	8438020987
9	11	12

* If 0 be multiplied by any number, or any number be multiplied by 0, the product will be 0.

CASE II.

When the multiplier consists of several figures.

RULE.

1. Set down the multiplier under the multiplicand, so that units shall stand under units, tens under tens, &c. as in Addition and Subtraction, and draw a line underneath.

2. Multiply the multiplicand by the right hand figure of the multiplier, and set down the product, as in Case first.

3. Multiply, in like maner, by the next figure in the multiplier, and place the right hand figure of the product exactly under the figure you multiply by, and place the other figures of the product to the left of this, in regular order, under the corresponding figures of the former product. In like manner multiply the multiplicand by each of the remaining figures of the multiplier, and place the right hand figure of each product directly under the figure you multiply by.

4. Draw a line below the several partial products thus found, and add them together, in the order in which they stand, and their sum will be the total product required.*

Note.—When there are ciphers between the significant figures in the multiplier; then, in multiplying, omit or neglect those ciphers; but be careful in multiplying by the significant figures, to set down the right hand figure of each partial product exactly below the figure of the multiplier by which it is produced.

**Demonstration.*—After having found the product of the multiplicand by the first figure of the multiplier, as in the former Case, the multiplier is supposed to be divided into parts, and the product is found for the second figure in the same manner: but as this figure stands in the place of tens, the product must be ten times its simple value; and therefore the first or right hand figure of this product must be set in the place of tens; or, which is the same thing, directly under the figure multiplied by. And by proceeding in this manner separately with all the figures of the multiplier, it is evident that we shall multiply all the parts of the multiplicand, by all the parts of the multiplier, or the whole of the multiplicand by the whole of the multiplier; therefore these several products being added together, will be equal to the whole required product; as in the example annexed.

```
    5207
     245
   -----
   26035=  5 times 5207.
  20828 = 40 times ditto.
 10414  =200 times ditto.
 -------  ---
 1275715=245 times ditto.
```

EXAMPLES.

1. Multiply 5207 by 245.

```
Multiplicand,   5207
Multiplier,      245
               -----
               26035
              20828
             10414
             -------
Product,     1275715
```

In performing the work of this example, I first set down the given factors as directed in the Rule, and draw a line below them. Then I multiply the multiplicand by 5, as in Case first, and the product is 26035. I next multiply by 4, and the product is 20828, the right hand figure of which I place under the 4 in the multiplier. I then multiply by 2, and the product is 10414, the right hand figure of which I place under the 2 in the multiplier. Lastly, I draw a line under these partial products, and add them together; and I find their sum to be 1275715; which is the total product required.

```
            (2)                           (3)

Multiply     43027          Multiply      7028356
   by         3205             by           10054
           -------                      ---------
            215135                       28113424
           86054                        35141780
         129081                       7028356
         ---------                    -----------
Prod.    137901535          Product,  70663091224
```

TO PROVE MULTIPLICATION.

There are three different methods of proving Simple Multiplication, which are as follows:

First Method.

Multiply the given multiplier by the multiplicand, and the total product thus found, will be the same as that found before, if the work be right.*

* When two numbers are to be multiplied together, either of them may be made the multiplier or multiplicand, and the product will be the same. Hence the reason of this method of proof is obvious.—It may be proper further to inform the learner, that when three or more numbers are

This method is illustrated by the following example.

(4)	Then, to prove the work,
Multiply 463	Multiply 48
by 48	by 463
3704	144
1852	288
	192
Prod. 22224	Prod. 22224 Proof.

Second method of Proof, by casting out the nines.

1. Cast the 9's out of the sum of the figures in the multiplicand, thus: Add together the figures of the multiplicand; but, in adding, omit all the 9's, and reject 9 from the sum as often as it amounts to 9 or more, always retaining the excess over 9; and the last excess thus found, will be the excess of 9's in the sum of all the figures; which excess set down a little to the right hand of the multiplicand.

2. In like manner find the excess of 9's in the multiplier, and place it against the multiplier.

3. Multiply the excess of 9's in the multiplicand, by the excess in the multiplier, and find, as before, the excess over 9 in the sum of the figures of this product, which excess place against the answer to the question, or the total product of the two given factors, which you wish to prove. Then cast the 9's out of the product last mentioned, as before; and if the excess is the same with that standing against the product, the work may be supposed to be right; but if these excesses differ, the work is erroneous.*

to be multiplied together, it is not material in what order the several factors are multiplied together: e. g. the product of the numbers 2, 5, and 8, is 80; and if the product of any two of them be multiplied by the remaining number, the final product will be 80; for $2\times5\times8=80$, and $2\times8\times5=80$, and $8\times5\times2=80$.

* This rule for proving multiplication depends upon a peculiar property of the number 9, which, except the number 3, belongs to no other digit whatever; viz. that "any number divided by 9, will leave the same remainder as the sum of its figures divided by 9."—The rule may be dem-

This method of proof is illustrated by the two following examples.

	(5)			(6)	
		Excess of 9's.			Excesses.
Multiply	53693	- - - 8	Multiply	9207	- - - 0
by	349	- - - 7	by	137	- - - 2
	483237			64449	
	214772			27621	
	161079			9207	
Product,	18738857	- - - 2	Prod.	1261359	- - - 0

I prove the work of example 5th thus: I begin at the left hand figure of the multiplicand, and say 5 and 3 are 8, and 6 are 14; which sum being more than 9, I cast 9 out of it, and 5 remains: then, (omitting the 9 in the multiplicand,) I say 5 and 3 are 8; which sum, being the excess of 9's in the multiplicand, I set down a little to the right. Then I proceed to the multiplier, and say 3 and 4 are 7; which sum I set down to the right hand of the multiplier.

---

onstrated as follows: Let A and B denote the number of 9's in the two factors to be multiplied, and a and b what remain; then $9A+a$ and $9B+b$ will be the factors themselves, and their product is $(9A\times9B)+(9A\times b)+(9B\times a)+(a\times b)$; but the first three of these products are each a precise number of 9's, because their factors are so, either one or both: these therefore being cast away, there remains only $a\times b$; and if the 9's also be cast out of this, the excess is the excess of 9's in the total product: but a and b are the excesses in the factors themselves, and $a\times b$ is their product; therefore the rule is true.

This method of proof, however, is not infallible; for although the work will always be wrong when it proves so according to this rule, yet it *may not* be right when it proves so; because the right figures may stand in the product, and not stand in the right order; or two or more wrong figures may amount to the same, when added together, as the right ones would. But as the method will usually detect mistakes, and is shorter than the other methods, it is thought useful to be retained.

I then multiply these two excesses together, and their product is 56. The two figures of this product, viz. 5 and 6, I add together, and their sum is 11, which exceeds 9 by 2; and this excess I place against the total product in the example. Lastly, I cast the 9's out of this product thus: I begin at the left hand, and say, 1 and 8 are 9; which I cast away: then, 7 and 3 are 10; I cast away 9, and 1 remains: then, 1 and 8 are 9, which I cast away: then, 8 and 5 are 13, which is 4 over 9—I cast away 9: then, 4 and 7 are 11, which is 2 over 9; and as this excess is the same as the excess standing against the product, I conclude the work is right.

Third method of Proof.

Multiplication may be proved by Division; viz. thus: Divide the product by either of the factors, and the quotient will be equal to the other factor, if the work be right. This is the safest method of proving multiplication, but it cannot be practised till the rule of Division is learned.

More Examples in Simple Multiplication.

7. Multiply 224676 by 474. Ans. 106496424.
8. Multiply 423801 by 651. Ans. 275894451.
9. Multiply 829921 by 911. Ans. 756058031.
10. Multiply 8192 by 4096. Ans. 33554432.
11. Multiply 80546 by 80052. Ans. 6447868392.
12. Multiply 40353607 by 16807. Ans. 678223072849.
13. Multiply 85173 by 78542. Ans. 6689657766.
14. Multiply 4897685 by 40003. Ans. 195922093055.
15. Multiply the number 8763 by itself. Ans. 76790169.
16. Required the continued product of the numbers 4096×512×64. Ans. 134217728.
17. Required the product of 1296×216×36×6. Ans. 60466176.

CONTRACTIONS IN SIMPLE MULTIPLICATION.

There are several methods of contraction which may be used in particular cases, in finding the products of numbers, by which the operations may be performed in a shorter manner than by the foregoing rules. The most useful of these contractions are the following.

I. *When there are ciphers on the right hand of one or both of the factors:* Set down the factors in the same manner as if there were no ciphers on the right hand: then, in multiplying, neglect those ciphers, and multiply together the other figures as usual; and annex to the product as many ciphers as there are on the right hand of both the factors.*

EXAMPLES.

1. Multiply 543000 by 520.

```
      543000
        520
      ------
      1086
     2715
     --------
Prod. 282360000
```

2. Multiply together the two numbers 4027 and 28000.

```
      4027
        28000
      -------
      32216
      8054
      ---------
Prod. 112756000
```

3. Multiply 359260 by 3040. Ans. 1092150400.

4. Multiply 9826000 by 82530. Ans. 810939780000.

Note.—To multiply any number by 10, *or* 100, *or* 1000, *&c.*: Annex to the given multiplicand as many ciphers as there are in the multiplier, and it will then be the product required.

So $54 \times 10 = 540$
And $342 \times 100 = 34200$
And $120 \times 1000 = 120000$

II. *When the multiplier exceeds* 12, *and is a composite number*† *which can be produced by multiplying together two or more small numbers not greater than* 12: Then multiply the given multiplicand by one of those small numbers, or component parts of the multiplier, and that product by another of those numbers, or parts, and so on, till you

---

* The reason of this method of contraction will easily appear from the demonstration of the rule for Case 2d, and from the nature of notation.

† A *composite number* is one that can be produced by multiplying together two or more smaller numbers; or, in other words, a composite number is one that can be divided by some smaller number besides unity, without a remainder. The factors which produce any composite number, are called the *component parts* of the number. Those numbers which are not composite, are called *prime numbers.*

have multiplied by all of them; and the last product will be the total product required.*

EXAMPLES.

1. Multiply 6423 by 35.

Here, because 7×5=35, I multiply by 7, and then by 5, as follows:

```
 6423
    7
 -----
 44961
     5
 ------
224805 Ans.
```

2. Multiply 8321 by 1728.

Here, because 12×12×12=1728, I multiply three times by 12, as follows:

```
    8321
      12
   ------
   99852
      12
  -------
 1198224
      12
 --------
14378688 Ans.
```

3. Multiply 1504 by 42.
7×6=42, and hence 1504×7×6=63168, the Ans.

4. Multiply 8647 by 252.
4×7×9=252; hence 8647×4×7×9=2179044, the Ans.

5. Multiply 64321 by 81. Ans. 5210001.

6. Multiply 712836 by 96. Ans. 68432256.

7. Multiply 3742 by 14400. Ans. 53884800.

III. *When all the figures of the multiplier are 9's:* Annex as many ciphers to the multiplicand as there are figures of 9 in the multiplier; then, from this number subtract the given multiplicand, and the remainder will be the product required.†

* The reason of this rule is obvious: for any number, multiplied by the component parts of another number, must give the same product as though it were multiplied at once by the whole number, or product of those parts: Thus, in example first, 5 times the product of 7 multiplied into 6423, makes 35 times that number, as plainly as 5 times 7 makes 35.

† It is evident that if any number be multiplied by 9, the product will be 9 tenths of the product of the same number multiplied by 10; and as the annexing of a cipher to the multiplicand increases it ten fold, it is evident that if the given multiplicand be subtracted from the tenfold multiplicand, the remainder will be ninefold the said given multiplicand, equal to the product of the same by 9; and the same will hold true of any number of 9's.

EXAMPLES.

1. Multiply 7528 by 99.

From	752800
subtract	7528
Ans.	745272

2. Multiply 475286 by 999. Ans. 474810714.
3. Multiply 86720 by 9999. Ans. 867113280.

APPLICATION OF MULTIPLICATION.

Note.—When the value, or weight, &c. of any article or thing is given, to find the value, &c. of any number of like articles: Multiply together the value of one article and the number of them, and the product will be the answer.*—Either of the two factors may be made the multiplier, but it will generally be the most convenient to multiply by the less number.

EXAMPLES.

1. A merchant sold 8 pieces of cloth, at 12 dollars a piece; how much money did he receive for the whole?
Here 12×8=96 dollars, Ans.

2. If 27 barrels contain each 196 pounds of flour, how much flour do they all contain? Ans. 5292 pounds.

3. What do 210 hats come to, at 3 dollars each?
Ans. 630 dollars.

4. What do 52 firkins of butter come to, at 7 dollars a firkin? Ans. 364 dollars.

5. My orchard contains 15 rows of trees, with 23 trees in each row; how many trees are there in it? Ans. 345.

6. What is the value of 526 acres of land, if each acre be worth 15 dollars? Ans. 7890 dollars.

7. What do 13 tons of hay come to, at 15 dollars a ton?
Ans. 195 dollars.

8. If seventeen casks contain each thirty gallons of cider, how much cider do they all contain?
Ans. Five hundred and ten gallons.

9. How many panes of glass will it take to make 22 win-

---

* The reason of this rule is very obvious; for it is plain that if one yard of cloth is worth 3 dollars, 2 yards are worth twice 3 dollars, or 6 dollars, and 3 yards are worth 3 times 3 dollars, or 9 dollars, and so on; and it is evident that the rule will hold true in all similar calculations.

dows, with 24 panes in each window? Ans. 528.

10. A merchant bought 342 bales of linen; each bale contained 56 pieces of cloth, and each piece 25 yards: how many pieces, and how many yards of cloth were there?

Ans. 19152 pieces, and 478800 yards.

QUESTIONS ON THE FOREGOING.

1. What is Multiplication? 2. What is the number to be multiplied called? 3. What is the number by which we multiply called? 4. What is the general name for both numbers? 5. What is *simple* multiplication? 6. What is the first case in simple multiplication? 7. How do you set down the factors? 8. Where do you begin to multiply; and how do you proceed? 9. When do you carry, and to what do you add the number carried? 10. What is the second case? 11. How do you set down the factors in this case? 12. How many figures of the multiplier do you use at a time? 13. In multiplying by each figure of the multiplier, how do you set down the product? 14. When you have multiplied by all the figures of the multiplier, what do you do next? 15. What is the first method of proving multiplication? the second? the third? 16. How do you perform multiplication when there are ciphers at the right hand of the factors? 17. How do you multiply any number by 10, or 100, &c.? 18. What is a composite number? 19. How may two numbers be multiplied together when the multiplier is a composite number? 20. What method of contraction may be used when all the figures of the multiplier are 9's? 21. How do you find the value of any number of similar articles when the value of one of them is known?

DIVISION,

Teaches how to separate any given number, or quantity, into any number of equal parts assigned; or to find how often one number is contained in another; and is a concise method of performing several subtractions.

The number to be divided is called the *Dividend;* the number to divide by is called the *Divisor;* and the number of times the dividend contains the divisor, is called the *Quotient.* If there is anything left after the operation is performed, it is called the *Remainder.* Sometimes there is a remainder, and sometimes none: when there is any, it is less than the divisor, and of the same denomination as the dividend.

SIMPLE DIVISION,

Is the dividing of one number by another, without regard to their values; as 20, divided by 5, produces 4 in the quotient; that is, 5 is contained 4 times in 20.

Note.—It will be proper for the scholar, before proceeding to perform any operations in division, to learn the multiplication table in an inverted order, as follows: Say 2 in 4, twice; 2 in 6, three times; 2 in 8, four times, &c.: then, taking the next column, say 3 in 6, twice; 3 in 9, three times; 3 in 12, four times, &c.; and in like manner proceed through the whole table.

CASE I.

When the divisor does not exceed 12.

RULE.

1. Place the divisor at the left hand of the dividend, with a curve line between them, and draw a line under the dividend.

2. See how many of the left hand figures of the dividend must be taken to make a number that will contain the divisor; that is, a number as large at least as the divisor; and the number expressed by those figures, call the *dividual.**

3. Find how many times the dividual contains the divisor, and also what remains over: The number of times the divisor is contained in the dividual will be the first figure of the quotient; which figure set down below the right hand figure of the dividual; and the remainder, (if any,) suppose to be prefixed to the next figure in the dividend, for a new dividual; but if there is no remainder, then take the said next figure of the dividend for a new dividual.

4. Find another quotient figure, as before, and so proceed, until the work is done.

* The dividual is a partial dividend, or so much of the dividend as is taken to be divided at a time, and which produces one quotient figure.

Note 1.—When any dividual after the first, is less than the divisor, place a cipher in the quotient, and suppose the dividual to be joined to the next figure of the dividend, for a new dividual; after which proceed as before.

Note 2.—If, after the division is performed, there is a remainder left, it may be placed a little to the right hand of the quotient, as in the 1st, 2d, and 3d examples; or, if you wish to make the quotient complete, then draw a short horizontal line from the quotient towards the right, and place the remainder above, and the given divisor below this line; which will form a fraction that will express the true value of the remainder. See examples 7th, 8th, &c.

To prove Division.

Multiply the quotient by the divisor; to the product add the remainder, if any; and the result will be equal to the dividend, if the work be right.*

EXAMPLES.

1. Divide 374027 by 4.

```
              Dividend.              93506 Quotient.
Divisor, 4)374027                        4 Divisor.
           ──────                   ──────
Quotient,  93506 - - 3 Rem.         374024
                                        +3 Rem.
                                    ──────
                                    374027 Proof.
```

In performing the work of this example, I proceed thus: I first set down the given numbers as directed in the Rule; then I examine the dividend, and find that the two left hand figures are the fewest that will contain the divisor; and therefore I take these two figures, which make the number 37, for the first dividual. This number contains the divisor 9 times, and 1 remains over; for 9 times 4 is

* Multiplication and Division being the reverse of each other, the reason of this method of proof is evident. There are also several other methods sometimes used for proving division; two of which may be briefly explained as follows: If the remainder be subtracted from the dividend, the difference will be the product of the divisor and quotient, if the work be right. Therfore, if this difference be found, the work may then be proved by either of the two last methods laid down for proving Multiplication.

36, and 1 makes 37. The 9 I place underneath, for the first quotient figure; and the 1 remainder, I suppose to be prefixed to the next figure of the dividend, viz. 4; and then I have 14, for a new dividual. This dividual contains the divisor 3 times, and 2 remains over; for 3 times 4 is 12, and 2 makes 14. The 3 I place underneath, and the 2 which remains, I suppose to be prefixed to the next figure of the dividend, which makes 20, for a new dividual. This dividual contains the divisor just 5 times; for 5 times 4 is 20. The 5 I set down; and as there is no remainder, I take the next figure of the dividend, viz. 2, for a new dividual; but as this is less than the divisor, I set down a cipher below, and then suppose the 2 in the dividend to be joined to the next figure, which makes 27, for a dividual. This dividual contains the divisor 6 times, and 3 remains over. The 6 I set down, and as there are no more figures in the dividend, I set down the remainder a little to the right hand of the quotient, which finishes the work.—Then, to prove the work, I multiply the quotient by the divisor, and add the remainder to the product; and as the result is equal to the dividend, I conclude the work is right.

	(2)	(3)	(4)
	5)470152	2)783002637	3)860250
Quotient,	94030 - - 2	391501318 - - 1	286750
	5	2	3
Proof,	470152	783002637	860250

I perform the work of example 2d thus: I say 5 in 47, 9 times, and 2 over—I set down 9: then, 5 in 20, 4 times—I set down 4: then, 5 in 1, no times, but 1 remains—I set down 0: then, 5 in 15, 3 times—I set down 3: then, 5 in 2, 0 times, but 2 remains—I set down 0, and place the 2 remainder near the quotient, at the right hand, which finishes the work.

(5)	(6)	(7)
4)3024004	5)82370055	6)196231
756001	16474011	$32705\frac{1}{6}$

(8)	(9)	(10)
7)32800256	8)860275	9)3003400
$4685750\frac{6}{7}$	$107534\frac{3}{8}$	$333711\frac{1}{9}$

11. If 758967 be divided by 11, what will the quotient be? Ans. 68997.
12. What is the quotient of 4765862 divided by 11? Ans. $433260\frac{2}{11}$
13. What is the quotient of 1072687 divided by 12? Ans. $89390\frac{7}{12}$
14. What is the quotient of 9876543210 divided by 12? Ans. $823045267\frac{6}{12}$
15. Divide 14738 by 8.
16. Divide 8300600 by 9.
17. Divide 102460175 by 12.

CASE II.

When the divisor exceeds 12.

RULE.*

1. After having set down the dividend, mark off a curve line on each side of it, and place the divisor on the left hand.

---

* This is a general rule for the division of whole numbers. The method by the rule for Case first, is a contraction of the process by the general rule; the subtrahends in the former case being made out and subtracted mentally. In performing division by either of these rules, we resolve the dividend into parts, and find the number of times the divisor is contained in each of those parts. Each dividual is just so much of the dividend, as is necessary to be divided in order to obtain one quotient figure; and the only thing which remains to be proved, is, that the several figures of the quotient, taken as one number, according to the order in which they are placed, are the true quotient of the dividend by the divisor; which may be thus demonstrated.

Dem.—The local or complete value of the first part of the dividend, or first dividual, is, by the nature of notation, 10, or 100, or 1000, &c. times the simple value of what it is taken in the operation; according as there are 1, 2, or 3, &c. more figures in the dividend; and consequently the true value of the quotient figure belonging to that part of the dividend, is also 10, 100, or 1000, &c. times its simple value; but the local or complete value of the quotient figure, belonging to that part of the dividend, found by the rule, is 10, 100, or 1000, &c. times its simple value; for there are as many more quotient figures as there are figures remaining in

2. Take as many of the left hand figures of the dividend as may be necessary, for a dividual, as in Case first. Find how many times the dividual contains the divisor, and place

---

the given dividend : therefore, the first quotient figure, taken in its complete value from the place it stands in, is the true quotient of the divisor in the complete value of the first part of the dividend. For the same reason, all the rest of the figures of the quotient, taken according to their places, are, each, the true quotient of the divisor, in the complete value of the several parts of the dividend belonging to each; consequently, taking all the quotient figures in the order in which they are placed by the rule, they make one number, which is equal to the sum of the true quotients of all the several parts of the dividend; and is, therefore, the true quotient of the whole dividend by the divisor.

That no obscurity may remain in the demonstration, it is illustrated by the following example.

Divisor=24)85701 Dividend.
1st dividual=85000(3000 the 1st quotient.
1st subtrahend=24×3000=72000
1st remainder=13000
+700
2d dividual=13700(500 the 2d quotient.
2d subtrahend=24×500=12000
2d remainder=1700
+00
3d dividual=1700(70 the 3d quotient.
3d subtrahend=24×70=1680
3d remainder= 20
+1
4th dividual=21(0 the 4th quotient.
4th subtrahend=24×0= 0
3570 the sum of all the
4th, and last remainder=21 quotients; or, the Answer.

Explanation.—It is evident that the dividend is resolved into these parts, 85000+700+00+1 ; for though the first part of the dividend, or the first dividual, is considered only 85, yet it is truly 85000; and therefore its quotient is not simply 3, but is really 3000, and the remainder is 13000 ; and so of the rest ; as may be seen in the operation.

When there is no remainder to a division, the quotient is the absolute

the result at the right hand of the dividend, for the first figure of the quotient.

3. Multiply the divisor by that quotient figure, and place the product under the dividual, for a subtrahend.

4. Subtract the subtrahend from the dividual, and bring down the next figure of the dividend, and annex it to the remainder, for a new dividual. Then find another quotient figure as before; and so proceed until the work is finished.

Note 1.—If it be necessary to bring down more than one figure to any remainder, to make it as large as the divisor, then a cipher must be put in the quotient for every figure so brought down more than one.

Note 2.—When the divisor is a large number, each quotient figure must be found by trial; and, in making these trials, the learner should always bear in mind, that the subtrahend must not be greater than the dividual, nor so much less as to leave a remainder as great as the divisor.

Note 3.—When division is performed by the rule for Case first, it is called *Short Division;* and when performed by the rule for Case second, *Long Division.*

EXAMPLES.

Divisor. Divid. Quotient.

14)33785(2413	
28···	
57 the 2d dividual.	2413 Quotient.
56 the 2d subtrahend.	14 Divisor.
18 the 3d dividual.	9652
14 the 3d subtrahend.	2413
45 the 4th dividual.	33782
42 the 4th subtrahend.	+3 Rem.
3 the last remainder.	33785 Proof.

and perfect answer to the question; but when there is a remainder, it may be observed, that it goes so much towards another time as it approaches the divisor: thus, if the remainder be equal to half the divisor, it will go half a time more, and so on. In order, therefore, to complete the quotient, put the last remainder to the end of it, above a line, and the divisor below it. Hence the origin of vulgar fractions, which are treated of hereafter.—[The preceding Note has been mostly copied from Pike's Arithmetic.]

In performing the work of example 1st, I find that the first dividual, viz. 33, contains the divisor twice; and therefore I set down 2 for the first quotient figure. I then multiply the divisor by this quotient figure, and the product is 28, which I place under the dividual, for a subtrahend. I then subtract the subtrahend from the dividual, and the remainder is 5, to which I annex the next quotient figure, and then I have 57, for a new dividual. This dividual contains the divisor 4 times, and therefore I set down 4 for the second quotient figure. Then I make out a subtrahend, &c. as before, and continue the operation in like manner until I have brought down all the figures of the dividend.

Note.—When a figure is brought down from the dividend a dot should be made under it, to show that the figure has been brought down; as in the foregoing example.

(2)

```
24)614472(25603
   48
   --
   134
   120
   ---
    144
    144
    ---
      72
      72
      --
      00
```

(3)

```
124)78300(631 56/124
    744
    ---
     390
     372
     ---
     180
     124
     ---
      56
```

4. If 645678 be divided by 13, what will the quotient be? Ans. $49667\frac{7}{13}$.

5. If 71640 be divided by 72, what will the quotient be? Ans. 995.

6. How many times is 422 contained in 253622? Ans. 601 times.

7. How many times is 456 contained in 378480? Ans. 830.

As learners sometimes find it difficult, when the divisor is a large number, to find the number of times each dividual contains the divisor, I would advise them, in such cases, to use the following method of finding the quotient figures.

After having made out a dividual, begin at the right hand figure of it, and count off as many figures lacking two as there are in the divisor; then find, by trial, how many times the part of the dividual remaining to the left, contains the number expressed by the two left hand figures of the divisor; and the result will be either the required quotient figure, or the next greater figure. Then, by making out a subtrahend, as usual, you may know whether the quotient figure thus found is the right one or not; and if it should prove to be too great, the next less figure will be the one required.

(8)

```
157496)53277693607(338279 Quot.
       472488
       ------
        602889
        472488
        -------
        1304013
        1259968
        -------
          440456
          314992
          -------
          1254640
          1102472
          -------
           1521687
           1417464
           -------
            104223
```

In this example, the divisor contains six figures; therefore, in seeking each quotient figure, I reject the four right hand figures of the dividual, and find how many times the number expressed by the remaining figures contains the number expressed by the two left hand figures of the divisor, viz. 15. Thus, the first dividual is 532776; but I reject the four right hand figures, and the remaining part is 53, which contains 15, (the two left hand figures of the divisor,) 3 times. I set down 3 for the first quotient figure; and then by making out a subtrahend as usual, I find that 3 is the right quotient figure. I then make out the 2d dividual; and by rejecting the four right hand figures of it, there remains 60, which contains the two left hand figures of the divisor 4 times. But, by making out a subtrahend, I find that the whole dividual does not contain the whole divisor 4 times; and therefore, instead of 4, I put down in the quotient the next less figure, viz. 3, which proves to be the right quotient figure. In like manner the other quotient figures are found.

9. Divide 823547 by 2401.	Quot. 343	Rem.	4
10. Divide 282475249 by 117649.	2401		0
11. Divide 10077698 by 1296.	7776		2
12. Divide 1249614 by 216.	....		54
13. Divide 7836416937 by 46656.	....		28521
14. Divide 761858465 by 90001.	....		0
15. Divide 119184693 by 38473.	....		33812
16. Divide 13060694016 by 279936.	....		0

CONTRACTIONS IN SIMPLE DIVISION.

I. *When there are ciphers at the right hand of the divisor;* cut off those ciphers from the divisor, and cut off the same number of figures from the right hand of the dividend; then divide the remaining part of the dividend by the remaining part of the divisor, as usual. If there is a remainder after the division, place the figures thus cut off from the dividend, to the right hand of it, and the whole will be the true remainder; otherwise, the figures cut off will be the remainder.*

---

* The reason of this method of contraction is easy to conceive; for cutting off the same number of figures from each, is the same as dividing by 10, or 100, or 1000, &c according to the number of figures cut off; and it is evident, that as often as the whole divisor is contained in the whole dividend, so often must any part of the divisor, be contained in the like part of the dividend. This method is used to avoid a needless repetition of ciphers, which would happen in the common way, as may be seen by working one of the examples in the common way, without cutting off the ciphers.

EXAMPLES.

1. Divide 48205 by 300.

3,00)482,05

Quot. $160\frac{205}{300}$

2. Divide 7052708 by 25000.

25,000)7052,708(282 $\frac{2708}{25000}$
50
205
200
52
50
2708 true rem.

3. Divide 20408 by 850. Ans. $24\frac{8}{850}$.

4. Divide 42140028 by 4900. Ans. $8600\frac{28}{4900}$

Divisor.	Dividend.		Remainder.
23000)	11980964(	Quotient.)	20964.
12500)	43625000(	)	0.
12000)	14952847(	)	847.
1000)	1854326 (	)	326.

Note.—To divide any number by 10, or 100, or 1000, &c. you have only to cut off as many of the right hand figures of the dividend, as there are ciphers in the divisor; then the remaining part of the dividend will be the quotient, and the figures cut off, the remainder; which remainder being placed over the given divisor, and annexed to the quotient, will make it complete.

So, $764 \div 10 = 76\frac{4}{10}$.
And $1800 \div 100 = 18$.
And $21507 \div 1000 = 21\frac{507}{1000}$.

II. *When the divisor exceeds* 12, *and is the exact product of two or more of the small numbers not greater than* 12; then, you may divide the dividend by one of those small numbers, or component parts of the given divisor; and the quotient thence arising, by another of them, and so on to the last; and the last quotient will be the one required.*

* This follows from the second contraction in Multiplication, being only the converse of it; for the half of the third part of any thing, is evidently the same as the sixth part of the whole; and so of any other numbers. The reason of

Note 1.—When only the *first* quotient has a *remainder* belonging to it, then this remainder will be the *true* or *total* remainder, the same as if the division had been performed all at once. But if there is any other remainder, then, to find the *total* remainder, proceed as follows: When only two divisors are used, multiply the last remainder by the first divisor; to the product add the first remainder, and the sum will be the total remainder. When more than two divisors are used, multiply the last remainder by the divisor used for finding the next foregoing quotient; to the product add the remainder, if any, belonging to that quotient; then multiply the sum by the next foregoing divisor, adding in the corresponding remainder, if any, and so on through all the divisors to the first; and the result will be the whole remainder required. After having found the total remainder, you may place it over the given divisor, and annex the fraction to the quotient, as usual.

EXAMPLES.

1. Divide 48322 by 35.

7×5=35; therefore I divide by 7, and then by 5; as follows:

```
  7)48322
  -------
   5)6903 - - - 1
   ------
     1380 - - - 3
```

Complete quotient, or Ans. $1380\frac{22}{35}$.

The total remainder is found thus:

```
  3=2d remainder.
  7=1st divisor.
 --
 21
 +1=1st remainder.
 --
 22=total remainder.
```

---

the method of finding the whole remainder from the several particular ones, will best appear from the nature of Vulgar Fractions. Thus, in the first example above, the first remainder being 1, when the divisor is 7, makes $\frac{1}{7}$: this must be added to the second remainder, 3, making $3\frac{1}{7}$ to be divided by the second divisor, viz. by 5. But $3\frac{1}{7}=\frac{3\times7+1}{7}=\frac{22}{7}$; and this divided by 5, gives $\frac{22}{7\times5}=\frac{22}{35}$, the value of the remainders.

2. Divide 97354 by 672.
8×7×12=672; therefore I divide by 8, then by 7, and then by 12.

```
 8)97354
   -----
 7)12169 - - - 2
   -----
12)1738 - - - 3
   -----
    144 - - - 10
```

Ans. $144\frac{586}{672}$.

```
10=last remainder.
 7=next preced. divisor.
--
70
+3=2d remainder.
--
73
 8=next preced. divisor.
---
584
+2=1st remainder.
---
586=total remainder.
```

Note 2.—To find out the component parts of a number greater than 144, proceed as follows; viz. Find, by trial, whether any of the numbers between 1 and 13 will divide the given number without a remainder; and if any of them will, then divide, and set down the quotient. Divide this quotient in like manner by some number between 1 and 13, if it can be done without leaving a remainder; and so proceed, till you find a quotient less than 13. Then the several divisors and the quotient last found, will be the component parts required; that is, their product will be equal to the given number. Thus, it may be readily found that the given divisor in example 2d is divisable by 8, and the quotient thus obtained, by 7; then the last quotient will be 12, which being less than 13, the division need not be continued any farther. So, 8×7×12=672.

3. Divide 5210015 by 81. Ans. $64321\frac{14}{81}$.
4. Divide 81799 by 96. Ans. $852\frac{7}{96}$.
5. Divide 538865 by 144. Ans. $3742\frac{17}{144}$.
6. Divide 52120 by 1728. Ans. $30\frac{280}{1728}$.
7. Divide 2179045 by 252. Ans. $8647\frac{1}{252}$.

APPLICATION OF DIVISION.

Note.—When the value, or weight, &c, of any number of articles is given, to find the value of each one of them, when they are all alike, or the average value of each, when they are unlike: Divide the value, &c. of the whole, by the number of articles, and the quotient will be the answer.*

* This rule is the converse of that given in the Note which precedes the practical questions in Multiplication; and the reason of it is very obvious.

EXAMPLES.

1. What is the value of a pound of flour, if 24 pounds be worth 96 cents? Here 96÷24=4 cents, the Ans.

2. If a farm containing 365 acres, be worth 8395 dollars, what is the value of each acre? Ans. 23 dollars.

3. A farmer butchered 25 hogs, which together weighed 6525 pounds: what was the average weight of each hog? Ans. 261 pounds.

4. The holy bible contains 1189 chapters. How many chapters must a man read in a day, to read the bible through in a year, or 365 days? Ans. $3\frac{94}{365}$ chapters.

5. The Hudson and Erie canal, in the State of New-York, is 363 miles long; and the whole expense of making it was about 7260000 dollars: What was the average expense for each mile? Ans. 20000 dollars.

6. Supposing a man labors a year for 132 dollars; how much has he a month? Ans. 11 dollars.

7. If an estate, valued at 25340 dollars, be divided equally among 7 heirs, what will be the value of each portion? Ans. 3620 dollars.

8. It is calculated that the number of square miles of land in the whole world is about fifty millions, and that the whole number of inhabitants is six hundred and fifty millions: What is the average number to each square mile of land? Ans. 13.

9. A man intending to go a journey of 1200 miles, would complete the same in 50 days: I demand how many miles he must travel each day? Ans. 24.

QUESTIONS ON THE FOREGOING.

1. What is division? 2. What is the dividend? divisor? quotient? 3. What is *simple* division? 4. What is the first case in simple division? 5. How are the dividend and divisor to be set down? 6. How many figures of the dividend must be taken for the first dividual? 7. Where is the first quotient figure to be placed? 8. What do you take for the second dividual? 9. When the number thus formed is less than the divisor, how do you proceed? 10. What is the second case? 11. How do you place the dividend and divisor in this case; and where do you place the quotient, when found? 12. When you have found a quotient figure, how do you make out a subtrahend; and from what do you subtract it? 13. How then do you make out a new dividual?

14. If, after annexing a figure to the remainder, it is still less than the divisor, how do you proceed? 15. How do you perform division when there are ciphers on the right hand of the divisor; and how do you find the true remainder? 16. How do you perform division when the divisor is a composite number; and how do you find the true or total remainder? 17. How do you prove division? 18. When the amount or value of several articles is given, how do you find the average value of each?

SUPPLEMENT TO MULTIPLICATION.*

Note.—A *Vulgar* or *Common Fraction*, is expressed by two numbers, placed one above the other, with a line between them; as $\frac{1}{2}$, $\frac{3}{4}$, &c.; the upper number being called the *numerator*, and the lower number the *denominator*.—A *mixed number*, is composed of a whole number and a fraction; as $4\frac{1}{2}$, $7\frac{5}{8}$, &c.

PROBLEM I.—*To multiply a whole number by a fraction.*

RULE.—Multiply the whole number by the numerator of the fraction, and divide the product by the denominator; and the quotient will be the required product of the whole number and fraction. When the numerator of the fraction is 1, the answer will be found by merely dividing the whole number by the denominator of the fraction.

EXAMPLES.

1. Multiply 425 by $\frac{2}{3}$; that is, find $\frac{2}{3}$ of 425.

$$\begin{array}{r} 425 \\ 2 \\ \hline 3)850 \\ \hline \text{Ans. } 283\frac{1}{3} \end{array}$$

2. What is $\frac{1}{4}$ of 72? Here $72 \div 4 = 18$, Ans.
3. What is the product of 34 and $\frac{13}{17}$? Ans. 26.
4. What is $\frac{1}{2}$ of 78? Ans. 39.

* The rules which are given in the Supplements to Multiplication and Division, properly belong to *Vulgar Fractions*; but, as the learner will have occasion to use them before he comes to the sections on fractions, I think it necessary to insert them here. The reason of these rules will appear from the rules for the multiplication and division of vulgar fractions.

PROBLEM II.—*To find the product of a whole number and a mixed number.*

RULE.—1. Find the product of the whole numbers, by multiplying them together as usual.

2. Multiply the factor which is a whole number by the fraction belonging to the other factor, as in Prob. I.; add the result to the product of the whole numbers, and the sum will be the total product required.

EXAMPLES.

1. Multiply 271 by $7\frac{3}{4}$.

```
   271                              271
     7                                3
  ----                             ----
  1897   Product by 7.            4)813
  203¾     Do.   by ¾.             ----
  ----                             203¾
Ans. 2100¾  Do.  by 7¾.
```

2. Multiply 84 by $12\frac{1}{2}$. Ans. 1050

3. Multiply 262 by $23\frac{4}{5}$. Ans. $6235\frac{3}{5}$

4. Multiply $65\frac{2}{7}$ by 8. Ans. $522\frac{2}{7}$

SUPPLEMENT TO DIVISION.

To divide a whole number by a mixed number.

RULE.

1. Multiply the integral part of the divisor by the denominator of the fraction, and take the sum for a new divisor.

2. Multiply the dividend by the fraction belonging to the given divisor, and take the product for a new dividend.

3. Divide the new dividend by the new divisor, and the quotient will be the answer sought.

EXAMPLES.

1. Divide 415 by $2\frac{1}{4}$.

Here $2\times4+1=9$, the new divisor; and $415\times4=1660$, the new dividend. Then $1660\div9=184\frac{4}{9}$, the Ans.

2. Divide 630 by $8\frac{2}{5}$. Ans. 75.

3. Divide 4518 by $125\frac{1}{2}$. Ans. 36.

4. Divide 140 by $5\frac{1}{9}$. Ans. $27\frac{18}{46}$.

Practical Questions, to exercise the learner in the preceding rules.

1. The population of the United States, in the year 1820, was 9625734; of which number, 7856269 were whites, 238029 free blacks, and the rest slaves: What was the number of slaves? Ans. 1531436.

2. The number of slaves transported from Western Africa in 25 years, ending in 1819, was stated to be such as would average 60000 a year: What was the whole number transported in that time? Ans. 1500000.

3. A bachelor, at his decease, left an estate worth 12426 dollars; and, in his will, ordered, that 1000 dollars should be given to his niece, and the remainder divided equally between his two nephews. What was the share of each nephew? Ans. 5713 dollars.

4. There are 16 bags of coffee, each weighing 120 pounds; and 8 bags more, weighing each 343 pounds: What is the weight of the whole? Ans. 4664 pounds.

5. There is an excellent well built ship just returned from the Indies. The ship only is valued at 27140 dollars; and one quarter of her cargo is worth 75274 dollars: What is the value of the whole ship and cargo?

Ans. 328236 dollars.

6. I received of A B and C a sum of money: A paid me 54 dollars, B paid me just three times as much as A, and C paid me twice as much as A and B both. Can you tell me how much money C paid me? Ans. 432 dollars.

7. What will $5\frac{1}{2}$ tons of hay come to, at 14 dollars a ton?

Ans. 77 dollars.

8. If 320 rods make a mile, and each rod contains $5\frac{1}{2}$ yards; how many yards are there in a mile?

Ans. 1760.

9. Sold a ship for 7500 dollars, and I owned $\frac{3}{4}$ of her: What is my part of the money? Ans. 5625 dollars.

10. If the number 42 be multiplied by 12, the product divided by 3, the quotient increased by 32, the amount divided by 4, and 49 subtracted from the last quotient, what will then remain? Ans. 1.

Note.—Some Authors on Arithmetic have given particular rules for the Addition, Subtraction, Multiplication, and Division, of *Federal Money*, immediately after the rules for performing the like operations in whole numbers. I do not think it would be useful to insert such rules in this part of the present Treatise, because the Federal currency is purely *decimal*, and most naturally falls in after Decimal Fractions. If any instructor should think it advisable for his pupils to acquire a knowledge of the method of reckoning in Federal money before they learn the rules of Reduction and the succeeding rules which precede Decimal Fractions, he can direct them to omit those rules until they have gone through with the first four rules in Decimals and the succeeding section on Federal Money.

REDUCTION,

Is the changing of numbers from one name or denomination to another, without altering their value. This is chiefly concerned in reducing money, weights, and measures.

When great names or denominations are brought into small, as pounds into ounces, feet into inches, &c., it is called *Reduction Descending*. When small names are brought into great, as ounces into pounds, inches into feet, &c., it is called *Reduction Ascending*. Reduction Descending is performed by Multiplication, and Reduction Ascending, by Division.

Before proceeding to the rules and questions of Reduction, it will be proper to set down the Tables of the denominations of money, weights, and measures. These tables the learner ought to commit to memory, as far as Table 16th, excepting the Supplemental Tables, which may be omitted.

TABLES OF MONEY, WEIGHTS, AND MEASURES.

1. *Federal Money.*

10 Mills, (marked m.) make	1 Cent,	c.
10 Cents,	1 Dime,	dm.
10 Dimes, or 100 cents,	1 Dollar,	D. or $.
10 Dollars,	1 Eagle,	E.

Supplement to Table 1st.

mills.	c.			
10=	1 dm.			
100=	10=	1 D.		
1000=	100=	10=	1 E.	
10000=	1000=	100=	10=	1

Federal Coins, or Coins of the United States.

Names of the Federal Coins.	*Value.*	*Standard Weight.*
GOLD COINS.	D. c.	Pwt. gr.
Eagle,	10.. 0	11 .. 6
Half-Eagle,	5.. 0	5 .. 15
Quarter-Eagle,	2.. 50	2 .. 19½
SILVER COINS.		
Dollar,	1.. 0	17 .. 8
Half-Dollar,	50	8 .. 16
Quarter-Dollar,	25	4 .. 8
Dime,	10	1 .. 17⅖
Half-Dime,	5	20⅘
COPPER COINS.		
Cent,	1	8 .. 16
Half-Cent,	½	4 .. 8

Note.—The proportional value of gold to silver, in all coins current by law in the United-States, is as 15 to 1; that is, 15 ounces of pure silver are equal in value to 1 ounce of gold.

2. *Sterling Money, and old Currencies of the several States.*

4 Farthings, (q.) make	1 Penny,	d.
12 Pence,	1 Shilling,	s.
20 Shillings,	1 Pound,	L. or l.

Supplement.

q.	d.		
4=	1 s.		
48=	12=	1 L.	
960=	240=	20=	1

Pence Table.

d.	s.	d.	d.	s.	d.	d.	s.	d.
20=	1 ..	8	60=	5 ..	0	100=	8 ..	4
30=	2 ..	6	70=	5 ..	10	110=	9 ..	2
40=	3 ..	4	80=	6 ..	8	120=	10 ..	0
50=	4 ..	2	90=	7 ..	6			

Note.—Farthings are usually written as fractional parts of a penny; 1 farthing being $\frac{1}{4}$, 2 farthings $\frac{1}{2}$, and 3 farthings $\frac{3}{4}$ of a penny.—People of business often write shillings at the left hand of a stroke, and pence at the right; thus, 15/4, is 15 shillings and 4 pence.

3. Troy Weight.

24 Grains, (gr.) make	1 Penny-weight,	pwt.
20 Penny-weights,	1 Ounce,	oz.
12 Ounces,	1 Pound.	lb.

Supplement.

gr.	pwt.		
24=	1 oz.		
480=	20=	1 lb.	
5760=	240=	12=	1

Note.—By this weight are weighed Gold, Silver, Jewels, Electuaries, and all liquors.

4. Refiners' Weight.

24 Blanks, make	1 Perrot.
20 Perrots,	1 Mite.
20 Mites	1 Grain.

Note 1.—What Refiners call a *Carat*, is the 24th part of any quantity, or weight.

Note 2.—The fineness of gold and silver is tried by fire. Gold that will abide the fire without loss, is accounted 24 carats fine: If it lose 2 carats in the trial, it is said to be 22 carats fine, &c. A pound of silver which loses nothing in the trial is said to be 12 ounces fine; but if it lose 5 penny-weights, it is 11oz. 15pwt. fine, &c.

Note 3.—*Alloy*, or *Allay*, is some base metal with which gold or silver is mixed to abate its fineness.*

* Gold and silver, in their purity, are so very soft and flexible, that they are not so useful, either in coin or otherwise, (except to beat into leaf gold or silver,) as when they are allayed, or mixed and hardened

5. *Apothecaries' Weight.*

20 Grains, (gr.) make	1 Scruple,	sc. or ℈	
3 Scruples,	1 Dram,	dr. or ʒ	
8 Drams,	1 Ounce,	oz. or ℥	
12 Ounces,	1 Pound.	lb.	

Supplement.

gr.	sc.			
20=	1 dr.			
60=	3=	1 oz.		
480=	24=	8=	1 lb.	
5760=	288=	96=	12=	1

Note 1.—Apothecaries make use of this weight in compounding their medicines; but they buy and sell their drugs by Avoirdupois weight.

Note 2.—The Apothecaries' pound, ounce, and grain, are the same as the pound, ounce, and grain, Troy.

6. *Avoirdupois Weight.*

The denominations of Avoirdupois weight are usually reckoned as follows:—

16 Drams, (dr.) make	1 Ounce,	oz.
16 Ounces,	1 Pound,	lb.
28 Pounds,	1 Quarter of a hundred weight,	qr.
4 Quarters, or 112 lb.,	1 Hundred weight,	cwt.
20 Hundred wt., or 2240 lb.,	1 Ton,	T.

In some of the United-States, the denominations of Avoirdupois weight, are, by law, as follows:

16 Drams =1 Ounce.
16 Ounces =1 Pound.
25 Pounds =1 Quarter of a cwt.
20 Hundred wt. or 2000 lb. =1 Ton.

---

with copper or brass.—In England, 22 carats of pure gold and 2 carats of copper, melted together, is the standard for gold coin; the alloy being one twelfth part of the mixture: and 11 oz 2 pwt. of fine silver melted with 18 pwt. of copper, is the standard for silver coin. In the United-States, the standard for gold coin, is 11 parts pure gold and 1 part alloy; the alloy to be silver and copper mixed, not exceeding one half copper. The standard for silver coin, is 1485 parts fine silver, and 179 parts alloy; the alloy to be pure copper. The copper coins of the United States are to be pure copper.

Supplement.

If a quarter of a cwt. be 28 pounds; then,

dr.	oz.			
16=	1	lb.		
256=	16=	1	qr.	
7168=	448=	28=	1 cwt.	
28672=	1792=	112=	4=1	T.
573440=	35840=	2240=	80=20=	1

If 25 pounds make 1 quarter of a cwt.; then,

dr.	oz.			
16=	1	lb.		
256=	16=	1	qr.	
6400=	400=	25=	1 cwt.	
25600=	1600=	100=	4= 1	T.
512000=	32000=	2000=	80=20=	1

Note 1.—In the examples given in this Work, 1 quarter of a cwt. is allowed to be 28 pounds, and the higher denominations of Avoirdupois weight are reckoned accordingly, excepting where it is otherwise mentioned in the question.

Note 2.—By Avoirdupois weight are weighed all coarse and bulky articles ;such as provisions, groceries, hay, &c. and all metals except gold and silver.

Note 3.—The Avoirdupois pound is heavier than the Troy pound; but the Avoirdupois ounce is lighter than the Troy ounce. 144 pounds Avoirdupois are equal to 175 pounds Troy; and 192 ounces Avoirdupois, to 175 ounces Troy. Therefore, 1 lb. Avoirdupois is equal to 7000 grains, or 1 lb. 2 oz. 11 pwt. 16 gr. Troy; and 1 lb. Troy is equal to 13 oz. $2\frac{114}{175}$ dr. Avoirdupois.

Note 4.—There are some other denominations in Avoirdupois weight, which are made use of in many places, in weighing some particular kinds of goods, &c., viz. the following:

	lb.		lb.
A barrel of flour is	196	A faggot of steel is	120
Do. of pork is	200	A gallon of train oil is	$7\frac{1}{2}$
Do. of beef is	200	A stone is	14
Do. of gunpowder is	112	A todd is	28
A quintal of fish is	100	A weigh is	182
A fother of lead is		A sack is	364
$19\frac{1}{2}$ cwt. or	2184	A last is	4368

7. *Long Measure.*

3	Barley corns (b. c.) make	1 Inch,	in.
12	Inches,	1 Foot,	ft.
3	Feet,	1 Yard,	yd.
$3\frac{3}{10}$	Feet,	1 Pace,	pac.
$5\frac{1}{2}$ 5	Yards, or $16\frac{1}{2}$ feet, or Paces,	1 Rod, Perch, or Pole,	rd.
40	Rods,	1 Furlong,	fur.
8	Furlongs, or 320 rods,	1 Statute Mile,	m.
$69\frac{1}{5}$	Statute miles,	1 Degree on the Earth,	deg.
360	Degrees, or 24912 m.=The circumference of the Earth.*		

Supplement.

b. c.	in.					
3=	1	ft.				
36=	12=	1	yd.			
108=	36=	3=	1	rd.		
594=	198=	$16\frac{1}{2}$=	$5\frac{1}{2}$=	1	fur.	
23760=	7920=	660=	220=	40=	1	m.
190080=	63360=	5280=	1760=	320=	8=	1

Note 1.—The use of Long Measure, is to measure the distance of places, or any thing where length is considered without regard to breadth.

Note 2.—Distances are frequently measured by a chain, 4 rods, or 66 feet long, containing 100 links. In measuring the height of horses, 4 inches make 1 hand. In measuring depths, 6 feet make 1 fathom.

8. *Cloth Measure.*

$2\frac{1}{4}$	Inches, (in.) make	1 Nail,	na.
4	Nails, or 9 inches,	1 Quarter of a yard,	qr.
4	Quarters, or 36 inches,	1 Yard,	yd.
3	Quarters,	1 Flemish Ell,	Fl. e.
5	Quarters,	1 English Ell,	E. e.
6	Quarters,	1 French Ell,	Fr. e.

* In this Work, where calculations are made which have any relation to the magnitude of the earth, its circumference is supposed to be 24912 American or English miles, and its diameter 7930 miles.

Supplement.

na. qr.
4=1 yd.
16=4=1
12=3= $\frac{3}{4}$=1 Fl. e.
20=5=1$\frac{1}{4}$=1 E. e.
24=6=1$\frac{1}{2}$=1 Fr. e.

Note.—By this measure, are measured cloth, tapes, &c.

9. *Superficial, or Square Measure.*

144 Square Inches, (sq. in.) make	1 Square Foot,	sq. ft.
9 Square Feet,	1 Square Yard,	sq. yd.
30$\frac{1}{4}$ Square Yards, or 272$\frac{1}{4}$ Square Feet,	1 Square Rod, Perch, or Pole,	sq. rd.
40 Square Rods,	1 Rood,	R.
4 Roods, or 160 square rods,	1 Acre,	A.
640 Acres,	1 Square Mile,	sq. m.

Also,

10000 Square Links, or 16 Square Rods, make	1 Square Chain.
10 Square Chains,	1 Acre.
6400 Square Chains,	1 Square Mile.

Supplement.

sq. in. sq. ft.
144= 1 sq. yd.
1296= 9= 1 sq. rd.
39204= 272$\frac{1}{4}$= 30$\frac{1}{4}$= 1 R.
1568160=10890=1210= 40=1 A.
6272640=43560=4840=160=4=1

Note.—This measure is used to ascertain the quantity of any surface, or any thing which has length and breadth, without regard to thickness; as the floor of a room, or a piece of land, &c. The length and breadth being multiplied together, to make the area, or superficial content. Hence, 144 square inches make 1 square foot; because 12 inches in length multiplied by 12 inches in breadth, gives 144 square inches; and so of the other denominations.

10. *Solid, or Cubic Measure.*

1728 Cubic Inches, (cub. in.) make	1 Cubic Foot,	cub. ft.
27 Cubic Feet,	1 Cubic Yard,	cub. yd.
40 Feet of round timber, or 50 Feet of hewn timber,	1 Ton or Load,	T.
128 Cubic Feet of wood,	1 Cord,	Cd.

Note.—This measure is used to ascertain the quantity of any solid, or any thing which has length, breadth, and thickness; and to regulate all measures of capacity, of whatever form. A solid or cubic foot contains 1728 cubic inches; because it is 12 inches long, 12 inches broad, and 12 inches thick, and 12×12×12, or the cube of 12, is 1728.

11. *Wine Measure.*

4 Gills, (gil.) make	1 Pint,	pt.
2 Pints,	1 Quart,	qt.
4 Quarts,	1 Gallon,	gal.
31½ Gallons,	1 Barrel,	bar.
42 Gallons,	1 Tierce,	tier.
2 Barrels, or 63 gal.,	1 Hogshead,*	hhd.
2 Hogsheads, or 4 bar.,	1 Pipe, or Butt,	p. or b.
2 Pipes,	1 Tun,	Tun.

Supplement.

gil.	pt.			
4=	1	qt.		
8=	2=	1	gal.	
32=	8=	4=	1	bar.
1008=	252=	126=	31½=	1

Note.—In the United-States, the ordinary measure for all liquids is Wine Measure. A gallon of this measure contains 231 cubic inches.

12. *Ale and Beer Measure.*

2 Pints make	1 Quart.
4 Quarts,	1 Gallon.
36 Gallons,	1 Barrel.
3 Barrels,	1 Butt.
2 Butts,	1 Tun.

* In the United States, a hogshead is no determinate quantity. Casks that are called hogsheads, usually hold more than 100 gallons.

Note.—This measure is used, in some places, for beer and some other liquids. A gallon of this measure contains 282 cubic inches.

13. *Dry, or Corn Measure.*

2 Pints, (pt.) make	1 Quart,	qt.
4 Quarts,	1 Gallon,	gal.
8 Quarts, or 2 gal.,	1 Peck,	pk.
4 Pecks,	1 Bushel,*	bush.
36 Bushels,	1 Chaldron of coal,	Chal.

Supplement.

pt.	qt.			
2 =	1	gal.		
8 =	4 =	1	pk.	
16 =	8 =	2 =	1	bush.
64 =	32 =	8 =	4 =	1

Note.—This measure is used for corn, seeds, fruit, roots, salt, coals, oysters, &c.—Corn, seeds, salt, &c. are measured by striked measure; but fruit, roots, coals, oysters, &c., are heaped to a handsome rounding measure.--A gallon of this measure, contains 268⅘, or 268.8 cubic inches; and a bushel contains 2150⅖, or 2150.4 cubic inches.

14. *Time.*

60 Seconds, (sec.) make	1 Minute,	min.
60 Minutes,	1 Hour,	h.
24 Hours,	1 Day,	da.
7 Days,	1 Week,	w.
52 Weeks and 1 day, or 365 Days,	1 Common Year,	yr.
365¼ Days,	1 Julian Year.	
365 Days, 5h. 48 min. 48sec.=	1 Solar Year.	
100 Years,	1 Century.	

Note 1.—The Solar year is the true length of a year, or the time in which the Sun passes through the 12 signs of the Zodiac. The Julian year takes its name from Julius Cæsar, who ordered that 3 civil years in every 4 should consist of

* In England, 8 bushels are equal to 1 quarter; 5 quarters, to 1 wey; 2 weys, or 10 quarters, to 1 last.

365 days each, and every fourth year, of 366 days; which made the average length of each civil year 365¼ days, or 365 days, 6 hours. Each year of 366 days, was denominated *Bissextile* or *Leap year*, and the other years were called *common* years. This method of computing time is still in use; but as the true length of a year is 11 minutes and 12 seconds less than 365¼ days, when the *new style*, or method of reckoning time, was adopted in England, in the year 1752, it was determined that three leap years in every four succeeding centuries should be reduced to common years; which will very nearly balance the error in the former mode of reckoning.

Note 2.—By the calendar the year is divided into 12 months, as follows:

1st Month,	January,	hath 31 days.
2d Month,	February,	28 or 29.
3d Month,	March,	31
4th Month,	April,	30
5th Month,	May,	31
6th Month,	June,	30
7th Month,	July,	31
8th Month,	August,	31
9th Month,	September,	30
10th Month,	October,	31
11th Month,	November,	30
12th Month,	December,	31

In leap years, February has 29 days; in other years only 28.

The number of days in each month may be known by committing to memory the following lines:

The ninth, fourth, eleventh, and sixth,
Have thirty days to each affix'd,
And every other thirty-one,
Except the second month alone,
Which has but twenty-eight, in fine,
Till leap year gives it twenty-nine.

Or the following:

Thirty days hath September,
April, June, and November;
All the rest have thirty-one,

Except February alone,
Which hath four and twenty-four,
And every fourth year one day more.

Note 3.—*To know whether any given year is leap year, or not;* divide the year of our Lord by 4, and if the division terminates without a remainder, it is leap year; otherwise it is a common year. But, when the number of years has two or more ciphers at the right hand, then you must reject two ciphers at the right hand, before you divide by 4.

15. *Circular Measure, or Motion.*

60 Seconds, (″) make	1 Minute,	min., or ′
60 Minutes,	1 Degree,	deg. or °
30 Degrees,	1 Sign,	S.
12 Signs, or 360 degrees, =	The whole great circle of the Zodiac.	

Note 1.—This table is used by Astronomers, Navigators, &c.

Note 2.—The circumference of every circle is supposed to be divided into 360 equal parts, called *degrees;* each degree into 60 equal parts, called *minutes;* each minute into 60 *seconds,* and each second into 60 *thirds,* &c.

16. *Of Particulars.*

12 Particular things make	1 Dozen,	Doz.
20 Do.	1 Score,	
12 Dozen,	1 Gross,	
12 Gross, or 144 dozen,	1 Great Gross.	

17. *Paper.*

24 Sheets of paper make	1 Quire.
20 Quires	1 Ream.
10 Reams, or 200 quires,	1 Bale.

18. *Books.*

A sheet of paper folded into 2 leaves, is called	Folio,	Fol.
4 do.	Quarto,	4to.
8 do.	Octavo,	8vo.
12 do.	Duodecimo,	12mo.

The smaller books are called 18's, 24's, 32's, &c. according to the number of leaves in a sheet.

19. *Of the Moneys of various foreign countries.*

Of France.

Accounts were formerly kept in France in Livres, Sols or Sous, and Deniers.

12 Deniers, or pence,=1 Sol, or shilling.
20 Sols, or shillings,=1 Livre, or pound.

Note 1.—A livre is equal to 18½ cents, or $ .185.

The present money of account in France is Francs, and Centimes or hundredths.

10 Centimes=	1 Decime.
100 Centimes, or 10 decimes,=	1 Franc.
80 Francs=	81 Livres.

Note 2.—In the United-States, the franc is usually estimated at 18¾ cents, or $ .1875, and the five franc pieces at 93¾ cents.

Of the Netherlands.

Accounts are kept here in Guilders, Stivers, Grotes, and Phennings.

8 Phennings, make	1 Grote.
2 Grotes,	1 Stiver.
20 Stivers,	1 Guilder, or Florin.

Note.—A guilder is=40 cents.

Of Spain.

Accounts are kept in Spain in Piastres, Rials, and Marvadies.

34 Marvadies, make	1 Rial.
8 Rials,	1 Piastre, or piece of eight.

Note.—There are two kinds of money in Spain; viz. *money of vellon*, or *current* money, and *money of plate*, or hard money. Plate money is more valuable than vellon, in the ratio of 32 to 17. Thus, 17 rials of plate are equivalent to 32 rials vellon; and so of the other denominations of money. In the United-States 1 rial of plate=10 cents, and 1 rial vellon=5 cents.

Of Portugal.

In Portugal accounts are kept in Reas and Milreas, 1000

reas being equal to one milrea.—In the United-States, a milrea is valued at $1.24.

English gold and silver Coins, with their full weight, and value.

Gold Coins.	Value in sterling mon. L. s. d.	Value in fed. mon. $ c.	Weight. pwt. gr.
A Sovereign,	1 .. 0 .. 0	4 .. 44$\frac{4}{9}$	5 .. 3$\frac{1}{4}$
A Guinea,	1 .. 1 .. 0	4 .. 66$\frac{2}{3}$	5 .. 9$\frac{1}{2}$
Half-Guinea,	10 .. 6	2 .. 33$\frac{1}{3}$	2 .. 16$\frac{3}{4}$
Quarter do.,	5 .. 3	1 .. 16$\frac{2}{3}$	1 .. 8$\frac{1}{4}$
Seven Shillings,	7 .. 0	1 .. 55$\frac{5}{9}$	1 .. 19$\frac{1}{6}$
Silver Coins.			
A Crown,*	5 .. 0	1 .. 11$\frac{1}{9}$	19 .. 8$\frac{1}{2}$
Half-Crown,	2 .. 6	55$\frac{5}{9}$	9 .. 16$\frac{1}{4}$
Shilling,	1 .. 0	22$\frac{2}{9}$	3 .. 21
Sixpence,	6	11$\frac{1}{9}$	1 .. 22$\frac{1}{2}$

* The crown passes in the U. S. at 1 dollar and 10 cents, and the half-crown at 55 cents.

Gold Coins of other foreign countries.

	Value. $ c.	Weight. pwt. gr.
Johannes, or Joe,	16 .. 0	18 .. 0
Half do.,	8 .. 0	9 .. 0
Doubloon,	14 .. 93$\frac{1}{3}$	17 .. 0
Moidore,	6 .. 0	6 .. 18
Spanish Pistole,	3 .. 77$\frac{1}{3}$	4 .. 6
French Pistole,	3 .. 66$\frac{2}{3}$	4 .. 4
French Guinea,	4 .. 60	5 .. 5
Louis d'or,	4 .. 44$\frac{4}{9}$	

Silver Coins, &c.

	Cents.		Cents.
French Crown,	= 110	Ruble of Russia, =	100
Spanish Dollar,	100	Piastre of Turkey,	88$\frac{8}{9}$
Pistareen,	18	Sequin of Arabia,	164
Rix-dollar of Sweden, or Denmark,	100	Pagoda of India,	184
Mark Banco of Hamburgh,	33$\frac{1}{3}$	Rupee of Bengal,	50
		Tale of China,	148

Note 1.—The values of some of the foreign coins often vary, according to the rate of exchange.

Note 2.—The Act of Congress of April 29, 1816, regulating the currency within the United-States of the gold coins of Great-Britain, France, &c. enacted,

That, of the gold coins of Great-Britain and Portugal,

27 grains=$1, or 1 pwt.=88⅘ cents;

Of France, 27½ do. = do. =87¼ do.

Of Spain, 28½ do. = do. =84 do.

Crowns of France, weighing 449 grains,=110 cents, or 1 oz.=117 cents.

Five franc pieces, weighing 386 grains,=93.3 cents, or 1 oz.=116 cents.

The Spanish dollar, weighing not less than 415 grains,= 1 Federal dollar.

Note 3.—The standard price of gold in England, is 3*l.* 17s. 10½d. an ounce, and of silver 5s. 2d. an oz. The standard coin of France is to contain one-tenth of alloy, and the standard value of gold to silver is 15 to 1.

20. *Weights and Measures of several foreign cities and nations, compared with the American Weights and Measures.*

Weights.

lb.	American lb.	oz.	lb.	American lb.	oz.
100 of England, Scotland and Ireland,=	100	0	100 of Russia,=	88	4
100 of Amsterdam, Paris, Bordeaux, &c.	109	8	100 of Vienna,	123	0
100 of Antwerp,	104	3	100 of Genoa,	73	0
100 of Lyons,	94	3	100 of Leghorn,	77	4
100 of Rochelle,	110	9	100 of Venice,	65	11
100 of Marseilles,	88	11	100 of Naples,	64	10
100 of Geneva,	123	0	100 of Portugal,	77	4
100 of Hamburgh,	107	4	100 of Spain,	97	0
			A Spanish Arrobe,	24	4
			A Russian Pood,	35	8

Measures.

	American yards.
100 Aunes or Ells of England, - - - =	125
100 ——— of Holland, - - -	75

		American yards.
100	Aunes or Ells of France, - - - -	= 128½
100	——— of Hamburgh, and Frankfort,	62½
100	——— of Geneva, - - - -	124¾
100	——— of Sweden, - - -	65¾
100	Metres of France, - - - - - -	109⅓+
100	Canes of Marseilles and Montpelier, - -	214½
100	—— of Genoa, of 9 palms, - - -	245¼
100	—— of Rome, - - - - -	227¼
100	Varas of Spain, - -	93¾
100	—— of Portugal, - - - - -	123
100	Cavidos of Portugal, - - - - -	75
100	Brasses of Venice, - - - -	73½
100	—— of Florence and Leghorn, - -	64
1	English mile=1 American mile, - -	1760
1	Scotch mile, - - -	1984
1	Irish mile, - - - -	2240
1	German mile=4 American miles, - -	7040
1	Dutch and Polish mile=3½ American miles,	6160
1	Swedish and Danish mile=5½ American miles,	9680
1	Spanish league=3⅔ American miles, - -	6453⅓
1	French league=2¾ American miles nearly.	
1	Russian verst=¾ of an American mile,	1320
11	Irish miles=14 American miles.	
30	Scotch miles=31 do.	
121	Irish acres=196 American and English acres.	
48	Scotch acres=61 do. nearly.	
1	French are=119.6046+ American square yards.	
607	French ares=15 American acres, nearly.	

21. *Ancient Measures, Weights, and Coins, mentioned in the Bible, and in the writings of Josephus.*

MEASURES OF LENGTH.

		American ft.	in.
A cubit, the standard,		= 1 ..	9
A zereth, or long span,	= ½ of a cubit,	= 0 ..	10½
A short span	= ⅓ do.	= 0 ..	7
A hand's breadth	= ⅙ do.	= 0 ..	3½
A thumb's breadth	= 1/18 do.	= 0 ..	1⅙
A finger's breadth	= 1/24 do.	= 0 ..	⅞
A fathom	= 4 cubits,	= 7 ..	0

			American ft.	in.
Ezekiel's canneh, or reed,	= 6	cubits,	= 10 ..	6
Arabian canneh, or pole,	= 8	do.	= 14 ..	0
A schaenus,* line, or chain,	= 80	do.	= 140 ..	0
A stadium, or furlong,	= 400	do.	= 700 ..	0
A sabbath-day's journey	= 2000	do.	= 3500 ..	0
A day's journey	= 96000	do.	= 168000 ..	0
A mile	= 4000	do.	= 7000 ..	0
A parasang	= 12000	do.	= 21000 ..	0

MEASURES OF CAPACITY.

Measures of liquids.

	Wine pints.
A homer, or cor,	= 605.3
A bath = $\frac{1}{10}$ of a cor	= 60.5
A hin = $\frac{1}{60}$ do.	= 10.1
A log = $\frac{1}{720}$ do.	= .8
A firkin	= 7.2

Measures of things dry.

	Dry pints.
A homer	= 513.6
A lethech = $\frac{1}{2}$ homer	= 256.8
An ephah = $\frac{1}{10}$ do.	= 51.4
A seah = $\frac{1}{30}$ do.	= 17.1
An omer, or Assaron = $\frac{1}{100}$ do.	= 5.1
A cab = $\frac{1}{180}$ do.	= 2.9

WEIGHTS.

	Troy wt. lb.	oz.	Avoirdupois wt. lb.	oz.
A shekel	= 0 ..	0.455 =	0 ..	$\frac{1}{2}$
A maneh=60 shekels	= 2 ..	3.321 =	1 ..	14
A talent=3000 do.	=113 ..	10.071 =	93 ..	12

MONEY.

			$
A shekel of silver		=	0.507
A bekah	= $\frac{1}{2}$ shek.	=	.253
A zuza	= $\frac{1}{4}$ do.	=	.127
A gerah	= $\frac{1}{20}$ do.	=	.025
A taneh, or mina,	= 60 do.	=	30.41$\frac{2}{3}$
A talent of silver	=3000 do.	=	1520.83$\frac{1}{3}$
A shekel of Gold		=	8.51$\frac{1}{9}$
A talent of Gold	=3000 shek.	=	25533$\frac{1}{3}$
A piece of silver (drachm)		=	.144

* The schaenus was the Egyptian line for land measure used to divide inheritances: they were of different lengths, but the shortest and most useful was 80 cubits.

				$
Tribute money (didrachm)	=	2 drac.	=	.287
A piece of silver (stater)	=	4 do.	=	.574
A pound (mina)	=	100 do.	=	14.352
A penny (denarius)	=	1 do.	=	.144
A farthing (assarium)	=	$\frac{1}{20}$ do.	=	.007
A farthing (quadrans)	=	$\frac{1}{40}$ do.	=	.004
A mite	=	$\frac{1}{80}$ do.	=	.002

REDUCTION DESCENDING.

RULE.*

1. *When the given quantity consists of several denominations;* multiply the number of the highest denomination by the number of units which it takes of the next lower denomination to make 1 of that higher; add to the product the given number, if any, of this lower denomination, and set down the amount. In like manner reduce this amount to the next lower denomination, adding in the given number, if any, of this denomination; and so proceed until the given quantity is reduced to the denomination required.

2. *When the given quantity is of one denomination only;* then it may be either reduced gradually from one denomination to another, as directed above; or, it may be reduced at once to the denomination required, by multiplying it by the number of units which 1 of the denomination given makes of that required. The multipliers to be used in working by the latter method, will be found in the Supplemental Tables.

---

* The reason of this rule is very evident: For instance, the reduction of Sterling Money, as 1 pound is eqnal to 20 shillings, 1 shilling to 12 pence, and 1 penny to 4 farthings; therefore, pounds are reduced to shillings by multiplying them by 20, shillings to pence by multiplying by 12; and pence to farthings by multiplying by 4; and the contrary by division: and this reasoning will hold true in the reduction of numbers consisting of any denominations whatever. The rule for Reduction Ascending is simply the reverse of this, and equally evident.

EXAMPLES.

Money.

1. Reduce 27 *l.* 14 s. 6 d. to farthings.

Operation.

```
L.  s.  d.
27 .. 14 .. 6
 20 s.=1 l.
 ---
540
 14
 ---
554 s.
 12 d.=1 s.
 ----
6648
   6
 ----
6654 d.
   4 q.=1 d.
 -----
26616 q. Ans.
```

Explanation.—Here, because 20 shillings make 1 pound, I multiply the 27 pounds by 20, to bring them into shillings; and the product is 540 shillings, to which I add the 14 shillings of the given sum, and the amount is 554 shillings. Then, as 12 pence make 1 shilling, I multiply those 554 shillings by 12, to reduce them to pence, and the product is 6648 pence, to which I add the 6 pence of the given sum, and the amount is 6654 pence. Then, as 4 farthings make 1 penny, I multiply those 6654 pence by 4, to reduce them to farthings, and have 26616 farthings; and there being no farthings to add to these, the work is done. So the answer is 26616 farthings.

2. Reduce 498 *l.* to pence.

```
L.
498
 240 d.=1 l.
 ------
19920
996
 ------
119520 d. Ans.
```

Here, the given quantity, or sum, is of one denomination only; and because 240 pence make 1 pound, I multiply the 498 pounds by 240, to reduce them to pence, and have 119520 pence for the answer.

N. B. The number of pence in a pound may be found in the Supplement to Table 2d.

Note.—In reducing quantities which consist of several denominations; when the numbers of the lower denominations are small, they may be added in *mentally*, (*i. e. in the mind,*) in multiplying; as in the next example.

3. Reduce 4*l.* 17s. 4d. to pence.

L. s. d.	
4 .. 17 .. 4	Here, in multiplying the 4 pounds by 20, I add in the 17 shillings; and, in multiplying the 97 shillings by 12, I add in the 4 pence.
20	
—	
97s.	
12	
—	
1168d. Ans.	

4. Reduce 31*l.* 11s. 10d. 1q. to farthings. Ans. 30329q.

5. Reduce 64*l.* 0s. 7d. to pence. Ans. 15367d.

6. In 18s. 9d. how many pence, and farthings? Ans. 225d., 900q.

7. In 48 guineas, each 21 shillings, how many shillings and pence? Ans. 1008s., 12096d.

Note.—From the foregoing Rule, and the first Contraction in Simple Multiplication, it appears that *Federal Money* may be reduced from higher to lower denominations by annexing as many ciphers as there are places from the denomination given to that required; or, if the given sum be of different denominations, by annexing the several figures of all the denominations in their order, and continuing with ciphers, if necessary, to the denomination required; as in the following examples.

8. In 5 dollars how many cents?

Here, $5\times100=500$ cents, the Ans.

Or, annexing two ciphers to the number of dollars, (which is the same as multiplying by 100,) gives 500, the number of cents, as before.

9. In 4 dollars, 8 dimes, how many dimes?

Annexing the 8 dimes to the 4 dollars (=40 dimes) gives 48 dimes for the answer.

10. Reduce 7 eagles and 5 dimes to cents. Ans. 7050 c.

11. Reduce 14 dollars, 2 dimes, to mills. Ans. 14200 m.

12. Reduce 8 dollars and 4 cents to cents. Ans. 804 c.

Troy Weight.

13. Reduce 3lb. 10oz. 12pwt. to penny-weights.

```
lb.  oz. pwt.
 3 .. 10 .. 12
12 oz.=1 lb.
—
46 oz.
  20 pwt.=1 oz.
———
932 pwt. Ans.
```

Here, in multiplying the 3lb. by 12, to reduce them to ounces, I add in the 10 ounces; and in multiplying the 46 oz. by 20, to reduce them to penny-weights, I add in the 12 penny-weights.

14. Reduce 4lb. 7 oz. 0 pwt. 8gr. to grains.
Ans. 26408 gr.

15. In 8oz. how many grains?
1 oz.=480 grains; therefore 480×8=3840 gr. the Ans.

Apothecaries' Weight.

16. In 6oz. 4dr. 1 sc. how many scruples?

```
oz. dr. sc.
 6 .. 4 .. 1
 8 dr.=1 oz.
—
52 dr.
  3 sc.=1 dr.
———
Ans. 157 scruples.
```

17. In 9 lb. 8oz. 1 dr. 2sc. 19gr. how many grains?
Ans. 55799 gr.

Long Measure.

18. In 42 miles, 5 furlongs, 8 rods, how many rods?

```
m. fur. rd.
42 .. 5 .. 8
  8 fur.=1 mile.
———
341
  40 rd.=1 fur.
———
Ans. 13648 rods.
```

19. Reduce 5 leagues, 2 miles, 4 furlongs, to rods.
Ans. 5600 rd.

20. In 18 rods, 2 yards, how many yards?

```
  rd.  yd.
 2)18 .. 2
   5½ yd.=1 rd.
  ----
   9=½ of 18 rd.
  90
  ----
  99
  +2
  ----
 101 yd. Ans.
```

Here the multiplier contains a fraction, and I perform the multiplication according to the rule for Prob. 2d in the Supplement to Simple Multiplication.

21. Reduce 22 rods to feet. Ans. 363 ft.

22. Reduce 2 miles, 7 fur. 30 rd. to feet. Ans. 15675 ft.

23. Reduce 1520 miles, 1 fur. 36 rd. 1 yd. to yards. Ans. 2675619 yd.

24. How many barley-corns will reach round the earth, supposing its circumference to be 24912 miles?

1 mile=190080 b. c.; therefore, 24912×190080= 4735272960 b. c. the Ans.

Cloth Measure.

25. Reduce 4 yards, 3 qr. to quarters. Ans. 19 qr.

26. Reduce 75 English ells to nails. Ans. 1500 na.

27. Reduce 39 French ells to inches. Ans. 2106 in.

Square Measure.

28. In 29 acres, 3 roods, 19 square rods, how many square rods? Ans. 4779.

29. In 5 square yards, 8 sq. feet, 140 sq. inches, how many square inches? Ans. 7772.

30. The area of the State of Connecticut is about 4764 square miles: reduce these to acres. Ans. 3048960 A.

31. St. Peter's church, (so called,) in the city of Rome, is 730 feet long, 520 wide, and 450 high; and it occupies about 8 acres and 2 roods of ground: Reduce these acres and roods to square feet.

8 acres, 2 roods,=34 roods; and 1 rood=10890 square feet: Hence, 10890×34=370260 sq. ft. the Ans.

Cubic Measure.

32. In 5 cords and 47 cubic feet of wood, how many cubic feet? Ans. 687.

33. Reduce 2 cords to cubic inches. Ans. 442368.
34. Reduce 15 tons of round timber to feet. Ans. 600.

Wine Measure.

35. In 5 gallons, 2 qt. 1 pt. of wine, how many gills? Ans. 180.

36. It is said that 20000 pipes of Port wine are annually exported from Oporto in Portugal: how many pints are in that quantity? Ans. 20160000.

37. How many pint bottles may be filled from a hogshead of cider? Ans. 504.

38. How many half gills in a gallon? Ans. 64.

Dry Measure.

39. Reduce 75 bushels, 3 pecks, to quarts. Ans. 2424.
40. Reduce 4 bushels and 1 quart to pints. Ans. 258.
41. Reduce 10 chaldrons to bushels. Ans. 360.

Time.

42. How many seconds in a solar year; it being 365 days, 5 hours, 48 min. 48 sec.? Ans. 31556928.

43. The time which elapses between two successive changes of the Moon, (called a synodic revolution of the Moon,) is, at an average, 29 days, 12 hours, 44 min. 3 sec.: How many seconds are in that time? Ans. 2551443.

44. Reduce 52 weeks to days. Ans. 364.
45. Reduce 8 years to calendar months. Ans. 96.
46. In 18 centuries and 31 years, how many years? Ans. 1831.

47. How many days from the birth of Christ to Christmas, A. D. 1832, allowing each of the 1832 years to consist of 365¼ days? Ans. 669138.

Circular Measure.

48. The Moon moves each day, at an average, through 13° .. 10′ .. 35″ of the Zodiac: Reduce these degrees, &c. to seconds. Ans. 47435.

49. Reduce 9 signs, 13° .. 25′ to seconds. Ans. 1020300.

50. Reduce 5 degrees and 48 sec. to seconds. Ans. 18048.

REDUCTION ASCENDING.

RULE.

Divide the given quantity by the number which it takes of that denomination to make 1 of the next higher, and the quotient will be of the said higher denomination, and the remainder, if any, will be of the same denomination as the dividend. In like manner reduce the quotient, thus found, to the next higher denomination, setting down the remainder, if any; and so proceed until the quantity is reduced to the denomination required: then, the last quotient, together with the several remainders, (if any,) of the lower denominations, will be the answer.

Or, instead of reducing the given quantity gradually from one denomination to another, as directed above, you may divide it by a number which will reduce it at once to the denomination required; that is, divide by the number which it takes of the denomination given to make 1 of that required. The divisors to be used in working by this last rule, will be found in the Supplemental Tables.

Note 1.—The remainder is always of the same denomination as the dividend.

Note 2.—When the divisor contains a fraction, the division must be performed according to the rule given in the Supplement to Simple Division. If, after dividing in this manner, there is a remainder, it must be divided by the denominator of the fraction belonging to the given divisor, and the quotient, thus found, will be the true remainder, or number of that denomination, to be set down in the answer.

PROOF.—Reduction Ascending and Descending are the reverse of each other, and hence they reciprocally prove each other. The following questions are the reverse of the corresponding questions in Reduction Descending, and the answers to the latter may serve as proofs of those to the former.

EXAMPLES.

Money.

1. Reduce 26616 farthings to pounds.

Operation.
4)26616 q.

12)6654 d.

2,0)55,4 s. 6d.

27*l.* 14s.

Ans. 27*l.* 14s. 6d.

Explanation.—Here, because 4 farthings make 1 penny, I divide the 26616 farthings by 4, to bring them into pence; and the quotient is 6654 pence. Then, because 12 pence make 1 shilling, I divide those 6654 pence by 12, to reduce them to shillings; and the quotient is 554 shillings, and 6 pence remain, which I set down a little to the right hand of the quotient. Then, because 20 shillings make 1 pound, I divide the 554 shillings by 20, to reduce them to pounds; and the quotient is 27 pounds, and 14 shillings remain. So, the answer is 27*l.* 14s. 6d.—The operation may be proved by reducing the answer to farthings, by Reduction Descending.

Reduce		L.	s.	d.	q.
119520 pence to pounds.	Ans.	498 ..	0 ..	0 ..	0
1168 pence to pounds.	Ans.	4 ..	17 ..	4 ..	0
30329 farthings to pounds.	Ans.	31 ..	11 ..	10 ..	1
15367 pence to pounds.	Ans.	64 ..	0 ..	7 ..	0
900 farthings to shillings.	Ans.		18 ..	9 ..	0

7. In 12096 pence, how many guineas at 21 shillings each? Ans. 48.

Note.—From the foregoing Rule, and the 1st Contraction in Simple Division, it appears that *Federal Money* is reduced from lower to higher denominations by merely cutting off as many places as the given denomination stands to the right of that required; the figures cut off belonging to their respective denominations.

8. How many dollars are equal to 500 cents?

Here, 500÷100=5 dollars, the Ans.

Or, what amounts to the same thing, by cutting off the two ciphers at the right hand, 5 remains, which is the number of dollars, as before.

9. Reduce 48 dimes to dollars. Ans. 4D. 8dm.
10. Reduce 7050 cents to eagles. Ans. 7E. 0D. 5dm.
11. Reduce 14200 mills to dollars. Ans. 14D. 2dm.
12. Reduce 804 cents to dollars. Ans. 8D. 4c.

Troy Weight.

13. Reduce 932 pennyweights to pounds.

2,0)93,2 pwt.

12)46 oz. 12 pwt.

Ans. 3 lb. 10 oz. 12 pwt.

Here, I divide the 932 pwt. by 20, to reduce them to ounces, and the ounces by 12, to reduce them to pounds.

14. Reduce 26408 grains to pounds.
Ans. 4 lb. 7 oz. 0 pwt. 8 gr.

15. Reduce 3840 grains to ounces. Ans. 8 oz.

Apothecaries' Weight.

16. Reduce 157 scruples to ounces.
Ans. 6 oz. 4 dr. 1 sc.

17. In 55799 grains, how many pounds?
Ans. 9 lb. 8 oz. 1 dr. 2 sc. 19 gr.

Long Measure.

18. Reduce 13648 rods to miles.
Ans. 42 m. 5 fur. 8 rd.

19. Reduce 5600 rods to leagues.
Ans. 5 Le. 2 m. 4 fur.

20. Reduce 101 yards to rods.

$5\frac{1}{2}$ yards=1 rod.

$5\frac{1}{2}\times2=$ 11, new divisor.

101 $\times2=202$, new dividend.

11)202

18 rods - 4 rem.

Then, 4÷2=2 yards, the true remainder.

Ans. 18 rods and 2 yards.

Here the divisor, $5\frac{1}{2}$, contains a fraction; and the reduction is performed according to the 2d Note after the Rule.

21. The river Niagara, at the great cataract, falls nearly 163 feet down a perpendicular precipice; and the depth of the river at the foot of the falls is supposed to be at least 200 feet; which makes the whole height of the precipice about 363 feet. Reduce this number of feet to rods.
Ans. 22 rods.

22. The highest mountain in Europe is Mont Blanc, in Switzerland; the summit of which is elevated about 15675 feet above the level of the sea. Reduce this number of feet to miles. Ans. 2 m. 7 fur. 30 rd.

23. The quantity of linen imported into the United States from Ireland, in the year 1806, was 2675619 yards. How many miles in length was the whole?

Ans. 1520 m. 1 fur. 36 rd. 1 yd.

24. Reduce 4735272960 barley-corns to miles,

4735272960÷190080=24912 miles, Ans.

Cloth Measure.

25. Reduce 19 quarters to yards. Ans. 4 yd. 3 qr.
26. Reduce 1500 nails to English ells. Ans. 75.
27. Reduce 2106 inches to French ells. Ans. 39.

Square Measure.

28. Reduce 4779 sq. rods to acres.

Ans. 29 A. 3 R. 19 sq. rd.

29. Reduce 7772 sq. inches to sq. yards.

Ans. 5 sq. yd. 8 sq. ft. 140 sq. in.

30. Reduce 3048960 acres to sq. miles. Ans. 4764.
31. Reduce 370260 sq. feet to acres.

370260÷10890=34 roods=8 acres, 2 roods, Ans.

Cubic Measure.

32. Reduce 687 cubic feet to cords. Ans. 5 C. 47 cub. ft.
33. Reduce 442368 cubic inches to cords. Ans 2.
34. Reduce 600 feet of round timber to tons. Ans. 15.

Wine Measure.

35. Reduce 180 gills to gallons. Ans. 5 gal. 2 qt. 1 pt.
36. Reduce 20160000 pints to pipes. Ans. 20000.
37. Reduce 504 pints to hogsheads. Ans. 1.
38. Reduce 64 half gills to gallons. Ans. 1.

Dry Measure.

39. Reduce 2424 quarts to bushels. Ans. 75 bush. 3 pk.
40. Reduce 258 pints to bushels. Ans. 4 bush. and 1 qt.
41. Reduce 360 bushels to chaldrons. Ans. 10.

Time.

Reduce

31556928 seconds to days. Ans. 365 da. 5 h. 48 m. 48 sec.

2551443 seconds to days. Ans. 29 da. 12 h. 44 m. 3 sec.

364 days to weeks. Ans. 52.

96 calendar months to years. Ans. 8.

Reduce
1831 years to centuries. Ans. 18 cent. and 31 years.
669138 days to Julian years. Ans. 1832.

Circular Measure.

48. Reduce 47435″ to degrees. Ans. 13° .. 10′ .. 35″.
49. Reduce 1020300″ to signs. Ans. 9 S. .. 13° .. 25′.
50. Reduce 18048″ to degrees. Ans. 5° and 48″.

QUESTIONS ON REDUCTION.

1. What is Reduction? 2. What is Reduction Descending? 3. What is Reduction Ascending? 4. How is Reduction Descending performed when the given quantity consists of several denominations? 5. How is it performed when the given quantity consists of one denomination only? 6. How is Reduction Ascending performed? 7. How do you proceed when the divisor contains a fraction? 8. How are operations in Reduction proved?

Note.—It will be useful for Instructors to put various other questions to their pupils respecting the most usual denominations of money, weight and measure; viz. such as these: How many farthings make a penny? How many ounces in a pound, Avoirdupois? How are feet reduced to inches? How are square rods reduced to acres? &c. And learners ought not to leave Reduction until they are able to answer such questions correctly.

COMPOUND ADDITION,

Is the addition of numbers of different denominations, but of the same general nature.

RULE.*

1. Set down the given numbers in such a manner that

* The reason of this rule is evident, from what has been said in Simple Addition and Reduction: For instance, the addition of Sterling money, as 1 in the pence is equal to 4 in the farthings; 1 in the shillings, to 12 in the pence; and 1 in the pounds to 20 in the shillings; therefore, carrying as directed, is nothing more than providing a method of placing the money arising from each column properly in the scale of denominations: and this reasoning will hold good in the addition of compound numbers of any denominations whatever.

those of the same denomination may stand directly under one another; placing the lowest denomination of each quantity at the right hand, the next higher next, and so on; and then draw a line underneath.

2. Add together the numbers of the lowest denomination, as in Simple Addition, and divide the amount by a number which will (according to the Rule for Reduction Ascending) reduce it to the next higher denomination: Set down the remainder under the column* added, (or, if nothing remains, set down a cipher,) and carry the quotient to the column of the next denomination. Then add up the numbers of this denomination, together with the number (if any) carried from the first column, and reduce the amount to the next higher denomination; setting down the remainder under the proper column, and carrying the quotient to the next column, as before; and so proceed through all the denominations, setting down the whole amount of the last column. Then the amount of the last column, together with the several remainders (if any) of the lower denominations, will be the answer, or whole amount sought.

PROOF.—The method of proof is the same as in Simple Addition.

EXAMPLES.

Money.

1. What is the total sum of 47*l.* 18s. 8d.+18*l.* 19s.+17*l.* 10d.+15s. 9d. and 11d.?

	L.	s.	d.
	47 ..	18 ..	8
	18 ..	19 ..	0
	17 ..	0 ..	10
		15 ..	9
			11
Ans.	84 ..	15 ..	2
Proof,	84 ..	15 ..	2

Explanation.—I first write down the given quantities according to the Rule. Then I add up the column of pence, as in Simple Addition, and find the amount to be 38 pence. This sum I divide by 12, to reduce it to shillings, and the quotient is 3 shillings, and 2 pence remain: The remainder I set down under the column of pence, and the quotient I carry to the column of shillings. I next add up

* By a *column*, is here meant a *column of numbers of the same denomination*; not a row of single figures, as in Simple Addition.

the column of shillings, together with the 3 shillings carried from the pence, and the sum is 55 shillings. This sum I divide by 20, to reduce it to pounds; and the quotient is 2 pounds, and 15 shillings remain; which remainder I set down below the column of shillings, and carry the quotient to the pounds. I then add up the column of pounds and the 2 pounds carried from the shillings, and the sum is 84 pounds, the whole of which I set down below the column of pounds, because there are no more columns to add, and then the work is done. So, the answer, or sum total, is 84*l.* 15s. 2d.—Then, to prove the work, I add all the columns downwards, proceeding in other respects as before, and as I find the same sum total as before, I conclude the work is right.

(2)

	L.	s.	d.	q.
	14	17	0	3
	8	0	8	2
	17	3	11	0
		18	8	1
Sum,	41	0	4	2

(3)

	s.	d.	q.
	15	11	2
	18	7	1
	6	9	1
	4	10	0
L. 2..	6	2	0

Note.—When the highest denomination mentioned in the question is not the highest of its kind, if the amount of that denomination in the answer be large, it may be reduced to a higher denomination, if necessary. Thus, in example 3d, the amount of the shillings is 46s., which being brought into pounds, is 2*l.* 6s.

4. Required the amount of 365*l.* 14s.+18s. 9d.+36*l.* 12s.+76*l.* 1s. 8d.+10d. 2q.

Ans. 479*l.* 7s. 3d. 2q.

Troy Weight.

5. What is the sum total of 5lb. 9oz. 13pwt. 8gr.+17lb. 8oz. 6pwt.+4oz. 18pwt. 9gr. +16lb. 14pwt.+8oz. 15gr.?

	lb.	oz.	pwt.	gr.
	5	9	13	8
	17	8	6	0
		4	18	9
	16	0	14	0
		8	0	15
Ans.	40	7	12	8

Here I divide the amount of the grains by 24, to reduce them to penny-weights; the amount of the penny-weights by 20, to reduce them to ounces; and the amount of the ounces by 12, to reduce them to pounds.

(6)

lb.	oz.	pwt.	gr.
55	0	14	6
7	6	19	8
18	1	5	0
	10	17	8

(7)

lb.	oz.	pwt.	gr.
8	7	14	16
	8	15	14
14	6	19	0
21	0	0	0

Avoirdupois Weight.

N. B. 28 lb.=1 qr. of a cwt.

(8)

	Cwt.	qr.	lb.
	21	1	12
	4	0	26
	5	0	6
	7	2	25
Sum,	38	1	13

(9)

Cwt.	qr.	lb.	oz.	dr.
3	3	27	14	12
2	0	18	10	0
	1	0	8	6
5	2	25	15	7

Long Measure.

(10)

	Rods.	ft.	in.
	15	14	11
	12	15	8
	8	2	0
	4	7	7
		15	0
Sum,	42	$5\frac{1}{2}$	2
Or,	42	5	8

In this example, the amount of the feet is 55, which I reduce to rods, according to the 2d Note after the rule for Reduction Ascending.

(11)

Leagues.	m.	fur.	rd.	yd.
24	2	7	20	4
7	0	4	28	4
2	1	0	4	0
	1	5	17	2
Sum, 35	0	1	30	4½

Or, 35 le. 0 m. 1 fur. 30 rd. 4 yd. 1 ft. 6 in.

(12)

Rods.	ft.	in.	b. c.
12	14	10	1
8	15	0	0
26	10	2	2
8	10	7	2
57	½	8	2

Or, 57 rd. 1 ft. 2 in. 2 b. c.

Cloth Measure.

Yards.	qr.	na.
24	2	2
17	1	1
12	2	0
10	0	1
64	2	0

E. e.	qr.	na.
18	4	1
4	2	2
1	0	0
13	3	3

Fl. e.	qr.	na.
21	2	1
8	0	3
4	2	0
12	0	3

Square Measure.

Square rods.	sq. yd.	sq. ft.
4	28	2
2	30	6
8	25	7
2	21	0
8	15	8
28	0	5

Acres.	R.	sq. rods.	sq. ft.	sq. in.
25	2	24	8	120
14	1	8	7	86
19	0	29	8	128
8	2	37	0	108
12	0	8	1	45

Cubic Measure.

Cords.	cub. ft.	cub. in.
18	102	1725
15	32	825
21	97	1200
12	48	867
68	25	1161

Cords.	cub. ft.	cub. in.
7	114	1540
5	100	287
4	82	1078
5	0	52

Wine Measure.

(20)

Bar.	gal.	qt.	pt.	gil.
2	9	3	1	2
3	21		0	1
5	8	3	1	3
6	28	2	1	0
18	6	0	0	2

(21)

Tuns.	p.	hhd.	bar.
24	1	1	0
5	0	0	1
8	1	1	1
10	0	1	1

Dry Measure.

Bush.	pk.	qt.	pt.
16	2	7	1
4	2	6	0
8	0	5	1
7	2	0	1
37	0	3	1

Bush.	pk.	qt.	pt.
12	3	[illegible]	1
9	0		0
5	2		1
20	1		0

Time.

Weeks.	da.	h.
2	6	21
2	8	18
1		17
1		20
9	0	4

Days.	h.	min.	sec.
20	17	45	27
5	12	22	18
16	12	53	47
	21	50	0

Circular Measure.

Signs.	°	′	″
4	28	54	47
2	8	17	20
1	12	16	52
	24	47	40
	17	45	46
10	2	2	25

S.	°	′	″
2	12	46	48
	5	18	54
	24	45	12
4	0	0	8
1	2	12	34

PRACTICAL QUESTIONS.

1. Bought a Geography for 10s. 6d.; an English Reader for 4s.; an Arithmetic for 4s. 6d.; a slate for 2s. 8d., and a penknife for 2s. 9d.: what do they all amount to?
Ans. 1*l.* 4s. 5d.

2. The national debt of England, at the Revolution, in 1689, was 1054925*l.* 12s. 6d.; at the close of the American war, in 1783, it was 238232247*l.* 19s. 11d.; and in 1827 it amounted to about 900000000*l.* Required the sum of these several debts? Ans. 1139287173*l.* 12s. 5d.

3. Bought 5 cheeses, weighing as follows; viz. the first cheese, 17lb. 15oz.; the second, 22lb. 14oz.; the third, 19lb; the fourth, 24lb. 7oz.; and the fifth, 21lb. 12oz. Required the weight of the whole? Ans. 106lb.

4. A wall-maker built four stone walls for a farmer: the first wall was 94 rods, 12 feet, in length; the second, 42 rods, 10 feet; the third, 37 rods, 9 feet; and the fourth, 28 rods. What length of fence did the four walls make?
Ans. 202 rods, 14½ feet.

5. A landlord has 4 farms: the first contains 120 acres, 2 roods; the second 150 acres; the third 215 acres, 1 rood, 28 square rods; and the fourth 96 acres, 2 roods, 22 square rods. How much land has he in all?
Ans. 582 A. 2R. 10sq. rd.

6. Bought three loads of wood: the first contained 124 cubic feet; the second 108cub. feet, 710cub. inches; and the third 95cub. ft. 1018cub. in. Required the contents of all the loads? Ans. 2 cords, 72cub. ft.

7. In the year 1820, there was raised in the county of Otsego, New-York, 125 bushels, 4 quarts, of Indian corn, on one acre; 120 bush. 2 pecks on another; 118 bush. 4 quarts on another; 117bush. on another; 111 bush. on another; 95 bush. 4 quarts on another; and 90 bush. 2 pecks, 6 quarts, on another: how much on the seven acres?
Ans. 777 bush. 2pk. 2qt.

8. The periods of time in which the primary planets revolve round the Sun, and round their own axes, are as follows:—

Planets.	*Rotation on their axis.*	*Time of moving round the sun.*
	days. h. m.	days. h. m.
Mercury	1..00.. 5	87..23..16
Venus	0..23..21.	224..16..49
The Earth	0..23..56½	365.. 6.. 9
Mars	1..00..39	686..23..31
Jupiter	0.. 9..56	4332..14..19
Saturn	0..10..16	10758..23..17
Herschel		30688..17.. 6
Find the sums,	4..20..13½	47145.. 4..27

COMPOUND SUBTRACTION,

Is the subtraction of a less compound quantity, or number, from a greater of the same generic kind.

RULE.*

1. Set down the subtrahend, or less quantity, under the minuend, or greater quantity, in such a manner that those numbers which are of the same denomination may stand directly under each other, as in Compound Addition.

2. Beginning at the right hand, subtract successively, if possible, the lower number in each denomination from the upper, and write the remainder underneath; and the several remainders taken together will be the whole difference sought.

3. But if the lower number of any denomination be greater than the upper, borrow and add to the upper number as many of that denomination as make 1 of the next higher; then subtract the lower number from the upper one thus increased, and set down the remainder. Then carry 1, (as an equivalent to the number borrowed,) and add it to the next superior denomination in the lower quantity, and proceed as before.

* The reason of this Rule will easily appear from what was said in Simple Subtraction; for the borrowing depends upon the same principle, only the number borrowed is not always 10, as in Simple Subtraction, because the denominations of compound numbers do not all increase in a tenfold ratio.

PROOF.—Add the total remainder, or answer, to the subtrahend, by the rule for Compound Addition; and, if the work be right, the amount will be equal to the minuend.

EXAMPLES.

Money.

1. From 572*l.* 15s. 11d. take 284*l.* 8s. 6d.

	L.	s.	d.
Minuend,	572 ..	15 ..	11
Subtrahend,	284 ..	8 ..	6
Remainder,	288 ..	7 ..	5
Proof,	572 ..	15 ..	11

Here I first subtract the 6 pence from the 11 pence, as in Simple Subtraction, and the remainder is 5 pence, which I set down underneath. I subtract, in like manner, the 8 shillings from the 15 shillings, and the 284 pounds from the 572 pounds, and set down the remainders, 7 and 288. So the whole remainder, or answer, is found to be 288*l.* 7s. 5d.—Then, to prove the work, I add the whole remainder to the subtrahend, (by Compound Addition,) and the sum being the same as the minuend, I conclude the work is right.

(2)

	s.	d.	q.
From	18 ..	5 ..	2
take	4 ..	7 ..	3
Rem.	13 ..	9 ..	3

Here, as I cannot subtract 3 farthings from 2 farthings, I borrow 4 farthings, (which are equal to 1 penny,) and add them to the 2 farthings, and the sum is 6 farthings, from which I subtract the 3 farthings, and set down the remainder. Then I carry 1 penny, (as an equivalent to the 4 farthings borrowed,) and add it to the 7 pence, and the sum is 8 pence. Then, as I cannot subtract 8 pence from 5 pence, I add 12 pence (which make 1 shilling) to the 5 pence, and from the amount I subtract the 8 pence, and set down the remainder. Then I add 1 to the 4 shillings, and subtract the sum from the 18 shillings, and set down the remainder, and the work is done.

Note.—In Compound Subtraction, when it is necessary to borrow, instead of adding the number borrowed to the upper number, and then subtracting the lower number from the amount, as directed in the foregoing Rule, you may, if you choose, subtract the lower number from the number borrowed, and add the remainder to the upper number, and set down the amount below; then carry 1 to the next superior denomination in the subtrahend; and so proceed.

(3)

	L.	s.	d.	q.
From	47 ..	0 ..	0 ..	0
take		15 ..	0 ..	2
Rem.	46 ..	4 ..	11 ..	2

(4)

L.	s.	d.	q.
7 ..	14 ..	0 ..	1
2 ..	17 ..	2 ..	2

5. From 18*l.* 12s. 4d. take 15*l.* 0s. 2d.
Ans. 3*l.* 12s. 2d.

6. Find the difference between 217*l.* 6s. and 178*l.* 18s. 5d. 1q.
Ans. 38*l.* 7s. 6d. 3q.

Troy Weight.

(7)

	lb.	oz.	pwt.	gr.
From	8 ..	4 ..	14 ..	5
take	7 ..	7 ..	19 ..	8
Rem.		8 ..	14 ..	21
Proof,	8 ..	4 ..	14 ..	5

(8)

lb.	oz.	pwt.	gr.
5 ..	9 ..	18 ..	20
2 ..	10 ..	6 ..	21

Avoirdupois Weight.

(9)

	Tons.	cwt.	qr.	lb.	oz.	dr.
From	7 ..	10 ..	0 ..	18 ..	12 ..	0
take	2 ..	15 ..	2 ..	12 ..	15 ..	8
Rem.	4 ..	14 ..	2 ..	5 ..	12 ..	8

(10)

lb.	oz.	dr.
8 ..	14 ..	11
4 ..	14 ..	15

Long Measure.

(11)

	Le.	m.	fur.	rd.	yd.
From	6	2	5	18	1
take	2	0	7	20	4
Rem.	4	1	5	37	$2\frac{1}{2}$

(12)

yd.	ft.	in.	b. c.
4	2	8	0
2	2	8	2

Cloth Measure.

(13)

	yd.	qr.	na.
From	7	1	1
take	4	2	2
Rem.	2	2	3

(14)

E. e.	qr.	na.
7	0	1
5	2	0

Square Measure.

(15)

	A.	R.	sq. rd.
From	5	1	27
take		2	28
Rem.	4	2	39

(16)

sq. yd.	sq. ft.	sq. in.
8	7	100
5	4	140

Cubic Measure.

(17)

	Cd.	cub. ft.	cub. in.
From	8	114	1540
take	2	122	1720
Rem.	5	119	1548

(18)

Cd.	cub. ft.	cub. in.
5	100	0
4	127	1

Wine Measure.

(19)

	T.	p.	hhd.	gal.
From	4	1	0	42
take	2	0	1	50
Rem.	2	0	0	55

(20)

gal.	qt.	pt.	gil.
27	2	1	2
18	1	0	3

Dry Measure.

(21)

	bush.	pk.	qt.	pt.
From	7	2	5	0
take		2	7	1
Rem.	6	3	5	1

(22)

bush.	pk.	qt.	pt.
5	1	0	1
2	2	2	0

Time.

(23)

	weeks.	da.	h.
From	4	5	20
take	2	6	12
Rem.	1	6	8

(24)

da.	h.	min.	sec.
4	14	40	50
	18	48	58

Circular Measure.

(25)

	S.	°	′	″
From	10	24	42	17
take	4	28	14	25
Rem.	5	26	27	52

(26)

S.	°	′	″
8	0	0	0
5	1	1	1

PRACTICAL QUESTIONS.

1. Bought a piece of cloth for 2*l.* 19s.; and sold the same for 3*l.* 18s. 6d.: what did I gain by the bargain?
Ans. 19s. 6d.

2. The value of the domestic and foreign produce exported from the United-States to foreign countries, in the year 1800, was 15968650*l.* 10s. sterling. The value of the exports from the city of London, in the same year, was 25428922*l.* 16s. 7d. How much did the value of the exports from London exceed that of the exports from the United-States? Ans. 9460272*l.* 6s. 7d.

3. A Silversmith had 14 lb. 9 oz. 10 pwt. of silver; he melts 7lb. 15pwt. 10 gr.; how much has he left?

Ans. 7 lb. 8 oz. 14 pwt. 14 gr.

4. The highest mountain in the known world is Dewalageri, the loftiest peak of the Himmaleh mountains, in Asia; and the next highest is Chimborazo, in South-America. The summit of Dewalageri is about 5 miles, 1 furlong, 37 rods, and that of Chimborazo 4 miles and 18 rods, above the level of the sea. How much does the height of Dewalageri exceed that of Chimborazo?

Ans. 1 m. 1 fur. 19 rd.

5. Bought several casks of cider, containing 152 gallons, 1 quart; and disposed of one which contained 41 gallons, 2 quarts, 1 pint: How much remained in the other casks?

Ans. 110 gal. 2 qt. 1 pt.

6. Napoleon Bonaparte was born in the town of Ajaccio, on the island of Corsica, the 15th of 8th Month (August) 1769; and he died on the island of St. Helena, the 5th of 5th Month (May) 1821. How old was he when he died?

	yr. mo. da.	
	1821 .. 5 .. 5	subsequent date.
	1769 .. 8 .. 15	prior date.
Ans.	51 .. 8 .. 20	difference.

Note.—To find the interval or space of time between any two given dates; subtract the years, months, and days, of the prior date, from those of the subsequent date, and the difference will be the time sought. When the number of days in the subtrahend exceeds the number in the minuend, subtract the former number from the number of days that the month next before that mentioned in the minuend contains, if this subtraction can be made; add the remainder to the number of days in the minuend, and set down the sum, for the number of days in the answer: but if this subtraction cannot be made, then the number of days in the minuend will be the number to set down in the answer. Always when the number of days in the subtrahend is greater than the number in the minuend, add 1 to the number of months in the subtrahend. In other respects proceed according to the rule for Compound Subtraction. Thus, in the foregoing example, the number of days in the subtrahend is greater than the number in the minuend, and I subtract the former number from 30, which is the number of days the 4th month contains, and add the remainder to the 5 days in the minuend, and the sum is 20, which I set down for the number of days in the answer. Then I carry 1 to the months in the subtrahend, and proceed as usual in compound subtraction.

7. How many months and days from the 31st of 1st Mo. (Jan.) to the 15th of 7th Mo. (July)?

	mo. da.
	7 .. 15
	1 .. 31
Ans.	5 .. 15

Here the number of days in the subtrahend cannot be subtracted from the number of days which the 6th month contains, and therefore I set down the number of days in the minuend for the days of the answer. Then, I add 1 to the number of months in the subtrahend, and subtract as usual.

8. The celebrated William Penn, the first governor of Pennsylvania, and the founder of the city of Philadelphia, was born in the city of London, 11th Mo. (Nov.) 4th, 1644; and he died at Rushcomb, in Buckinghamshire, England, 8th Mo. (Aug.) 10th, 1718: how old was he when he died?

Ans. 73 yr. 9 mo. 6 days.

9. What is the difference between the Julian and the Solar year; the former consisting of 365 days, 6 hours, and the latter of 365 da. 5 h. 48 min. 48 sec.?

Ans. 11 min. 12 sec.

10. The latitude of the city of London is 51 degrees, 31 minutes, north; and that of Quebec, in Lower-Canada, is 46 degrees, 50 minutes, north: What is the difference between the latitudes of these two cities? Ans. 4° .. 41′.

COMPOUND MULTIPLICATION,

Is when the multiplicand consists of different denominations.

RULE.

1. Set down the multiplier under the lowest denomination of the multiplicand.

2. Multiply the lowest denomination of the multiplicand by the multiplier, as in Simple Multiplication; divide the product by such a number as will reduce it to the next higher denomination; set down the remainder, and carry the quotient to the product of the next denomination.

3. Multiply the next denomination of the multiplicand by the multiplier; add to the product the number (if any) carried from the denomination below; then reduce this sum to the next higher denomination, setting down the remainder, and carrying the quotient to the product of the next denomination, as before.

4. Proceed in this manner through all the denominations to the highest, and the product or amount of that denomination, together with the several remainders (if any) of the lower denominations, will be the answer, or whole product required.*

EXAMPLES.

Money.

	L.	s.	d.
1. Multiply	8 ..	17 ..	11
by			4
Product,	35 ..	11 ..	8

Here, I first multiply the 11 pence by 4, and the product is 44 pence; which I reduce to shillings, and have 3 shillings, and 8 pence remain. The 8d. remainder I set down below the pence of the multiplicand, and the 3 s. I carry to the product of the shillings. I then multiply the 17 shillings by 4, and the product is 68s.; to which I add the 3s. carried from the pence, and the sum is 71 s.; which I reduce to pounds, and have 3*l*., and 11 s. remain. The 11 s. I set down below the shillings of the multiplicand, and the 3*l*. I carry to the product of the pounds. Lastly, I multiply the 8*l*. by 4, and add to the product the 3*l*. carried from the shillings, and the amount

* The product of a number, consisting of several parts or denominations, by any simple number whatever, will be expressed by taking the product of the simple number and each part of the compound number, by itself, as so many distinct questions: Thus 8*l*. 17s. 11 d. multiplied by 4, will be 32*l*. 68 s. 44 d. = (by taking the shillings from the pence, and the pounds from the shillings, and placing them in the shillings and pounds respectively,) 35*l*. 11 s. 8d; and this will be true when the multiplicand is any compound number whatever.

is 35*l.*, which I set down, and the work is done. So the whole product, or answer, is 35*l.* 11s. 8d.

(2)

	L.	s.	d.	q.
Multiply	18	16	0	2
by				8
Prod.	150	8	4	0

(3)

	s.	d.	q.
	6	4	1
			15
L. 4	15	3	3

4. Multiply 1*l.* 17s. 5d. by 6. Ans. 11*l.* 4s. 6d.
5. Multiply 18*l.* 7s. by 8. Ans. 146*l.* 16s.
6. Multiply 9s. 7d. 2q. by 12. Ans. 5*l.* 15s. 6d.
7. Multiply 4*l.* 19s. 10d. 2q. by 365.
Ans. 1822*l.* 14s. 4d. 2q.

Weights and Measures.

(8)

	lb.	oz.	dr.
Multiply	14	15	12
by			4
Product,	59	15	0

(9)

yd.	ft.	in.
2	2	8
		5
14	1	4

(10)

m.	fur.	rd.
20	7	25
		6
125	5	30

(11)

	A.	R.	sq. rd.
Multiply	1	0	15
by			7
Product,	7	2	25

(12)

sq. yd.	sq. ft.	sq. in.
5	8	140
		8
47	8	112

(13)

Cd.	cub. ft.
7	120
	9
71	56

(14)

	gal.	qt.	pt.	gil.
Multiply	7	2	0	3
by				12
Product,	91	0	1	0

(15)

bush.	pk.	qt.	pt.
5	2	7	1
			20
114	2	6	0

	(16)	(17)
	w. da. h. min. sec.	S. ° ′ ″
Multiply	2 .. 6 .. 8 .. 20 .. 40	1 .. 8 .. 7
by	25	275
Product,	72 .. 4 .. 16 .. 36 .. 40	10 .. 12 .. 12 .. 5

CONTRACTIONS.

I. *When the multiplier exceeds* 12, *and is a composite number*, you may multiply successively by its component parts, instead of the whole number at once.

EXAMPLES.

1. Multiply 27 lb. 12 oz. 8 dr. Avoirdupois, by 42.
7×6=42.

lb.	oz.	dr.
27	12	8
		6
166	11	0
		7
Ans. 1166	13	0

2. Multiply 8*l.* 12 s. 4 d. 2 q. by 32.
Ans. 275*l.* 16 s.

3. Multiply 8 acres, 3 roods, 28 sq. rods, by 144.
Ans. 1285 A. 0 R. 32 sq. rd.

4, Multiply 8 bush. 2 pk. by 640.
8×8×10=640. Ans. 5440 bush.

II. *When the multiplier is large, and is not a composite number*, you may reduce the multiplicand to the lowest denomination of which it consists, and then multiply it by the multiplier, as in Simple Multiplication; and the product will be the answer in the denomination that the multiplicand is reduced to, which may be brought into a higher denomination, if necessary.

EXAMPLES.

1. Multiply 8*l.* 17 s. 4 d. by 57.

```
L.  s.  d.        d.
8 .. 17 .. 4  =  2128
20                 57
----             -----
177              14896
 12             10640
----            ------       L.   s.
2128 d.         121296 d.=505 .. 8  Ans.
```

Explanation.—Here, I first reduce the given multiplicand, viz. 8*l.* 17s. 4d., to pence, and find it to be 2128 pence. I then multiply those 2128 pence by 57, the given multiplier, and the product is 121296 pence; which I reduce to pounds, and have 505*l.* 8s. for the answer.

2. Multiply 19s. 8d. 2q. by 149.
Ans. 146*l.* 16s. 6d. 2q.

3. Multiply 17 lb. 8 oz. Avoirdupois, by 71.
Ans. 1242 lb. 8 oz.

4. Multiply 22 yards, 3 qr. 1 na. by 173.
Ans. 3946 yd. 2 qr. 1 na.

PRACTICAL QUESTIONS.

1. What do 4 yards of broadcloth amount to, at 2*l.* 7s. 10d. per yard?

```
     L. 2 ..  7 .. 10   value of 1 yard.
                   4   number of yards.
     ----------------
Ans. L. 9 .. 11 ..  4   value of 4 yards.
```

Questions.	L. s. d.		Answers. L. s. d.
4 lb. of tea, at	0 .. 7 .. 8	a lb.	1 .. 10 .. 8
10 cwt. of cheese, at	2 .. 17 .. 10	a cwt.	28 .. 18 .. 4
11 tons of hay, at	2 .. 1 .. 10	a ton.	23 .. 0 .. 2
12 bush. of apples, at	1 .. 9	a bush.	1 .. 1 .. 0
19 lb. of indigo, at	11 .. 6	a lb.	10 .. 18 .. 6

7. Find the amount of the following

BILL OF PARCELS.

New-York, Nov. 2d, 1831.

James Paywell, Bought of *John Grocer,*

[*See next page.*]

	L.	s.	d.
12 cwt. of sugar, at 2 *l.* 12 s. a cwt. =	31 ..	4 ..	0
35 lb. of loaf sugar, at 1 s. 1 d. a lb. =	1 ..	17 ..	11
28 lb. of rice, at 3 d. a lb. =		7 ..	0
4 cwt. of raisins, at 2 *l.* 5 s. a cwt.=	9 ..	0 ..	0
Amount,	L42 ..	8 ..	11

Received payment in full. *John Grocer.*

8. How much cloth in 25 pieces, each measuring 27 yd. 1 qr.? Ans. 681 yd. 1 qr.

9. How much land in 9 fields, each containing 14 A. 1 R. 25 sq. rd.? Ans. 129 A. 2 R. 25 sq. rd.

10. How many bushels of apples in 144 casks, each containing 3 bush. 3 pk.? Ans. 540 bush.

11. Find the difference between 100 Julian, and 100 Solar years. Ans. 18 h. 40 min.

COMPOUND DIVISION,

Is when the dividend consist of different denominations.

RULE.

1. Set down the divisor at the left hand of the dividend, as in Simple Division.

2. Divide the highest denomination of the dividend by the divisor, as in Simple Division, and the quotient will be a part of the answer, of the same denomination.

3. If there is any remainder after the division of the highest denomination, reduce that remainder to the next lower denomination, and add to it the given number (if any) of this denomination in the dividend. Then divide this amount by the divisor, as before; and so proceed till the lowest denomination has been divided; and the several numbers of the quotient, taken together, will be the answer.*

* To divide a number consisting of several denominations by any simple number, is the same as dividing all the parts or members of which the compound number is composed, by the simple number. And this will be true when any of the parts are not an exact multiple of the divisor; for, by conceiving the number, by which it exceeds that multiple, to have its proper value by being placed in the next lower denomination, the dividend will still be divided into parts, and the true quotient found, as before: Thus 41 *l.* 17 s. 6d. divided by 6, will be the same as 36 *l.* 114 s. 42 d. divided by 6, which is equal to 6 *l.* 19 s. 7 d, as by the Rule.

Note 1.—Each particular or partial quotient, and each remainder, will be of the same denomination as the dividend. When the lowest denomination of the dividend is not the lowest of its kind; then, if there is a remainder after the division of this denomination, it may be reduced to the next lower denomination, and then divided by the given divisor; and so on.

PROOF.—Compound Multiplication and Division prove each other.

EXAMPLES.

Money.

1. Divide 244 *l.* 17 s. 8 d. by 14.

Operation.

```
    L.    s.   d.  L.   s.   d.
14)244..17 .. 8(17 .. 9 .. 10 Quotient.
   14                     14
   ---             -----------
   104              244 .. 17 .. 8 Proof.
    98
   ---
Rem. 6
    20
   ---
   120
    17
   ---
   137(9 s.
   126
   ---
Rem. 11
    12
   ---
   132
     8
   ---
   140(10d.
   14
   ---
     0
```

Explanation.—I first divide the 244 pounds by 14; and the quotient is 17 *l.*, and 6 *l.* remain. The remainder I multiply by 20, to reduce it to shillings, and add to the product the 17s. of the given dividend, and the sum is 137 s. This sum I divide by 14, and the quotient is 9 s. and 11 s. remain. The remainder I multiply by 12, to reduce it to pence, and add to the product the 8 d. of the dividend, and the sum is 140 d. This sum I divide by 14, and the quotient is 10 d.; and as this is the lowest denomination of the given dividend, and nothing remains, the work is done. So the whole quotient, or answer, is 17 *l.* 9 s. 10 d.—Then, to prove the work, I multiply the quotient by the divisor; and the product being the same as the dividend, I conclude the work is right.

Note 2.—When the divisor does not exceed 12, the division may be performed after the manner of Short Division; as in the 2d and 3d examples.

(2)		(3)
L. s. d.		L. s. d. q.
4)14 .. 9 .. 10		5)2 .. 19 .. 1 .. 1
q.		
Quot. 3 .. 12 .. 5 .. 2		11 .. 9 .. $3\frac{2}{5}$

4. Divide 73*l.* 16s. 9d. by 7. Ans. 10*l.* 10s. 11d. $2\frac{2}{7}$q.

5. Divide 376*l.* 11s. 4d. by 11. Ans. 34*l.* 4s. 8d.

6. Divide 961*l.* 4s. by 37. Ans. 25*l.* 19s. 6d. 3q.+

7. Divide 97*l.* 6s. 8d. by 365. Ans. 5s. 4d.

Weights and Measures.

(8)		(9)
lb. oz. dr.		yd. ft. in.
4)59 .. 15 .. 4		5)14 .. 1 .. 8
		b. c.
Quot. 14 .. 15 .. 13		2 .. 2 .. 8 .. $2\frac{2}{5}$

10. Divide 280 miles, 4 furlongs, by 16. Ans. 17 m. 4. fur. 10 rd.

11. Divide 570 yd. 2 qr. 1 na. by 47. Ans. 12 yd. 0 qr. 2 na.+

12. Divide 2 acres and 10 sq. rods by 6. Ans. 1 R. 15 sq. rd.

13. Divide 15 sq. ft. 120 sq. in. by 12. Ans. 1 sq. ft. 46 sq. in.

14. Divide 53 cords, 16 cub. ft. 1585 cub. in. by 7. Ans. 7 Cd. 75 cub. ft. 967 cub. in.

15. Divide 69 gallons, 3 qt. by 8. Ans. 8 gal. 2 qt. 1 pt. 3 gil.

16. Divide 51 bush. 3 pk. 6 qt. 1 pt. by 25. Ans. 2 bush. 2 qt. 1 pt.

17. Divide 5 bushels by 8. Ans. 2 pk. 4 qt.

18. Divide 365 days, 6 hours, by 12. Ans. 30 da. 10 h. 30 min.

19. Divide 4 S. 5° .. 45′ by 9. Ans. 13° .. 58′ .. 20″.

CONTRACTIONS.

I. *When the divisor exceeds* 12, *and is a composite num-*

ber, you may divide successively by its component parts, instead of the whole divisor at once.

EXAMPLES.

1. Divide 54 lb. 3 oz. Avoirdupois, by 24.

6×4=24.

```
     lb. oz.
   6)54 .. 3
   ---------   dr.
   4)9 .. 0 .. 8
   -------------
Ans. 2 .. 4 .. 2
```

2. Divide 275 *l.* 16 s. by 32.
Ans. 8 *l.* $12\frac{1}{3}\frac{2}{2}$ s., or 8 *l.* 12 s. 4 d. 2 q.

3. Divide 1285 acres and 32 sq. rd. by 144.
Ans. 8 A. 3 R. 28 sq. rd.

4. Divide 11426 bush. 1 pk. by 1320.
12×11×10=1320. Ans. 8 bush. 2 pk. 5 qt.

II. *When the divisor is large, and is not a composite number*, you may reduce the dividend to the lowest denomination of which it consists, and then divide it by the divisor; and the quotient will be the answer; which may be brought into a higher denomination, if necessary.

EXAMPLES.

1. Divide 657 *l.* 6 s. 11 d. by 79.

657 *l.* 6 s. 11 d.=157763 d.; and 157763÷79=1997 d., and 1997 d.=8 *l.* 6 s. 5 d. Ans.

Here, I first reduce the given dividend to pence, and find it to be 157763 d. I then divide those 157763 d. by 79, and the quotient is 1997 d., which I reduce to pounds, and have 8 *l.* 6 s. 5 d. for the answer.

2. Divide 505 *l.* 8 s. by 57. Ans. 8 *l.* 17 s. 4 d.

3. Divide 154 lb. 8 oz. 7 dr. Avoirdupois, by 379.
Ans. 6 oz. 8 dr.+

4. Divide 506 yd. 2 ft. 9 in. by 711. Ans. 2 ft. 1 in. 2 b.c.

PRACTICAL QUESTIONS, *to exercise the learner in Compound Division and the preceding rules.*

1. An estate, valued at 8615 *l.* 12 s. is to be divided equally among 4 heirs: What is each share?

```
      L.     s.
   4)8615 .. 12
   ------------
Ans. 2153 .. 18
```

2. A certain house is worth 850 *l.*, and I own $\frac{3}{4}$ of it: What is my part worth?

L 850
3
4)2550
s.
Ans. L 637 .. 10

3. If 75 bushels of wheat be worth 28 *l.* 2 s. 6 d., what is the value of 1 bushel? Ans. 7 s. 6 d.

4. If 8 cheeses weigh 162 lb. 8 oz., what is the average weight of each? Ans. 20 lb. 5 oz.

5. A farmer has three farms: the first contains 125 acres, 3 roods; the second 175 acres and 10 sq. rods; and the third 200 acres, 2 roods, 18 sq. rods. He intends to divide these farms equally between his two sons: What will be the share of each son?

Ans. 250 A. 2 R. 34 sq. rd.

6. Bought 110 lb. of cheese, for 2 *l.* 10 s. 8 d., and 140 lb. more, for 3 *l.* 14 s. 4 d.: What did the whole cost me a lb.? Ans. 6 d.

7. If a man drinks 4 barrels of cider, (each $31\frac{1}{2}$ gallons,) in a year, or 365 days, how much does he drink in a day? Ans. 1 qt. 3 gil.+

8. In 48 *l.* 6 s. how many guineas, at 21 shillings each?

48 *l.* 6 s.=966 s., and 966÷21=46 guineas, Ans.

9. How many moidores at 36 shillings each; guineas at 28 s., and pistoles at 22 s., are there in 283 *l.* 16 s., and the number of each equal?

283 *l.* 16 s.=5676 s., and 36+28+22=86: then, 5676÷86=66, the Ans.

10. If 11 s. 3 d. sterling be equal to $2\frac{1}{2}$ dollars, how much sterling money is equal to 1 dollar?

s. d.
11 .. 3
2
$2\frac{1}{2}\times2=5$)22 6
Ans. 4 .. 6

Here the divisor contains a fraction, and I perform the division according to the rule given in the Supplement to Simple Division.

11. The sun is about 95000000 miles from the earth, and a cannon ball, at its first discharge, flies about a mile in $7\frac{1}{2}$ seconds: How long would a cannon ball be, at that rate,

in flying from here to the sun?

Ans. 22 yr. 216 da. 12 h. 40 min.

12. Light moves with the wonderful velocity of about 200,000 miles in one second of time. How long a time must it require for light to pass from Sirius, (one of the nearest of the fixed stars,) to the earth, supposing the distance of Sirius from the earth to be 20,000,000,000,000 miles?

Ans. 3 years, 62 da. 9 h. 46 min. 40 sec.

QUESTIONS ON THE FOREGOING.

1. What is Compound Addition? 2. How are the numbers to be written down? 3. Which column is to be added first? 4. What is to be done with its sum? 5. How then do we proceed? 6. How is Compound Addition proved? 7. What is Compound Subtraction? 8. How are the numbers to be written down? 9. How do we subtract when each part of the lower quantity is less than the corresponding part of the upper? 10. How do we proceed when the lower number is greater than that above it? 11. How is Compound Subtraction proved? 12. What is Compound Multiplication? 13. Where must the multiplier be placed? 14. Which denomination of the multiplicand must be multiplied first? 15. What is to be done with that product? 16. How must the rest of the work be performed? 17. What is Compound Division? 18. How are the dividend and divisor to be placed? 19. Which denomination of the dividend must be divided first? 20. If, after the division of this denomination, there is a remainder, what is to be done with it? 21. How must the rest of the work be performed? 22. How are Compound Multiplication and Division proved?

FRACTIONS.

A *fraction* is an expression of some part or parts of a unit, or of something considered as *a whole*. When the unit, or integer, is divided into two equal parts, the parts are called *halves*; when into three parts, *thirds*; when into four parts, *fourths*, &c.

Fractions arise naturally from the operations of Division,

when the divisor is not contained some number of times *exactly* in the dividend. For the *remainder* after the division is performed, is a part of the dividend which has not been divided; the *divisor* being the number of parts into which the integer is divided, and the *remainder* showing the number of those parts expressed by the fraction. Thus, 4 is contained in 9, *two and one-fourth times*, and the quotient cannot be fully expressed, except by a whole number and a fraction.

Fractions are divided into two kinds, viz. *Vulgar* and *Decimal.*

INTRODUCTION TO VULGAR FRACTIONS.

Note.—I have thought proper to give, in this part of the Treatise, only some general definitions, and a few of the most useful problems in vulgar fractions. The other rules for vulgar fractions will be given in a subsequent part of the Treatise.

A *vulgar fraction* is represented by two numbers, placed one above the other, with a line between them; as $\frac{1}{2}$, which signifies *one-half;* and $\frac{3}{4}$, which denotes *three-fourths.*

The upper number, or part, of any vulgar fraction, is called the *numerator,* and the lower number, the *denominator,* and both numbers, or parts, are called the *terms* of the fraction.*

The *denominator,* (which is the divisor in division,) shows how many parts the integer is divided into; and the *numerator,* (which is the remainder after division,) shows how many of those parts are meant by the fraction. Thus, $\frac{1}{3}$, (read *one-third,*) shows that the unit is divided into *three* parts, and that *one* of those parts is to be taken; and $\frac{4}{5}$, (read *four-fifths,*) denotes that the unit is divided into *five* parts, and that *four* of those parts are to be taken. The *denominator* is so called because it gives the *name* or *denomination* to the parts; that is, it shows whether they are *halves,* or *thirds,* &c.; and the *numerator* is so called because it *numbers* the parts, or shows how many of them are to be taken.

* The denominator, instead of being put under the numerator, is sometimes written after it, separated by a hyphen; thus, 1-2 signifies *one-half,* and 3-4 *three-fourths,* &c.

A vulgar fraction is said to be in its *least*, or *lowest terms*, when it is expressed by the least numbers possible. Thus, $\frac{4}{8}$, when reduced to its lowest terms is $\frac{1}{2}$; and $\frac{15}{20}$ is equal to $\frac{3}{4}$.

A *mixed number* is composed of a whole number and a fraction; as $4\frac{1}{2}$, or $7\frac{2}{3}$, &c.

PROBLEM I.

To abbreviate or reduce fractions to their lowest terms.

RULE.

Divide the terms of the given fraction by any number that will divide each of them without a remainder, and set down the quotients in the form of a fraction. Then divide the terms of the new, or reduced fraction, in the same manner; and so proceed until it appears that there is no number greater than 1 that will divide both the terms without a remainder; and the fraction will then be in its lowest terms.*

---

* It will not alter the value of a fraction to multiply both terms by a common multiplier, or divide them by a common divisor, because both parts will be increased or diminished in the same proportion; and hence the reason of this rule is evident.

Note 1.—Any number ending with an even number, or a cipher, is divisible, or can be divided, by 2.

2 Any number ending with 5, or 0, is divisible by 5.

3. If the two right hand figures of any number be divisible by 4, the whole is divisible by 4; and if the three right hand figures be divisible by 8, the whole is divisible by 8; and so on.

4 If the sum of the digits, constituting any number, be divisible by 3, or by 9, the whole is divisible by 3, or by 9.

5. If a number cannot be divided by some number less than the square root thereof, that number is a *prime*.

6. All *prime* numbers, except 2 and 5, have 1, 3, 7, or 9 in the place of units: all other numbers are *composite*.

7. When numbers, with the sign of addition or subtraction between them, are to be divided by any number, then each of those numbers must be divided by it. Thus, $\frac{20+8-4}{4}=5+2-1=7-1=6$; or $\frac{20+8-4}{4}=\frac{24}{4}=6$.—But if the numbers have the sign of multiplication between them, then only one of them must be divided. Thus, $\frac{20\times8\times4}{4}=5\times8\times4=160$; or $\frac{20\times8\times4}{4}=\frac{640}{4}=160$.

EXAMPLES.

1. Reduce $\frac{144}{192}$ to its lowest terms.

$2)\frac{144}{192}=\frac{72}{96}$; then, $8)\frac{72}{96}=\frac{9}{12}$; then, $3)\frac{9}{12}=\frac{3}{4}$, the Ans.

Here, I first divide both terms of the given fraction by 2, and it gives $\frac{72}{96}$. I then divide both terms of this fraction by 8, and it gives $\frac{9}{12}$; which I divide in like manner by 3, and it gives $\frac{3}{4}$; and as the terms of this fraction cannot both be divided by any number greater than 1, the fraction is in its lowest terms, as required.

2. Reduce $\frac{21}{42}$ to its lowest terms. Ans. $\frac{1}{2}$.
3. Reduce $\frac{216}{288}$ to its lowest terms. Ans. $\frac{3}{4}$.
4. Reduce $\frac{44}{48}$ to its lowest terms. Ans. $\frac{11}{12}$.
5. Reduce $\frac{175}{700}$ to its lowest terms. Ans. $\frac{1}{4}$.
6. Reduce $\frac{40}{50}$ to its lowest terms. Ans. $\frac{4}{5}$.
7. Abbreviate $\frac{101}{100}$ as much as possible. Ans. $\frac{1}{10}$.
8. Express $\frac{770}{1008}$ by the least numbers possible. Ans. $\frac{55}{72}$.
9. Reduce $\frac{2472}{3502}$ to its lowest terms. Ans. $\frac{12}{17}$.

Note.—Throughout the rest of this Work, where the answers to questions contain vulgar fractions, they are reduced to their lowest terms.

PROBLEM II.

To find the value of a fraction of any of the higher denominations of Money, or Weight, &c., in whole numbers in the lower denominations.

RULE.

Multiply the numerator of the given fraction by the number which 1 of that denomination makes of the next lower, and divide the product by the denominator: If there is a remainder, reduce it in like manner to the next lower denomination, and then divide it by the denominator, and so on. Then the several quotients will be the answer.

EPAMPLES.

1. What is the value, or proper quantity, of $\frac{13}{18}$ of a foot?

Operation.
13 Numer.
12 in.=1 ft.

[*Carried up.*

[*Brought up.*]

```
Denom. ——
      18)156(8 in.
         144
         ———
          12 Rem.
           3 b. c.=1 in.
          ——
          36(2 b. c.
          36
```

Ans. 8 inches, 2 barley-corns.

2. What is the value of $\frac{2}{3}$ of a pound sterling? Ans. 13 s. 4 d.

3. What is the value of $\frac{3}{8}$ of a shilling? Ans. 4 d. 2 q., or $4\frac{1}{2}$ d.

4. Reduce $\frac{4}{5}$ of a lb. Avoirdupois, to its proper quantity. Ans. 12 oz. $12\frac{4}{5}$ dr.

5. Reduce $\frac{5}{6}$ of a mile to its proper quantity. Ans. 6 fur. 26 rd. 11 ft.

6. Reduce $\frac{5}{9}$ of an English ell to its proper quantity. Ans. 2 qr. $3\frac{1}{9}$ na.

7. Reduce $\frac{4}{5}$ of an acre to its proper quantity. Ans. 3 R. 8 sq. rd.

8. Reduce $\frac{2}{3}$ of a cord of wood to its proper quantity. Ans. 85 cub. ft. 576 cub. in.

9. How much is $\frac{6}{7}$ of a barrel of wine? Ans. 27 gal.

10. What is the value of $\frac{1}{4}$ of a week? Ans. 1 day, 18 h.

PROBLEM III.

To reduce any given quantity to the fraction of a higher denomination of the same kind.

RULE.

Reduce the given quantity to the lowest denomination or term mentioned in it, for a numerator; then reduce the integral part to the same term, for a denominaton; and you will have the fraction required.

EXAMPLES.

1. Reduce 5 furlongs, 8 rods, to the fraction of a mile.

Operation.

fur. rd.		1 m.
5..8		8
40		—
——		8 fur.
Numer. 208 rods.		40
		——
Ans. $\frac{208}{320}=\frac{13}{20}$ m.		Denom. 320 rods.

Here the lowest denomination in the given quantity is rods, and therefore I reduce the quantity to rods, for a numerator. Then, because the quantity is to be reduced to the fraction of a mile, 1 mile is the integer; which I reduce to rods, for a denominator. Then I place the numerator over the denominator, and the fraction is $\frac{208}{320}$; which I reduce to its lowest terms, by Prob. I., and have $\frac{13}{20}$ for the answer.

2. Reduce 13 s. 6 d. 2 q. to the fraction of a pound. Ans. $\frac{650}{960}=\frac{65}{96}$ *l.*

3. Reduce 4 d. 2 q. to the fraction of a shilling. Ans. $\frac{3}{8}$ s.

4. Reduce 8 oz. 12 dr. to the fraction of a pound, Avoirdupois. Ans. $\frac{35}{64}$ lb.

5. Reduce 2 ft. 9 in. 1 b. c. to the fraction of a yard. Ans. $\frac{25}{27}$ yd.

6. What part of a yard of cloth is 2 qr. 2 na.? Ans. $\frac{5}{8}$ yd.

7. What part of an acre is 3 roods, 8 sq. rods? Ans. $\frac{4}{5}$.

8. What part of a cord of wood is 85 cub. ft. 576 cub. in.? Ans. $\frac{2}{3}$.

9. What part of a week is 1 day, 18 hours? Ans. $\frac{1}{4}$.

10. What part of a degree is 20 min. 40 sec.? Ans. $\frac{31}{90}$.

QUESTIONS ON THE FOREGOING.

1. What is a fraction? 2. How are fractions produced? 3. How many kinds of fractions are there? 4. How is a vulgar fraction represented? 5. What are the upper and lower parts of any vulgar fraction called? 6. What does the denominator denote? 7. What does the numerator denote? 8. What is a mixed number? 9. How do we reduce a fraction to its lowest terms? 10. How do we find the value of a fraction of a higher denomination in whole num-

bers in a lower? 11. How do we reduce any quantity to the fraction of a higher denomination?

DECIMAL FRACTIONS.

A *Decimal Fraction*, is a fraction whose denominator is a unit (1), with a cipher or ciphers annexed; as $\frac{4}{10}$, $\frac{5}{100}$, $\frac{65}{1000}$, &c.

As the denominator of a decimal fraction is always 10, or 100, or 1000, &c. it need not be expressed; for the numerator only may be made to express the value of the fraction: For this purpose it is only required to write the numerator with a point before it at the left hand, to distinguish it from a whole number, when it consists of so many figures as the denominator has ciphers: So, $\frac{5}{10}$ is written thus .5, and $\frac{45}{100}$ thus .45. But if the numerator has not as many places as the denominator has ciphers, then ciphers must be prefixed to make up that number of places: So, $\frac{5}{100}$, must be written thus .05, and $\frac{7}{1000}$, thus .007, &c. Thus do these fractions receive the form of whole numbers.

Any decimal may be expressed in the form of a common fraction by writing under it its proper denominator, (viz. a unit with as many ciphers annexed as there are figures in the given decimal,) rejecting from the numerator the decimal point, and also the ciphers, if any, to the left hand of the significant figures. Thus, .75, expressed in the form of a common fraction, is $\frac{75}{100}$; and .027, is $\frac{27}{1000}$.

When a whole number and decimal parts are expressed together, in the same number, it is called a *mixed number*. Thus, 25.48 is a mixed number, 25., or all the figures on the left hand of the decimal point, being whole numbers, and .48, or all the figures on the right hand of the decimal point, being decimals.

The point prefixed to decimals is called the *separatrix*, or the *decimal point*. This point must never be omitted; because, without it, decimals and mixed numbers cannot be distinguished from whole numbers.

Decimals are numerated from left to right, (which is contrary to the way of numerating whole numbers;) and each figure takes its value by its distance from the unit's place: If it be in the first place after units, (viz. in the first place to the right hand of the decimal point,) it signifies *tenths*; if in the second place, *hundredths*; and so on. For in decimals, as well as in whole numbers, the values of the places increase towards the left hand, and decrease towards the right, both in the same tenfold proportion; as in the following

TABLE.

&c.	Hundreds.	Tens.	Units.	Tenths.	Hundredths.	Thousandths.	Ten-Thousandths.	Hundred-Thousandths.	Millionths.	Ten-Millionths.	&c.	
				.5								read 5 Tenths.
				.0	5							.. 5 Hundredths.
				.0	2	5						.. .. 25 Thousandths.
				.6	8	4	2					.. .. 6842 Ten-Thousandths.
			4	.0	0	0	9	8				.. .. 4, and 98 Hund. Thousandths.
		4	8	.1	2	6	8	4	9			.. .. 48, and 126849 Millionths.
	5	9	1	.0	0	0	0	0	0	8		.. .. 591, and 8 Ten-Millionths.

Integers. Decimal parts.

Ciphers placed at the right hand of a decimal do not alter its value, since every significant figure continues to possess the same place: So .5, .50, and .500, are all of the same value, each being equal to $\frac{5}{10}$, or $\frac{1}{2}$. Therefore, when there are ciphers at the right hand of any decimal fraction, they may be omitted.

But ciphers placed at the left hand of decimals, decrease their value in a tenfold proportion, by removing them farther from the decimal point: Thus .5, .05, .005, &c. are $\frac{5}{10}$, $\frac{5}{100}$, $\frac{5}{1000}$, &c. respectively. It is therefore evident that the value of a decimal fraction, compared with another, does not depend upon the number of its figures, but chiefly upon the value of its first left hand figure: for instance, a fraction beginning with any figure less than .9, such as .89978, &c, if extended to an infinite number of figures, will not equal .9 or $\frac{9}{10}$.

Decimals are read in the same manner as whole numbers, giving the name of the lowest denomination, or right hand figure, to the whole. Thus, .7854,(the lowest denomination, or right hand figure, being ten-thousandths,) is read, 7854 *ten-thousandths.*

ADDITION OF DECIMALS.

RULE.*

Place the numbers, (whether pure decimals or mixed numbers,) according to the values of their places, so that the decimal points shall stand exactly under one another, and then proceed as in addition of whole numbers, only taking care to put the decimal point in the sum exactly under those in the numbers added.

Note.—The methods of proving Addition, Subtraction, &c., of Decimals, are the same as in whole numbers.

EXAMPLES.

1. Required the sum of .4123+.25+27.5+548+.028?

```
   .4123
   .25
 27.5
548.
   .028
--------
Ans. 576.1903
```

(2)	(3)	(4)	(5)
31.47	.978	.86	.4
480.	.02	.415	5.782
1.76	2.5	.085	7.808
.007	4.	.97	.01
513.237	7.498	2.330	14.000

* As decimals increase from right to left, and decrease from left to right, in the same tenfold ratio as whole numbers, it is evident that Addition and Subtraction of Decimals may be performed in the same manner as in whole numbers, if care be taken to write down the decimals in such a manner that tenths shall stand under tenths, hundredths under hundredths, &c. Thus, the sum of 5 and .4 is evidently .9; and the difference between 8 and .3 is .5.

6. What is the sum of 429+21.37+355.1+1.07+1.7? Ans. 808.24

7. What is the sum of 972+20+1.75+.7164+65.4? Ans. 1059.8664

8. To .9 add one-tenth part of a unit. Ans. 1.

SUBTRACTION OF DECIMALS.

RULE.

Place the subtrahend under the minuend, so that the decimal points shall be one under the other, and then proceed as in subtraction of whole numbers, only putting the decimal point in the remainder under those of the other numbers.

EXAMPLES.

	(1)		(2)	(3)	(4)
From	52.18	From	.5075	67.21	42.
take	7.459	take	.42	.1875	.725
Rem.	44.721	Rem.	.0875		

N. B. There being no figure above the 9, from which to subtract, I *suppose* a cipher.

5. From 270.2 take 75.4075 Ans. 194.7925
6. From 27.8 take 27.75 Ans. .05
7. From 5.4 take 1.4 Ans. 4.
8. From a unit, or 1, take the hundredth part of itself. Ans. .99

MULTIPLICATION OF DECIMALS.

RULE.*

Place the factors, and multiply them together, as in

* The reason of this rule will be easily understood after the student has become acquainted with the method of multiplying vulgar fractions together. In Multiplication of Vulgar Fractions, we multiply together the numerators, for a numerator, and the denominators, for a denomin-

whole numbers. Then point off in the product just as many decimal places as there are in both factors; and if the product has not so many figures, supply the defect by prefixing ciphers.

EXAMPLES.

1. Multiply .0127 by .25

Operation.	
.0127	Here I multiply 127, the significant figures of the multiplicand, by 25, and get 3175 for the figures of the product; but, as there are four decimal places in the multiplicand, and two in the multiplier, there must be six decimal places in the product; and therefore I prefix two ciphers to the product, to make up that number of places.
.25	
635	
254	
Prod. .003175	

2. Multiply 24.72 by 1.4 — Ans. 34.608
3. Multiply .785 by .0027 — Ans. .0021195
4. Multiply 284 by .07 — Ans. 19.88
5. Multiply .0582 by 21. — Ans. 1.2222
6. Multiply 7.528 by 120. — Ans. 903.36
7. Multiply .0025 by .03 — Ans. .000075
8. Multiply 26.4 by 1.25 — Ans. 33.

CONTRACTION I.

When it is required to multiply a decimal, or a mixed number, by 10, *or* 100, *or* 1000, *&c.*, it may be performed by merely removing the decimal point in the multiplicand as many places farther to the right hand as there are ciphers in the multiplier; annexing ciphers to the multiplicand, when necessary.

So, $475.87 \times 10 = 4758.7$
And $21.7 \times 100 = 2170.$
And $87.5 \times 1000 = 87500.$

ator; and the fraction, thus obtained, is the product required. In Multiplication of Decimals, the operation is, in effect, the same, only the denominators of the fractions are omitted; and it is evident that the product of the numerators, or given decimals, ought to contain just as many places of decimals as there are in both factors, because the product of the denominators would contain that number of ciphers.

CONTRACTION II.

To contract the operation, so as to retain only as many decimal places in the product as may be necessary, when the whole product would contain several more places.

RULE.—Count off, after the decimal point in the multiplicand, (annexing ciphers if necessary,) as many figures of decimals as it is necessary to have in the product. Below the last of these, write the units figure of the multiplier, and set down its other figures in a contrary order to what they are usually placed in. Then, multiply by each significant figure of the multiplier thus inverted, neglecting all the figures of the multiplicand to the right hand of the multiplying digit, except to find what is to be carried; and place all the partial products so that their right hand figures may stand in the same column. Lastly, add together these partial products, and point off the assigned number of decimal places in the sum, and you will have the product required.*

☞ In carrying from the rejected figures of the multiplicand, always take what is nearest the truth, whether it be too great or too small; that is, when the product or amount is from 5 to 14 inclusive, carry 1; when it is from 15 to 24, carry 2; when from 25 to 34, carry 3; &c.

EXAMPLES.

1. Multiply 27.149863 by 92.4105, so as to retain only four decimal places in the product.

---

* The reason of this method of contraction will appear by multiplying together, in the common way, the factors given in the first example, and then comparing the operation with the contraction; as follows:—

```
    27.149863
      92.4105
    ---------
     135|749315
    2714|9863
  108599|452
  542997|26
24434876|7
--------|------
2508.9324|147615
```

The figures which are here cut off at the right hand, by the perpendicular line, are omitted in the contracted way; and the last product here is the first there; and hence the reason of inverting the order of the figures of the multiplier, and of placing the several partial products as directed in the rule, is obvious.

Operation.
27.149863
5014.29

24434877
542997
108599
2715
136

2508.9324 Ans.

Expla.—Under the multiplicand I write the multiplier with the order of its figures inverted; taking care to place them so that the units figure of the multiplier, (viz. 2,) stands under the fourth decimal figure of the multiplicand. Then, I multiply the figures 27.14986 by 9; the figures 27.1498 by 2; the figures 27.149 by 4; the figures 27.14 by 1, and the figures 27 by 5; carrying from the figures rejected, and setting down the several partial products as directed in the rule. Lastly, I add these partial products together, and point off four decimal places in the sum.

2. Multiply .6479 by .07658, and retain only three decimal places in the product.

.6479
85670.

45 Prod. by 7.
4 Do. by 6.

.049 Ans.

3. Multiply 56.7534916 by 5.37692, and retain only five decimal places in the product. Ans. 305.15899

4. Multiply .8274 by 5.214, and retain only three decimal places in the product. Ans. 4.313

DIVISION OF DECIMALS.

RULE.

Place the divisor and dividend as in division of whole numbers. Then see whether the dividend contains as many places of decimals as the divisor; and if it does not contain as many, supply the deficiency by annexing ciphers; and always annex as many ciphers to the dividend as may be necessary to make it contain the divisor. Then divide as in whole numbers; and point off in the quotient as many places for decimals as the decimal places in the dividend exceed (in number) those in the divisor; taking care when the quotient does not contain so many figures to supply the defect by prefixing ciphers.

Note 1.—When there is a remainder after the division

of the dividend, the operation may be carried on as much farther as may be necessary, by annexing ciphers to the remainders, and dividing as usual.* It will be best always to place the decimal point in the quotient immediately after dividing the given dividend, or that first set down; and then, if there is a remainder, the quotient may be carried on farther, if necessary.

Note 2.—The reason of pointing off so many decimal places in the quotient as those in the dividend exceed those in the divisor, will easily appear; for, since the quotient multiplied by the divisor gives the dividend, therefore the number of decimal places in the dividend is equal to those in the divisor and quotient, taken together, by the nature

* By proceeding in this manner, the division will sometimes soon terminate without a remainder; but, in many cases, there will be a remainder left if the quotient is extended to ever so many places of decimals.

A decimal which cannot be exactly expressed, but which may be continued to an unlimited number of figures, is called an *interminate decimal*, to distinguish it from others, which in respect of it, are called *terminate*. An interminate decimal which is expressed either by the continual repetition of the same figure, or of the number expressed by two or more figures, is called a *periodical* or *circulating decimal;* and the figure, or number, so repeated, is called the *period* Thus, the decimals .333, &c., .5454, &c are periodical decimals, the period in the former consisting of one figure, and that in the latter of two figures. A periodical decimal is said to be *mixed*, if it consists of one or more figures prefixed to a periodical part; others are called *pure*. Thus, the decimals .333, &c., .5454, &c, are pure periodicals; and the decimals 833, &c., .12436436, &c. are mixed. For the sake of brevity, in writing decimals of this kind, it will often be sufficient to write the period but once, and to denote its continuation by placing a point or dot over the first figure of the period, and another over the last figure, or one over the repeating figure, if there be but one figure in the period. Thus, .4733, &c., may be expressed by $.47\dot{3}$, and .5637637, &c., by $.5\dot{6}3\dot{7}$

As periodical decimals often occur in the division of decimals, it may be proper to show the learner how the exact values of them may be determined, when necessary.

To find the value of a pure periodical decimal; take the period for a numerator, and as many nines as there are figures in the period, for a denominator; and you will have a vulgar fraction equivalent to the given decimal.

Thus, .8181, &c.$=\frac{81}{99}=\frac{9}{11}$; $.\dot{2}9\dot{7}=\frac{297}{999}=\frac{11}{37}$; $.\dot{3}=\frac{3}{9}=\frac{1}{3}$, and $.\dot{9}=\frac{9}{9}=1$.

To find the value of a mixed periodical decimal; from the number expressed by the finite part with the period annexed, subtract the finite part for the numerator; and, for the denominator, to as many nines as

of multiplication; and consequently the quotient itself must contain as many as the dividend exceeds the divisor.—When the dividend contains just as many decimal places as the divisor, the quotient will evidently be a whole number.

EXAMPLES.

1. Divide .0008625 by .0345

Operation.

```
.0345).0008625(.025 Ans.
        690
        ———
        1725
        1725
```

Here I divide 8625, the significant figures of the dividend, by 345, the significant figures of the divisor, and the quotient is 25. Then, to find where the decimal point must be placed in the quotient, I count the decimal places of the divisor, and find them four; and then count the decimal places of the dividend, and find them seven, that is three more than those of the divisor. There

---

there are figures in the period, annex as many ciphers as there are figures in the finite part.

Thus, to find the value of $.8\dot{3}$; from 83 take 8, and there will remain 75, and the required fraction is $\frac{75}{90}$, or $\frac{5}{6}$. In like manner, to find the value of $.26\dot{3}0\dot{1}$, we have for numerator 26301—26=26275, and for denominator 99900; and hence the required fraction is $\frac{26275}{99900}$, or $\frac{1051}{3996}$.

By the method thus shown, interminate decimals may be reduced to vulgar fractions, and subjected to the rules for managing such quantities. Unless when complete precision is required, however, this is not necessary: and indeed in all useful cases, their values, instead of being found with entire accuracy, are to be approximated, by carrying the decimals out to as many figures as may be necessary in a particular case.

In the application of decimals to practical purposes, it is generally known, from the nature of the case under consideration, to how many places it is necessary that the result should be true When a result is thus required to be true to an assigned number of places of decimals, it is proper to carry the decimals which consist of more places, to at least one place beyond the assigned number, and to reject the last figure. In this case, it is proper to observe, that when a decimal is not carried out to its full length, the last figure of the part retained should be increased by 1 if the succeeding figure be 5 or greater than 5 Thus, if we would reject the two last figures of the decimal .47263, it is proper to increase the third decimal figure by 1; which being done, and the two last figures rejected, the decimal is .473, which is nearer the value of the given decimal than .472 is.

must, therefore, be three decimal places in the quotient; and so I prefix a cipher to the quotient to make up that number of places, and place the decimal point before it.

2. Divide 8.6 by 2.718

```
2.718)8.600(3.164+ Quotient.
      8154
      ----
       4460
       2718
       ----
       17420
       16308
       -----
        11120
        10872
        -----
          248
```

Here, I annex two ciphers to the given dividend, to make up as many decimal places in it as there are in the divisor. After the division of this dividend, there is a remainder, and I carry on the operation farther, according to Note 1.

3. Divide 17.1 by 8.

```
  8)17.1
  ------
Quot. 2.1375
```

Here, after dividing 17.1 by 8, there is a remainder, and I *suppose* ciphers to be annexed to the dividend, and continue the operation till nothing remains.

4. Divide 8564.825 by 63.21 Ans. 135.49+
5. Divide 56.7 by .7 Ans. 81.
6. Divide 246.1 by 6.0427 Ans. 40.72+
7. Divide 7.25406 by 957. Ans. .00758
8. Divide 76 by .7438 Ans. 102.17+
9. Divide 65.8 by 1.2 Ans. 54.833,&c.
10. Divide 27 by .05 Ans. 540.

CONTRACTION I.

When the divisor is an integer, with any number of ciphers on the right hand; cut off those ciphers, and remove the decimal point in the dividend as many places farther to the left as there are ciphers cut off from the divisor; prefixing ciphers to the dividend, if necessary: then divide the dividend by the remaining part of the divisor, as usual.

EXAMPLES.

1. Divide 7.14 by 7400.

74,00).0714(.00096+ Quotient.
666

480
444

36

2. Divide 485.2 by 1780. Ans. .272+
3. Divide 45.5 by 2100. Ans. .021+
4. Divide 7.8 by 8000. Ans. .000975

Note.—When the divisor is 10, or 100, or 1000, &c., the quotient may be found by merely removing the decimal point in the dividend as many places farther to the left as the divisor has ciphers; prefixing ciphers, if necessary. So 21.4÷10=2.14; and .54÷100=.0054; &c.

CONTRACTION II.

To contract division when there are many figures in the divisor, and it is required to find only a certain number of figures in the quotient.

RULE 1.

1. Take as many of the left hand figures of the divisor as will be equal to the number of figures (both integers and decimals) required to be found in the quotient, and find how many times they may be had in the first figures of the dividend, as usual.

2. Let each remainder be a new dividual, and for every such dividual reject one figure more from the divisor; and in making out each subtrahend, carry from the figures cut off from the divisor, as in the 2d contraction in Multiplication of Decimals.

Note 1.—When there are not as many figures in the given divisor as are required to be in the quotient, begin the operation with all the figures, as usual, and continue it until the number of figures in the divisor and those remaining to be found in the quotient are equal; after which use the contraction.

Note 2.—To know where to place the decimal point in the quotient; find, by the general rule for dividing decimals, the value, or place, of the first quotient figure, and you will then know where the decimal point must be placed.

EXAMPLES.

1. Divide 2508.9324 by 92.4105, so as to have only three decimal places in the quotient, in which case the quotient will contain five figures.

Operation.

```
92.410,5)2508.93,24(27.149 Quotient.
         184821
         ------
  92.41)66072
        64687
        -----
   92.4)1385
         924
        ----
     92)461
        370
        ---
      9)91
        83
        --
         8
```

N. B. I have set down every divisor, in order to explain the work; but you need only put a dot over every figure rejected, as you proceed, to show that it is omitted.

2. Divide 4109.2351 by 230.4091, so as to have six figures in the quotient. Ans. 17.8345

3. Divide 37104.36 by 57.1396, so that the quotient may contain seven figures. Ans. 649.3633

4. Divide 913.08 by 21372, so that the quotient may contain four places of decimals. Ans. .0427

RULE 2.

Take, for a defective divisor, one or two more of the left hand figures of the given divisor than the number of figures required to be found in the quotient, and divide the dividend by this defective divisor, as usual.

Note.—When any of the figures rejected from the divisor are integers, then you must remove the decimal points in the divisor and dividend, or suppose them to be removed, as many places farther to the left as there are integral figures rejected from the given divisor.

EXAMPLES.

1. Divide 721175.62 by 222574.12, so as to have only hree figures in the quotient.

Operation.

22257,4.12)721175.62(

```
22257)72117.562(3.24 Quotient.
      66771
      -----
       53465
       44514
       -----
        89516
        89028
        -----
          488
```

Here, because I cut off one integral figure from the divisor, I remove the decimal point in the dividend one place to the left.

2. Divide 250.8928 by 92.41052, and find only three figures in the quotient. Ans. 2.71

3. Divide 12.169825 by .031415926, so as to have four figures in the quotient. Ans. 387.3

FEDERAL MONEY.

Having explained the nature of decimal fractions, and given rules for adding, subtracting, multiplying, and dividing decimals, I shall now proceed to show the application of those rules to the currency of the United-States, usually denominated *Federal Money.*

The denominations of federal money increase, from the lowest to the highest, in a tenfold ratio, like whole numbers and decimal fractions. The denominations are as follows:

10 Mills (m.)	= 1 Cent,	c. or ct.
10 Cents	= 1 Dime,	dm.
10 Dimes, or 100 cents,	= 1 Dollar,	D. or $.
10 Dollars	= 1 Eagle,	E.

There are coins of all these denominations, excepting that of mills, which is merely nominal.

As all the denominations of federal money increase in a tenfold ratio, like whole numbers, it is very obvious that the value of any sum of federal money consisting of several denominations, may be expressed in the lowest denomination mentioned, by writing the numbers of the several denominations one after another, in regular order, beginning with the highest, and placing the figures so that they may be read as one whole number. Thus, 17 eagles, 5 dollars, 6 dimes, 2 cents, and 8 mills, are equal to 175628 mills. In writing down sums of federal money in this manner, if any denomination between the highest and lowest, mentioned in the given sum, be wanting, a cipher must be written in its place: Thus, 6 dollars and four cents, are equal to 604 cents; and 7 dollars and 5 mills, are equal to 7005 mills.—It is also evident, that if the number of any one denomination be considered so many units, the lower denominations will be decimal parts of a unit, and may therefore be expressed like other decimal fractions. So, 8 eagles, 4 dollars, 6 dimes, 7 cents, and 5 mills, are equal to 8.4675 eagles,=84·675 dollars,=846.75 dimes,=8467.5 cents,=84675 mills.

In reckoning in federal money, dollars are considered as units or integers, and parts of a dollar, as decimal parts of a unit: So, 5 dollars, 8 dimes, 4 cents, 2 mills, are written thus, $5.842; and 7 eagles, 2 dollars, and 4 cents, thus, $72.04; &c. Hence, Addition, Subtraction, Multiplication, and Division of Eederal Money, are performed by the foregoing rules for decimal fractions.

Eagles are usually considered as tens of dollars, and dimes as tens of cents; and the names of eagles and dimes are seldom mentioned; accounts being kept in dollars, cents and mills, or more frequently in dollars and cents only, the mills being usually considered of too little value to be retained. So, $27.57, is read thus, *twenty-seven dollars and fifty-seven cents*, or *twenty-seven dollars and fifty-seven hundredths*; and $.254, thus, *twenty-five cents, four mills*, or *two hundred fifty-four thousandths of a dollar*.

☞ In writing down parts of a dollar in the decimal form, always remember to prefix a cipher to the cents when the number is less than 10, and two ciphers to the mills, when the fractional part of a dollar consists of any number of mills less than 10. So, write 7 dollars and 4 cents thus, $7.04; and 14 dollars and 5 mills thus, $14.005.

Examples in Addition of Federal Money.

1. Find the amount, or total sum, of 58D. 89c.+9D. 5c.+ 14D. 8m.+98c. 7m.

$58.89
9.05
14.008
.987

Ans. $82.935

2. Bought a hat for 5D.; a vest for 2D. 12c. 5m.; a coat for 15D.; and a pair of boots for 6D. 50c.: What did they all cost? Ans. $28.625

3. Bought a quantity of goods in New-York for 575D. 62c. 5m.; paid for carting the goods to the dock 1D. 25c.; for freighting the same to Bridgeport 5D 7c.; for carrying the same to New-Milford 5D. 25c.; and my own expenses were 10D. 12c. 5m.: How much do the goods stand me in at New-Milford? Ans. $597.32

4. Suppose I am indebted
To A twenty-six dollars, fifty cents,
B forty-eight cents,
C sixty dollars, four cents, five mills,
D nine dollars and five mills,
E two hundred dollars:
How much is the amount of my debts? Ans. $296.03

Examples in Subtraction of Federal Money.

1. From 85D. subtract 4D. 8c. 2m.

$85.
4.082

Ans. $80.918

2. A merchant bought a quantity of goods for 485D. 50c., and afterwards sold the same for 557D.: how much did he gain by the sale? Ans. $71.50

3. Suppose the effects of a bankrupt amount to $2000, and he owes to A 1250D.; to B 875D. 78c.; to C 271D. 18c.; and to D 40D. 75c.: What is the deficiency?
Ans. $437.71

4. From twenty-one cents take three mills.
Ans. $.207, or 20c. 7m.

5. A has an account against B, to the amount of one hundred dollars and ten cents; and B pays him forty-two dollars twelve and a half cents in cash, and agrees to give

his note for the remainder—for what sum must the note be drawn? Ans. $57.975

Examples in Multiplication of Federal Money.

1. Multiply 58 D. 5 c. by 7.

$58.05
7

Ans. $406.35

2. Multiply 72 D. by .12 Ans. $8.64

3. What do 20 barrels of flour amount to, at 5 D. 25 c. per barrel? $5.25×20=$105, Ans.

4. What do 174 bushels of wheat amount to, at 1 D. 20 c. a bushel? 174×$1.20=$208.80, Ans.

5. What is the value of 48.75 acres of land, if each acre be worth 15 D. 25 c.?

Ans. $743.4375, or 743 D. 43 c. 7.5 m.

Note.—When there are figures to the right hand of the place of mills, they are decimal parts of a mill, and of too little value to be retained in ordinary calculations.

Questions.			*Answers.*
What is the value of	$		$
97 pieces of cloth, at	9.87	a piece?	957.39
155 yards of muslin, at	.755	a yd.?	117.025
175 yards of broadcloth, at	5.08	a yd.?	889.
8.5 lb. of tea, at	.875	a lb.?	7.437+
27.25 lb. of butter, at	.125	a lb.?	3.406+
284 lb. of cheese, at	.065	a lb.?	18.46
180 oranges, at	.025	each?	4.50
48 pairs of shoes, at	1.625	a pair?	78.
4.567 cords of wood, at	2.80	a cord?	12.787+

15. Find the amount of the articles in the following bill.

New-York, ———

Jacob Wanzer, Bought of *John Merchant,*

	$	
24 pieces of Irish linen, at	7.65	a piece,
10 do. blue satin, at	7.	a piece,
23 yards of durant, at	.45	a yd.
14 do. book muslin, at	.78	a yd.
19 do. black calimanco, at	.39	a yd.

Amount, $282.28

Received payment. *John Merchant.*

Examples in Division of Federal Money.

1. If 254 dollars be divided equally among 4 men, how much will each man receive?

$ 4)254.

Ans. $ 63.5
Or 63 D. 50 c.

Note.—When there is only one decimal figure in the quotient, (found by dividing any number of dollars,) it is dimes, or tenths of a dollar; and, to reduce it to cents, a cipher must be annexed; as in the foregoing example.

2. Bought 250 lb. of cheese, for 16 D. 25 c.: what did it cost me a pound?

$ 16.25÷250=$ .065=6 c. 5 m.=6½ c. Ans.

3. If a contribution of $ 27000 is to be made up in equal shares, by 625 persons, how much must each contribute?
Ans. $ 43.20

4. Suppose a man labors a month for 12 D.; how much does he have a day; there being 26 working days in a month? Ans. 46 c. 1 m.+

5. The canal of Languedoc, in France, is 180 miles long, and cost $ 2400000: how much is that per mile?
Ans. $ 13333.33+

6. The salary of the President of the United-States is $ 25000 a year: how much is it a day, allowing 365.25 days to make a year? Ans. $ 68.446+

7. Bought a quantity of wheat, at $1.25 per bushel, which amounted to $375. How many bushels of wheat did I buy?
Ans. 300 bush.

REDUCTION OF DECIMALS.

CASE I.

To reduce a vulgar fraction to its equivalent decimal.

RULE.

Divide the numerator by the denominator, as in division of decimals, and the quotient will be the decimal fraction required.

EXAMPLES.

1. Reduce $\frac{1}{8}$ to its equivalent decimal. 8)1.000

Ans. .125

Note.—In dividing after the manner of short division, you may *suppose* ciphers to be annexed to the numerator, and perform the operation without setting down the ciphers; as in the next following example.

2. Reduce $\frac{3}{4}$ to a decimal. 4)3.

Ans. .75

3. Reduce $\frac{1}{2}$ to a decimal. Ans. .5
4. Reduce $\frac{1}{4}$ to a decimal. Ans. .25
5. What decimal is equal to $\frac{2}{3}$? Ans. .666, &c.
6. What decimal is equal to $\frac{1}{12}$? Ans. .0833, &c.
7. What decimal is equal to $\frac{3}{40}$? Ans. .075
8. What decimal is equal to $\frac{7}{2000}$? Ans. .0035
9. Reduce $\frac{5}{13}$, $\frac{11}{16}$, and $\frac{27}{5719}$, to decimals.

Answers, .3846+, .6875, and .00472+

CASE II.

To find the value of a decimal fraction of any of the higher denominations of Money, Weight, or Measure, in whole numbers in the lower denominations.

RULE.

Multiply the given decimal by the number which 1 of that denomination makes of the next lower, and point off the proper number of decimal places in the product, as in Multiplication of Decimals. In like manner reduce the *decimal part*, (if any,) of this product to the next lower denomination; and so proceed through all the inferior denominations, if necessary: then the last product, together with the integers, (or numbers at the left hand of the decimal points,) in the other products, will be the answer.

Note 1.—This Case is similar to Reduction Descending, and the following Case to Reduction Ascending, in whole numbers.

EXAMPLES.

1. Required the value of .775 of a pound sterling.

.775 *l.*
20 s.=1 *l.*
―――
15.500 s.
12 d.=1 s.
―――
6.000 d.
Ans. 15 s. 6 d.

2. Required to find the proper quantity of .722 of a lb., Avoirdupois.

.722 lb.
16 oz.=1 lb.
―――
11.552 oz.
16 dr.=1 oz.
―――
8.832 dr.
Ans. 11 oz. 8.832 dr.

Explanation of example 1*st.*—I first multiply the given decimal of a pound by 20, to reduce it to shillings, and the product is 15.500 shillings. I then multiply the decimal of a shilling, viz. .500, by 12, to reduce it to pence, and the product is 6 d.; and there being no decimal of a penny to reduce to farthings, the work is done. So the answer is 15 s. 6 d.

What is the value, or proper quantity,

3. Of .625 of a shilling? Ans. 7 d. 2 q., or 7½ d.
4. Of .03125 of a mile? Ans. 10 rods.
5. Of .125 of a foot? Ans. 1 in. 1.5 b. c.
6. Of .50625 of an acre? Ans. 2 roods, 1 sq. rd.
7. Of .1875 of a sq. foot? Ans. 27 sq. in.
8. Of .125 of a cord of wood? Ans. 16 cub. ft.
9. Of .182 of a gallon of wine? Ans. 1 pt. 1.824 gil.
10. Of .21 of a day? Ans. 5 h. 2 min. 24 sec.
11. Of .015 of a degree? Ans. 54″.

Note.—The addition and subtraction of decimals of different denominations, may be performed after the decimals are reduced to their proper quantities; as in the following examples.

12. What is the sum of .17 of a lb. Troy, and .87 of an ounce, reduced to their proper quantities?

	oz.	pwt.	gr.
.17 lb.=	2 ..	0 ..	19.2
.87 oz.=		17 ..	9.6
Ans.	2 ..	18 ..	4.8

13. What is the sum of .15 lb. Avoirdupois, and .25 oz.? Ans. 2 oz. 10.4 dr.

14. What is the difference between .17*l*. and .7s.?
Ans. 2s. 8d. 1.6q.

15. What is the difference between .41 of a day, and .16 of an hour?
Ans. 9h. 40min. 48sec.

CASE III.

To reduce any given quantity to a decimal of a higher denomination.

RULE.

I. *When the given quantity consists of several denominations*, proceed as follows:

1. Write the given numbers under each other, for dividends; proceeding orderly from the least denomination to the greatest.

2. Opposite to each dividend, on the left hand, place such a number, for a divisor, as will, (according to the rule for Reduction Ascending,) bring the dividend to the next superior denomination.

3. Divide each dividend, beginning at the top of the column, by its divisor, and write the quotient of each division as decimal parts on the right hand of the dividend next below it, and the last quotient will be the decimal sought.

Note 1.—In dividing in this manner, you must find as many decimal figures in each quotient as are required to be in the decimal sought, when the true quotients will contain so many figures.

II. *When the given quantity consists of one denomination only*, it may be reduced to a decimal as directed above; or it may be divided, at once, by such a number as will reduce it to the decimal required.

Note 2.—Cases 2d and 3d prove each other.

EXAMPLES.

1. Reduce 15s. 6d. to the decimal of a pound.

Operation.

12	6d.
20	15.5s.
Ans.	.775 *l*.

Here, I first set down the 15s. below the 6d. and place the proper divisors against these numbers. I next divide the 6d. by 12, to reduce them to the decimal of a shilling, and the quotient is .5s., which I annex to the

15s. I then divide the 15.5s. by 20, to reduce them to the decimal of a pound, and the quotient, .775*l.*, is the answer.

☞ The student must remember to prefix ciphers to the quotients, when necessary, to make up the proper number of decimal places, according to the rule for the division of decimals.

2. Reduce 11 oz. 8.832 dr. to the decimal of a pound, Avoirdupois. Ans. .722 lb.

3. Reduce 7d. 2q. to the decimal of a shilling. Ans. .625 s.

4. Reduce 10 rods to the decimal of a mile.

By the 2d part of the Rule thus, 10÷320 (the number of rods in a mile)=.03125 m. Ans.

5. Reduce 1 in. 1.5 b. c. to the decimal of a foot. Ans. .125 ft.

6. Reduce 2 roods, 1 sq. rod, to the decimal of an acre. Ans. .50625 A.

7. Reduce 27 sq. inches to the decimal of a sq. foot. Ans. .1875 sq. ft.

8. Reduce 16 cub. feet to the decimal of a cord. Ans. .125 Cd.

9. Reduce 1 pt. 1.824 gil. to the decimal of a gallon. Ans. .182 gal.

10. Reduce 5h. 2min. 24 sec. to the decimal of a day. Ans. .21 da.

11. Reduce 54 seconds to the decimal of a degree. Ans. .015 deg.

QUESTIONS ON THE FOREGOING.

1. What is a decimal fraction? 2. In what manner are decimal fractions written? 3. What is a mixed number? 4. How are decimals enumerated? 5. What does the first figure to the right hand of the decimal point denote? the second? the third? &c. 6. What effect have ciphers when placed at the right or left hand of the significant figures of a decimal? 7. What is the rule for the addition of decimals? 8. How is the subtraction of decimals performed? 9. How are decimals multiplied together; and what is the rule for placing the decimal point in the product? 10. What is the shortest method of multiplying a decimal, or mixed

number, by 10, or 100, &c.? 11. How is the division of decimals performed; and what is the rule for placing the decimal point in the quotient? 12. What is the shortest method of dividing a decimal, or mixed number, by 10, or 100, &c.? 13. What are the denominations of Federal Money? 14. In what ratio do these denominations increase? 15. If dollars be considered as integers, what will each of the lower denominations be? and what will eagles be? 16. Which of these denominations is usually considered the integer; and how are sums of federal money usually written? 17. Is it customary, in reckoning in federal money, to mention the names of eagles and dimes? 18. How do we write down, in the decimal form, a part of a dollar which consists of any number of cents less than 10, or any number of mills less than 10? 19. How are Addition, Subtraction, Multiplication, and Division of Federal Money performed? 20. What is the method of reducing a vulgar fraction to a decimal? 21. What is the rule for reducing a decimal fraction of any of the higher denominations of money, &c. to its value in the lower denominations? 22. What is the method of reducing a compound quantity to an equivalent decimal of a higher denomination? 23. How is a simple quantity reduced to a decimal of a higher denomination?

PROPORTION,

Is the relation which one quantity has to another.

Numbers are compared to each other in two different ways: One comparison considers the *difference* of the two numbers, and is named *Arithmetical Relation*, and the difference is sometimes called the *Arithmetical Ratio*: the other considers their *quotient*, which is called *Geometrical Relation*, and the quotient, the *Geometrical Ratio*.* So, of these two numbers, 8 and 2, the difference, or arithmetical ratio, is 8—2, or 6, but the geometrical ratio is 8÷2, or 4.

* It may be proper to inform the learner that, in this Work, where mention is made of the *ratio* of two quantities, the geometrical ratio is intended, excepting where it is otherwise expressed.

There must be two numbers to form a comparison: the number which is compared, being placed first, is called the *antecedent;* and that to which it is compared, the *consequent.* So, of the two numbers above, 8 is the antecedent, and 2 the consequent.

Note.—Arithmetical and Geometrical Proportions will be treated of at large in a subsequent part of this Treatise; but I shall here explain so much of the latter as is necessary to give the learner a correct idea of the nature of the following *Rule of Proportion,* called the *Rule of Three.*

A *geometrical ratio* is usually denoted by writing a colon between the two terms, (that is between the antecedent and the consequent,) of the ratio. Thus, 3 : 5, denotes the ratio of 3 to 5.

Four quantities are said to be in *geometrical proportion,* when the ratio of the first to the second is equal to the ratio of the third to the fourth. Thus, the four numbers 8, 4, 14, 7, are proportional, because the ratio 8 : 4 is equal to the ratio 14 : 7; that is, $8 \div 4$ is equal to $14 \div 7$, equal to 2. The equality or identity of two such ratios is usually denoted by writing a double colon between the ratios: Thus, 2 : 3 :: 6 : 9, denotes that the ratio of 2 to 3 is the same with, or equal to, the ratio of 6 to 9: read thus, *as* 2 *is to* 3, *so is* 6 *to* 9. Such a series, consisting of four terms in geometrical proportion, is called an *Analogy;* the first and last terms being called the *extremes,* and the other terms, the *means.*

In any analogy, the product of the extremes is equal to the product of the means; that is, the product of the first and fourth terms is equal to the product of the second and third terms.* Thus, in the analogy 2 : 4 :: 3 : 6, the pro-

* This proposition, or assertion, may be demonstrated as follows: Let a, b, and r, represent any three quantities whatever; then, $a : ar :: b : br$, will denote any analogy, or any four quantities in geometrical proportion; r being the ratio, or quotient, of the first and second, and of the third and fourth terms. Now, $a \times br = ar \times b = abr$; that is, the product of the extremes is equal to the product of the means. Q. E. D.

Corollary.—Hence, if the product of the two mean terms of any analogy be divided by either of the extremes, the quotient will be the other extreme; and, if the product of the extremes be divided by either of the mean terms, the quotient will be the other mean; for it is evident that if the product of any two numbers be divided by one of the num-

duct 2×6 is equal to the product 4×3, equal to 12.—Hence, if the product of the two mean terms of any analogy be divided by either of the extremes, the quotient will be the other extreme; which is the foundation of the following Rule of Simple Proportion, commonly called the Rule of Three.

The Rule of Proportion is divided into *Simple* and *Compound.*

Simple Proportion is a single analogy, or the equality of the ratio of two quantities to that of two other quantities; as 2 : 6 :: 8 : 24.

Compound Proportion is the equality of the ratio of two quantities to another ratio, the antecedent and the consequent of which are respectively the products of the antecedents and consequents of two or more ratios; as $\left.\begin{matrix}2:3\\9:7\end{matrix}\right\}$ :: 6 : 7, in which the ratio 6 : 7 is equal to a ratio whose antecedent is 2×9, or 18, and its consequent 3×7, or 21.

SIMPLE PROPORTION, OR

THE SINGLE RULE OF THREE.

The *Rule of Simple Proportion*, or *Rule of Three*, teaches how to find the fourth term of any analogy from three given terms. It is called the *Rule of Three*, because three terms or numbers are given to find a fourth; and, because of its great and extensive usefulness, it is sometimes called the *Golden Rule.*

In every question belonging to the Rule of Three, two of the three given terms, or numbers, are contained in a *supposition*, and the other in a *demand;* and hence the former are called the *terms of supposition*, and the latter the *term of demand.* One of the terms of supposition is always of the same kind or quality with the demanding term, and the

---

bers, the quotient will be the other number. Consequently, if any three of the terms of an analogy be given, or known, the other term may be found; which admits of four cases; viz 1st, when the two mean terms and the first term are given, to find the fourth: 2d, when the two mean terms and the fourth term are given, to find the first: 3d, when the extremes and the second term are given, to find the third: 4th, when the extremes and the third term are given, to find the second.

other term of supposition is of the same kind with the answer, or term sought. When the three numbers, or quantities, given in any such question, are arranged in the proper order, they form the first three terms of an analogy of which the answer, or number sought, is the fourth term; and the business of placing the three given terms in the proper order is called *stating* the question.

The Rule of Simple Proportion has usually been divided into *Direct*, and *Inverse Proportion*; and in all the old systems of Arithmetic a particular rule is given for solving questions in direct proportion, and another for questions in inverse proportion. This distinction, however, is unnecessary, and therefore I shall omit it.*

All questions in Simple Proportion may be solved by the following

GENERAL RULE.

1. State the question as follows; viz. Set down that quantity, or term of supposition, which is of the same kind with the answer, or number sought, for the third term of the proportion; that is, if the answer is to be money, the third term must be money; or, if the answer is to be weight or measure, then the third term must be weight or measure. Then consider, from the nature of the question, whether

---

* As the terms *Direct* and *Inverse Proportion*, are very frequently used by mathematicians, it may be proper, in this place, to explain their meaning.

One quantity is said to be *directly* as another quantity, when the one increases in the same ratio in which the other increases, or diminishes in the same ratio in which the other diminishes: For example, if I purchase cloth at 2 dollars a yard, the amount of the cost depends upon the quantity purchased; that is, the greater the quantity, the greater the amount of the cost; and therefore the amount is said to be directly as the quantity.—But, when one quantity increases in the same ratio in which the other diminishes, the one is said to be *inversely* as the other: e. g. if I have to ride a certain distance, the time requisite depends upon the speed employed; that is, the greater the speed, the less time will be requisite; and hence the time is said to be inversely as the speed.

The *reciprocal* of any geometrical ratio, is that ratio, (or an equivalent one,) with the order of its terms inverted. Thus, the ratio 5 : 8 is the reciprocal, or inverse, of the ratio 8 : 5.

When two ratios are equal, they, (or the numbers which express them,) are said to be in *direct* proportion. Such are the ratios 2 : 4 and 5 : 10. But when one ratio is equal to the reciprocal of another, the ratios are said to be in *inverse*, *indirect*, or *reciprocal* proportion. Such are the ratios 2 : 4 and 10 : 5.

the answer, (or fourth term,) is to be greater or less than the third term:* If the answer is to be greater than the third term, write the greater of the two other given numbers for the second term, and the less for the first; but, if the answer is to be less than the third term, then make the less remaining number the second term, and the greater the first term.

2. If the first and second terms be simple numbers of different names; or, one or both of them of divers denominations; reduce them both to the same denomination, viz. to the lowest denomination mentioned in either.† Then multiply the second and third terms together, and divide their product by the first term, and the quotient will be the fourth term, or answer, in the same denomination as the third term; which may be brought into any other denomination required.‡

Note 1.—When the third term consists of several denominations, you may multiply it by the second term, and divide the product by the first term, according to the rules for Compound Multiplication and Division: Or, you may

---

* In every case that can occur, the conditions of the question will show whether the fourth term, or number required, is to be greater or less than the third term.

† The first and second terms must always be of the same kind or quality; and when they are not both simple numbers of the same denomination, they must be made such, by reduction.

‡ The truth and reason of this Rule will easily appear from the demonstration of a proposition respecting four proportional numbers, given in a preceding note. The truth of the Rule, as applied to ordinary inquiries, may also be made evident by attending to principles which have been explained in a preceding part of the Work. It has been shown, in the application of Multiplication to practical purposes, that the value of one article, or yard, &c. multiplied by the whole quantity, or number of articles, is the value of the whole; and, in the application of Division, that the value of the whole, divided by the quantity, or number of articles, is the value of one. Now, in all cases of valuing goods, &c. where the first term of the proportion is 1, it is plain that the answer found by this Rule will be the same as that found by the rule which has been given for the application of Multiplication; and, where 1 is the second term, it will be the same as that found by the rule for the application of Division. In like manner, if the first term be any number whatever, it is plain that the product of the second and third terms will be greater than the answer required, by as much as the price in the third term exceeds the price of one, or as the first term exceeds a unit; consequently this product, divided by the first term, will give the true answer required.

reduce the third term to the lowest denomination of which it consists, and then multiply it by the second term, and divide the product by the first term; and the quotient will be the answer in the denomination which the third term was reduced to; which may be brought into a higher denomination, if necessary. The latter method will often be more convenient than the former, especially when the first and second terms are large numbers.—When the third term is Federal money, or contains any decimals, then multiply and divide according to the rules for the multiplication and division of decimals.

Note 2.—Sometimes two or more statements will be necessary, in order to solve a question; which may always be known from the conditions of the question.

Proof.—The method of proof is by inverting or varying the order of the question.

EXAMPLES.

1. If 5 yards of cloth cost $14, what would 80 yards cost, at the same rate?

yd. yd. $
Stated thus; As 5 : 80 :: 14 :
80
———
5)1120
———
Ans. $224

This statement may be read thus; As 5 yards is to 80 yards, so is $14 to $224, the answer.

Explanation.—In the foregoing question, the *supposition* is, *that* 5 *yards of cloth cost* $14; and the *demand* is, *to find what* 80 *yards would cost at the same rate:* Therefore, the terms of supposition are 5 yards and $14; and the term of demand is 80 yards. Now, it is evident that the answer to the question must be *money*; and therefore, in stating the question, I set down the $14, (the term of supposition which is of the same kind with the answer,) for the third term. It is also evident that 80 yards of cloth must cost more than 5 yards; and therefore I make the greater number of yards the second term, and the less the first term. Then, as the first and second terms are both whole numbers of the same denomination, there are no reductions to be performed, to prepare these terms for multiplying and dividing: so I proceed to multiply the second and third terms together, and then divide their product by the first term; and the quotient is the fourth term, or answer; being the value of 80 yards of cloth at the rate of $14 for 5 yards.

2. If 80 yards of cloth cost $224, what will 5 yards cost?

As 80 yd. : 5 yd. :: $224 :
5

8,0)112,0

Ans. $14.

N. B.—The second example is only the first *varied*, being given to show the learner how the answer to any question in Simple Proportion may be proved by inverting or varying the order of the question.*

3. If 100 lb. of butter cost $12, how much butter may be bought for $50?

$12 : $50 :: 100 lb.
100

12)5000

Ans. 416 lb. $10\frac{2}{3}$ oz.

Here, after the division of the product of the second and third terms by the first, there is a remainder of 8 lb., which I reduce to ounces, and then divide it by the divisor, as in Compound Division.

4. If 8 lb. of cheese cost 3 s. 5 d., what would 20 lb. cost?

Operation.

8 lb. : 20 lb. :: 3 s. 5 d.
12

41 d.
20

8)820

12)102 d. 2 q.

Ans. 8 s. 6 d. 2 q.

Or, otherwise thus:—

lb. lb. s. d.
8 : 20 :: 3 .. 5
20

8)68 .. 4

Ans. 8 .. $6\frac{1}{2}$

* We may, in like manner, form two other questions from Ex. 1. Thus; if $14 buy 5 yards of cloth, how many yards would $224 buy? Or, 2dly, if $224 will buy 80 yards of cloth, how many yards will $14 buy? And thus, whenever a question has been solved by the Rule

Remarks.—When the third term consists of several denominations, if the first and second terms are large numbers, or contain several significant figures, it will usually be most convenient to reduce the third term to the lowest denomination of which it consists; but, if the first and second terms are small numbers, the operation may be conveniently performed by Compound Multiplication and Division. Both these methods of operation are illustrated by example 4th.

The principal difficulty in solving questions in Simple Proportion, consists in *stating* them; and, in order to assist the learner, several of the following question are stated, which are not worked out at full length. In these abridged examples, where the first and second terms need reducing, all the terms are first set down in the denominations in which they are given in the question; and then the statement is given again, with the first and second terms properly reduced.

In each of the eighteen questions next following, (viz. questions 5th to 22d,) the first and second terms are whole numbers of the same name or denomination; and hence the answer is found by merely multiplying the second and third terms together, and then dividing their product by the first term; only it may be proper in some cases to reduce the third term to a simple number before multiplying it by the second term.

5. If 12 yards of cloth cost $52, what would 45 yards cost? Ans. $195.

6. Bought 4 pieces of cloth for 15 D. 25 c.: what would 50 pieces have cost at the same rate?

As 4 pieces :: 50 pieces : $15.25 : $190.625, Ans.

7. If a bag of coffee, weighing 110 pounds, cost 18 D. 15 c., what will 14 lb. cost? Ans. $2.31

8. How many yards of sarcenet may be purchased for $117, at the rate of $4 for 3 yd. 1 qr.?

As $4 : $117 :: 3 yd. 1 qr. : 95 yd. 0 qr. 1 na. Ans.

9. If 385 yards of linen cost $315, how many yards might be bought for $90? Ans. 110 yd.

---

of Three, the student may be profitably exercised in forming three other questions adapted to prove the truth of his answer, since we can find any one of the four terms of an analogy from having given the three others.

10. If a piece of linen, containing 20 yards, cost 2*l*. 10s., what would 155 yards cost at the same rate?

Ans. 19*l*. 7s. 6d.

11. If the yearly rent of a farm containing 182 acres be $273, what is the rent of a part of it containing 43 acres?

As 182 A. : 43 A. :: $273 : $64.50 Ans.

12. How many barrels of flour can I purchase for $744 when the price of flour is $6 per barrel?

As $6 : $744 :: 1 bar. : 124 barrels, Ans.

13. If 1 cwt. of beef cost $5.25, what will 39 cwt. cost?

Ans. $204.75

14. If 25 men can do a certain piece of work in 60 days, how many men would do the same in 100 days?

As 100 da. : 60 da. :: 25 men : 15 men, Ans.

Here, because it would not require as many men to do the work in 100 days as it would to do it in 60 days, I make the less number of days the second term, and the greater the first term.

15. If 6 men can build a certain wall in 192 days, how many men must be employed to build it in 24 days?

Ans. 48.

16. If the shilling loaf weighs 36 ounces when flour is $4 per barrel, how much must it weigh when flour is $6 per barrel?

As $6 : $4 :: 36 oz. : 24 oz. Ans.

It is evident that the higher the price of flour is, the lighter the shilling loaf must be; and, because the answer must be less than the third term, I make the less remaining number the second term, and the greater the first.

17. If 84 sheep can be grazed in a certain field for 12 days, how long may 112 sheep be grazed equally well in the same field?

Ans. 9 days.

18. If a person lend me $270 for 8 months; in return for his kindness, how much money ought I to lend him for 18 months?

Ans. $120.

19. Suppose a principal of $360 to produce a certain interest in 5 months; in what time, at the same rate, will $150 produce the same interest?

Ans. 12 months.

20. If $100 principal gain $6 interest in a year, how much will $49 gain in the same time, at that rate?

Prin. prin. int. int.
As $100 : $49 :: $6 : $2.94, Ans.

Here all the terms are *money*. The $6 interest is made the third term because the answer is interest; and the less principal is made the second term and the greater the first, according to the Rule.

21. If I sell goods to the amount of $400, and thereby gain $75.50, how much should I gain, at the same rate, by selling goods to the amount of $1000? Ans. $188.75

22. If a staff 4 feet high cast a shade on level ground 6 feet, what is the height of a tower whose shadow at the same time extends 100 feet?

Shade. shade. height. height.
As 6 ft. : 100 ft. :: 4 ft. : 66 ft. 8 in. Ans.

In each of the fourteen questions next following, (viz. questions 23d to 36th,) the first and second terms are of different denominations.

23. If 1 lb. of butter cost 1 s. 5 d., what would 8 lb. 6 oz. cost?

```
      lb.        lb. oz.   s.  d.
As    1 :        8..6 ::   1..5 :
     16          16        12
     ---         ---       ---
     16 oz.  :  134 oz. :: 17 d. :
                 17
                ---
                938
               134
               ---- Pence.   s.   d.
         16)2278( 142¼ = 11 .. 10¼, Ans.
            16
            ---
             67
             64
             ---
              38
              32
              ---
         Rem.  6 d.
               4
              ---
           16)24(1 q.
              16
              ---
   Last rem.  8 q.
```

Here, I reduce the first and second terms to ounces, to make them both simple numbers of the same denomination: I also reduce the third term to the lowest denomination of which it consists. I then multiply the second and third terms together, and divide their product by the first term; and the quotient is 142¼ pence; which I reduce to shillings, and have 11s. 10¼d. for the answer.

N. B. The learner ought to bear in mind, that the first and second terms must not only be reduced to simple numbers when they are compound numbers, but they must always be simple numbers of the *same name*, or *denomination.*

24. If 1 lb. of tobacco cost 1s. 2d. what will 5 lb. 14 oz. cost? Ans. 6s. $10\frac{1}{4}$d.

25. What is the value of 5 acres, 2 roods, of land, if one acre be worth 12*l.* 17s.?

As 1 A. : 5 A. 2 R. :: 12*l.* 17s. : the 4th term.
R. R. s. s. d. L. s. d.
Or, as 4 : 22 :: 257 : 1413 .. 6=70 .. 13 .. 6 Ans.

26. If the rent of 5 acres of land be 4*l.* 13s. 4d. how much land could be rented, at the same rate, for 70*l.* 10s. 6d.?

L. s. d. L. s. d. A.
As 4 .. 13 .. 4 : 70 .. 10 .. 6 :: 5 : the 4th term.
d. d. A. A. R. sq. rd.
Or, as 1120 : 16926 :: 5 : 75 .. 2 .. 10 Ans.

27. If 12 bushels of wheat cost 6*l.* 10s. 8d. how many bushels may be bought for 11*l.* 8s. 8d.? Ans. 21 bush.

28. How much will 250 bushels, 1 peck, of salt cost, at 52 cents per bushel?

As 1 bush. : 250 bush. 1 pk. :: $.52 : the 4th term.
Or, as 4 pks. : 1001 pk. :: $.52 : $130.13, Ans.

29. An eagle contains 11 pwt. 6 gr. of standard gold.* What is the value of 1 lb. of gold, at that rate?

As 11 pwt. 6 gr. : 1 lb. :: $10 : the 4th term.
Or, as 270 gr. : 5760 gr. :: $10 : $$213\frac{1}{3}$ Ans.

30. An American silver dollar contains 17 pwt. 8 gr. of *standard* silver.† What is the value of 1 lb. of silver, at that rate? Ans. $$13\frac{11}{13}$.

31. If the weight of an eagle be 11 pwt. 6 gr. Troy, and 175 oz. Troy be equal to 192 oz. Avoirdupois, what is the Avoirdupois weight of an eagle?

Troy wt. Troy wt. Avoir. wt.
As 175 oz. : 11 pwt. 6 gr. :: 192 oz. : the 4th term.
Or, as 84000 gr. : 270 gr. :: 192 oz. : $9\frac{153}{175}$ drams, Ans.

---

* The weight of an eagle is 11 pwt. 6 gr; but the quantity of *pure* gold in an eagle is only 10 pwt. 7 1-2 gr., the rest being *alloy.*

† An American dollar contains 15 pwt. 11 1-4 gr. of pure silver, and 1 pwt. 20 3-4 gr. of alloy.

32. An American silver dollar weighs 17 pwt. 8 gr. Troy: required the Avoirdupois weight? Ans. $15\frac{187}{875}$ drams.

33. If 7 tons of potashes cost $262.50, what would 9 cwt. 3 qr. cost? Ans. $18.281+

34. How much must be paid for 8cwt. 3qr. 25lb. of pig-iron, at the rate of $28 per ton? Ans. $12.5625

35. If the digging of a mile of canal cost $6500, what will the digging of 30 miles, 7 furlongs, 21 rods, cost? Ans. $201114.0625

36. If a person walk 17 miles in 5 hours, 12 min. 31 sec., how far would he walk, at the same rate, in 3 hours, 40 min. 36 sec.? Ans. 12 miles.

In each of the eleven questions next following, (viz. questions 37th to 47th,) there are decimals in one or both of the terms in the first and second places.

37. If a piece of cloth containing 25.5 yards be worth $12.75, what is the value of 3 yards of it? Ans, $1.50

38. If 1.47 cwt. of sugar be worth $20, what is 1.7 cwt. worth? Ans. $23.129+

39. If 9.75 bushels of wheat cost $15, what would .5 of a bushel cost? Ans. 76c. 9m.+

40. If a man lays out $121.25 in merchandize, and thereby gains $37.50, how much would he gain, at the same rate, by laying out $500?

As $121.25 laid out : $500 laid out :: $37.50 gained : $154.639+ gained, Ans.

Note 3.—When the third term is a compound quantity, (i. e. when it consists of several denominations,) and there are decimal figures in the first and second terms, or in either of them; if you do not wish to reduce the third term to a simple quantity, you may equalize the decimal places in the first and second terms, when they are unequal, by annexing to the term which has the fewest decimal places, as many ciphers as may be necessary to make up as many places of decimals as there are in the other term; and then reject the decimal points from both these terms, and consider them as whole numbers. Then multiply the third term by the second, and divide the product by the first, (by Compound Multiplication and Division,) and the quotient will be the answer.* This method may also be used when the third term is not a compound quantity, when it is more convenient than the common mode. By proceeding in this way, decimal fractions, which would sometimes be very troublesome, may be avoided in the operation.—It will be proper for the learner to solve questions 41st to 47th in this way.

41. If 12lb. 2oz. of butter cost $2, how much butter may be bought for $1.20?

As $2 : $1.20 :: 12lb. 2oz. : the 4th term.

By annexing two ciphers to the first term, and rejecting the decimal point from the second, (which is in effect the same as reducing the first and second terms to cents,) the statement is as follows:

As 200 : 120 :: 12lb. 2oz. : 7lb. 4$\frac{2}{5}$oz. Ans.

42. If 1.2 bushels of wheat cost 7s. 6d. what would 2.75 bushels cost?

As 1.2 bush. : 2.75 bush. :: 7s. 6d. : the 4th term.

Or, as 120 : 275 :: 7s. 6d. : 17s. 2$\frac{1}{4}$d. Ans.

43. If 2.75 yards of cloth cost 4*l.* 13s. 6d. what cost 12.25 yards? Ans. 20*l.* 16s. 6d.

44. If the rent of 7 acres and 1 rood of land be $12.75, how much land could be rented for $100?

Ans. 56A. 3R. 18sq. rd.+

45. If 25.75 bushels of rye cost 10*l.* what would 4.5 bushels cost?

As 25.75 bush. : 4.5 bush. :: 10*l.* : the 4th term.

Or, as 2575 : 450 :: 10*l.* : 1*l.* 14s. 11$\frac{1}{4}$d.+ Ans.

The same answer may be obtained without rejecting the decimal points from the first and second terms, viz. as follows:

As 25.75 bush. : 4.5 bush. :: 10*l.*

10

—— L.

25.75)45.0(1.7475+ Then, by reducing the decimal of a pound to shillings, &c., by Case II. Reduc-

* It is evident that by equalizing the decimal places in the multiplier and divisor, (i.e. in the first and second terms,) and then rejecting the decimal points, both numbers are increased in the same ratio; and hence the answer must evidently be the same as though the decimal points were retained, as usual.

tion of Decimals, the answer is 1*l.* 14s. 11¼d., as before.

46. If 2.5lb. of sugar cost 2s. what cost 85.25lb.?

Ans. 3*l.* 8s. 2¼d.+

47. If 11 yards of linen cost $9, how much may be bought for $78.75? Ans. 96 yd. 1 qr.

The following questions are promiscuously placed.

48. A cent weighs 208 grains Troy; and 7000 grains Troy are equal to 1 lb. Avoirdupois. What is the Avoirdupois weight of as many cents as are worth $1000?

Cent. cents. grains. grains.
First, As 1 : 100000 :: 208 : 20800000.

Secondly, As 7000 grains Troy : 20800000 grains Troy :: 1 lb. Avoir. : 2971$\frac{3}{7}$ lb. Avoir. Ans.

Question 48th is solved by two statements. The Troy weight of 100000 cents, (=$1000,) is found by the first statement, and the Avoirdupois weight by the second.

49. A factor bought a certain quantity of broadcloth and drugget, which, together, cost $270: The quantity of broadcloth was 50 yards, at $3 per yard; and for every 5 yards of broadcloth he had 9 yards of drugget. I demand the quantity of drugget, and what it cost per yard?

As 5 yards of broadcloth : 50 yards of broadcloth :: 9 yd. of drugget : 90 yd., the quantity of drugget.

yd. yd. $ $
As 1 : 50 :: 3 : 150, value of the 50 yd. of broadcloth.
$270—150=$120, value of the 90 yd. of drugget.
$120÷90=$1⅓, value of a yard of drugget.

Ans. 90 yards, at $1⅓ per yard.

50. A and B depart from the same place, and travel the same road; but A goes 5 days before B, at the rate of 15 miles a day—B follows at the rate of 20 miles a day: What distance must B travel to overtake A?

As 1 day : 5 days :: 15 miles : 75 miles, the distance A had travelled when B started.

20—15=5, the number of miles B gains of A in travelling 20 miles. Then,

Gained. gained. travelled. travelled.
As 5 miles : 75 miles :: 20 miles : 300 miles, Ans.

51. A harmless dove was soaring high,
To stretch her wings in space—
At length a hawk did her espy,
And gave the dove a chase:
Just forty chains were then between
The hawk and dove that flew—
While the poor dove flew seventeen,
The hawk just twenty-two:
The hawk pursued with all his strength,
As those who saw did say—
Pray tell the chains he flew in length
Before he caught his prey?
Ans. 176 chains.

52. Just fifteen pair of ladies' gloves
For sixty dimes had I;
How many pair of that same kind
Will forty eagles buy? Ans. 1000.

53. There are four pieces of cloth; the first contains 21 yards, the second 23, the third 24, and the fourth 27. What is the value of the whole, at $1.43 per yard?

$21+23+24+27=95$ yards, the whole quantity.

Then, as 1 yd. : 95 yd. :: $1.43 : $135.85, Ans.

54. If 5 barrels of flour cost $28.75, how many barrels may be bought for $5750? Ans. 1000.

55. Bought 4 casks of wine, each containing $65\frac{1}{4}$ gallons, at the rate of 2 gallons for $3: what did they cost? Ans. $391.50

56. After observing a flash of lightning, it was 12 seconds before the thunder was heard: required the distance of the cloud from whence it came? Ans. $2\frac{4}{7}$ miles.

Note.—Sound, if not interrupted, moves at the rate of 1142 feet in a second of time, or 1 mile in about $4\frac{2}{3}$ seconds. Therefore, the answer to the preceding question is found thus: as $4\frac{2}{3}$ sec. : 12 sec. :: 1 mile : $2\frac{4}{7}$ miles.

57. Perceiving a man at a distance hewing down a tree with an ax, I remarked that 6 of my pulsations passed between seeing him strike and hearing the report of the blow: what was the distance between us, allowing 70 pulses to a minute? Ans. $5873\frac{1}{7}$ feet.

58. In what time would wind move from the pole to the equator, at the rate of 3 miles an hour, the distance being 6228 miles? Ans. 86 da. 12 h.

59. The Earth describes its orbit round the Sun in 365 days, 5 h. 48 min. 48 sec.: through what space does it move each hour, at an average, the circumference of the orbit being 596902655 miles? Ans. 68094 m.+

60. How many yards, 3 quarters wide, are equal in measure to 24 yards, 5 quarters wide?

As 3qr. : 5qr. :: 24yd. : 40 yards, Ans.

61. How many yards of carpeting, which is 3 feet wide, will cover a floor which is 27 feet long and 20 feet wide?

Width. width. length. length.
As 3 ft. : 20 ft. :: 27 ft. : 180 ft.=60 yards, Ans.

62. How many yards of stuff, 3 quarters wide, will line a cloak that is 3yd. 2qr. in length, and 1yd. 3qr. wide?

Ans. 8yd. $2\frac{2}{3}$ na.

63. Supposing I have 200 yards of cloth, which cost me 90 cents per yard, but some damage having happened to it, I am willing to lose $20 by the whole; at what rate must I sell it per yard? Ans. 80 cents.

64. Bought a pipe of wine for $84, and afterwards found it had leaked out 12 gallons; I sold the remaining 114 gallons at 20 cents a quart: what did I gain or lose?

Ans. I gained $7.20

65. If the Legislature of a State grant a tax of 8 mills on the dollar, how much must that man pay whose list amounts to $1084.75? Ans. $8.678

66. If 30 bushels of corn, at 50 cents a bushel, will pay a certain debt, how many bushels at 75 cents a bushel would pay the same debt? Ans. 20.

67. If 50 gallons of water, in one hour, fall into a cistern that will hold 230 gallons, and by a pipe in the cistern 35 gallons run out in an hour; in what time will the cistern be filled? Ans. 15 hours, 20 min.

68. Bought 4 pieces of cloth, each containing 24 English ells, for $96: how much is that per yard?

Ans. 80 cents.

69. Bought 126 gallons of rum for $110: how much water must be added to reduce the cost to 75 cents per gallon? Ans. $20\frac{2}{3}$ gal.

70. The moon moves through 13 degrees, 10 min. 35 sec. of the zodiac in a day: in what time does it move through 360 degrees, or perform an entire revolution?

Ans. 27da. 7h. 43 min.+

CONTRACTIONS IN THE RULE OF THREE.

There are several methods of contraction, which may be used in particular cases, in working out the proportions. The following are some of them.

Note.—When the first and second terms of the given analogy are of different denominations, they must be reduced to the same denomination before any of the following methods of contraction can be used.

CONTRACTION I.

When the first term of the proportion is 1, multiply the second and third terms together, and the product will be the answer: And when either the second or the third term is 1, divide the other by the first term, and the quotient will be the answer, in the same denomination as the third term.*

Ex. 1. What cost 80 lb. of cheese, at $4\frac{1}{2}$ cents a lb.?

As 1 lb. : 80 lb. :: \$.045 : the 4th term.

Then, \$.045×80=\$3.60, Ans.

2. What is the value of 7 acres of land, if each acre be worth 8*l.* 5s.? Ans. 57*l.* 15s.

3. If 8 lb. of butter cost \$1.28, what cost 1 lb.?

As 8 lb. : 1 lb. :: \$1.28 : the 4th term.

Then, \$1.28÷8=\$.16=16 cents, Ans.

4. If 1 yard of broadcloth cost \$5, how many yards may be bought for \$70? Ans. 14 yd.

CONTRACTION II.

The work may often be much abbreviated by dividing either, the first and second, or the first and third terms, (but never the second and third,) by any number which will measure or divide them, and then using the quotients instead of the numbers so divided.†

---

* If any number be either multiplied or divided by 1, the product, or quotient, will evidently be the same as the given number, and hence, in such cases, the operation of multiplying or dividing may be omitted.

† The ratio of any two numbers, is the same as the ratio of the products or quotients obtained by multiplying or dividing both the given numbers by any one number: Or, which amounts to the same thing, if the divisor and dividend be both multiplied or divided by the same number, the quotient is not changed; and hence the reason of the 2d method of contraction is evident.

Note.—It will be best to take, for a divisor, one of the two terms which you would divide, when the other can be divided by it; for then one of the quotients will be 1, and the answer may be found by Contraction 1st.

Ex. 1. If 15 yards of cloth cost $17.43, what will 55 yards cost, at the same rate?

```
 yd.  yd.    $
5)15 : 55 :: 17.43
 --    --       11
  3 : 11   -------
          3)191.73
          --------
   Ans.  $63.91
```

Here I divide the first and second terms of the proportion by 5, and the quotients are 3 and 11, which I use instead of 15 and 55.

2. If 16 men eat 25 loaves of bread in a week, how many loaves will 24 men eat in the same time? Ans. $37\frac{1}{2}$.

3. Find a fourth proportional to 100, 256, and 700.

As 100 : 256 :: 700 : the 4th term.

Or, as 1 : 256 :: 7 : 1792, Ans.

Here I divide the first and third terms by 100; or, which is the same thing, I reject two ciphers from the right hand of each of these terms; and then I find the answer by Contraction 1st.

4. If 8 yards of broadcloth cost $32.80, what will 15 yards cost? Ans. $61.50

CONTRACTION III.

If by adding to, or subtracting from the first term, any part of itself, the sum or the remainder will be equal to the second term; then add to the third term, or subtract from it, (as the case shall require,) the like part of itself, and the result will be the fourth term, or answer.

Ex. 1. If 50 bushels of wheat be worth $60.25, what is the value of 75 bushels?

As 50 bush. : 75 bush. :: $60.25 : the 4th term.

```
   2)50                          2)60.25
   +25                           +30.125
   ---                           -------
   75=the 2d term.               $90.375  Ans.
```

Here, $\frac{1}{2}$ of the first term added to the first, is equal to the second term; and therefore, $\frac{1}{2}$ of the third term added to the third, gives the answer.

2. If 6 acres and 2 roods of land can be bought for $80, how much may be bought for $100?

Ans. 8 acres and 20 sq. rods.

3. How many pounds sterling are equal in value to 24 *l.* 15s. New-England currency, if 4s. 6d., or 54d., of the former currency be equal to 6s., or 72d., of the latter?

As 72d. : 54d. :: 24*l.* 15s. : the 4th term.

```
  4)72                     4)24 .. 15
  —18                      —6 ..  3 .. 9
  ——                       ————————————
  54=the 2d term.          L 18 .. 11 .. 3  Ans.
```

Here, the first term diminished by $\frac{1}{4}$ of itself is equal to the second term; and therefore the third term diminished by $\frac{1}{4}$ of itself is the answer.

4. How many pounds, &c. New-England currency, are equal in value to 8*l.* 16s. 4d. New-York currency; 6 shillings of the former currency being equal to 8 shillings of the latter? Ans. 6*l.* 12s. 3d.

Note.—The two last methods of contraction are very useful; for we may not only solve some particular questions by them, but also find *general rules*, for making, in the shortest manner possible, many numerical calculations which frequently occur in transacting business. The particular rules for the Reduction of Currencies, given in the 3d Problem in Exchange, are found by these methods of contraction.

Solution of questions in Simple Proportion by Analysis.

Questions in Simple Proportion may sometimes be easily solved by *analysis*, that is, by general principles, without the formality of *stating* the proportions. This method of solving such questions may be illustrated by a few examples.

Ex. 1. If 2 yards of cloth cost $4.20, what would 7 yards cost?

It is evident that if 2 yards cost $4.20, one yard would cost one-half of $4.20, viz. $2.10; and 7 yards would cost 7 times as much; that is, $2.10×7=$14.70, Ans.

2. If 8 sheep cost $10, what would 3 sheep cost?

Ans. $3.75

3. If a staff, 5 feet, 8 inches, in length, cast a shadow of

6 feet, how high is that steeple whose shadow measures 153 feet?

If 6 feet shadow require a staff of 5 ft. 8 in.=68 inches, one foot shadow will require a staff of $\frac{1}{6}$ of 68 inches, or $\frac{68}{6}$ inches, and 153 feet shadow will require 153 times as much; that is, $\frac{68\times153}{6}=\frac{10404}{6}=1734$ inches$=144\frac{1}{2}$ feet, Ans.

4. If 4 tons of hay will keep 3 horses through the winter, how many tons will keep 15 horses the same time?

Ans. 20.

PRACTICE,

Is a contraction of the Single Rule of Three when the first term is 1; and has its name from its frequent use in business, being a concise method of finding the value of any number of articles when the price of *one* is known.

The method of proof is by the Rule of Three, or by varying the order of the question.

Note 1.—One number is said to be an *aliquot part* of another, when the former will divide the latter without a remainder: So, 2 is an aliquot part of 8, and 5 is an aliquot part of 15. Therefore, to find whether any given number is an aliquot part of a greater number, divide the greater number by the less, and, if the division terminates without a remainder, the less number is an aliquot part of the greater: If the quotient be 2, the less number is equal to $\frac{1}{2}$ of the greater; if the quotient be 3, the less number is $\frac{1}{3}$ of the greater; and so on.

Note 2.—To find $\frac{1}{2}$, or $\frac{1}{3}$, or $\frac{1}{4}$, &c. of any given quantity or number, divide the quantity by the denominator of the fraction, viz. by 2, or 3, &c. and the quotient will be the part required: Thus, $\frac{1}{2}$ of 8 is$=8\div2=4$; and $\frac{1}{3}$ of 15 is $=15\div3=5$.

A Table of Aliquot or Proportional Parts of Money, Weight, and Measure.

Sterling Money, &c.

q.	d.	s.	L.
1	$=\frac{1}{4}$	$=\frac{1}{48}$	$=\frac{1}{960}$
2	$=\frac{1}{2}$	$=\frac{1}{24}$	$=\frac{1}{480}$
3	$=\frac{3}{4}$	$=\frac{1}{16}$	$=\frac{1}{320}$
	1	$=\frac{1}{12}$	$=\frac{1}{240}$
	$1\frac{1}{2}$	$=\frac{1}{8}$	$=\frac{1}{160}$
	2	$=\frac{1}{6}$	$=\frac{1}{120}$
	3	$=\frac{1}{4}$	$=\frac{1}{80}$
	4	$=\frac{1}{3}$	$=\frac{1}{60}$
	5		$=\frac{1}{48}$
	6	$=\frac{1}{2}$	$=\frac{1}{40}$
	8	$=\frac{2}{3}$	$=\frac{1}{30}$
	10		$=\frac{1}{24}$

s.	d.	L.
1	.. 0	$=\frac{1}{20}$
1	.. 3	$=\frac{1}{16}$
1	.. 4	$=\frac{1}{15}$
1	.. 8	$=\frac{1}{12}$
2	.. 6	$=\frac{1}{8}$
3	.. 4	$=\frac{1}{6}$
4	.. 0	$=\frac{1}{5}$
5	.. 0	$=\frac{1}{4}$
6	.. 8	$=\frac{1}{3}$
10	.. 0	$=\frac{1}{2}$

Federal Money.

c.	$
1	$=\frac{1}{100}$
$1\frac{1}{4}$	$=\frac{1}{80}$
$1\frac{2}{3}$	$=\frac{1}{60}$
2	$=\frac{1}{50}$
$2\frac{1}{2}$	$=\frac{1}{40}$
$3\frac{1}{3}$	$=\frac{1}{30}$
5	$=\frac{1}{20}$
$6\frac{1}{4}$	$=\frac{1}{16}$
$8\frac{1}{3}$	$=\frac{1}{12}$
$12\frac{1}{2}$	$=\frac{1}{8}$
$16\frac{2}{3}$	$=\frac{1}{6}$
20	$=\frac{1}{5}$
25	$=\frac{1}{4}$
$33\frac{1}{3}$	$=\frac{1}{3}$
50	$=\frac{1}{2}$

Avoirdupois Weight.
Here, 1 cwt.=112 lb.

lb.	cwt.	oz.	lb.
4	$=\frac{1}{28}$	1	$=\frac{1}{16}$
7	$=\frac{1}{16}$	2	$=\frac{1}{8}$
8	$=\frac{1}{14}$	$2\frac{2}{3}$	$=\frac{1}{6}$
14	$=\frac{1}{8}$	$3\frac{1}{5}$	$=\frac{1}{5}$
16	$=\frac{1}{7}$	4	$=\frac{1}{4}$
28	$=\frac{1}{4}$	$5\frac{1}{3}$	$=\frac{1}{3}$
56	$=\frac{1}{2}$	8	$=\frac{1}{2}$

Cloth Measure.

na.	yd.
1	$=\frac{1}{16}$
2	$=\frac{1}{8}$
qr.	
1	$=\frac{1}{4}$
2	$=\frac{1}{2}$

Liquid Measure.

gil.	qt.	gal.
$\frac{1}{2}$	$=\frac{1}{16}$	$=\frac{1}{64}$
1	$=\frac{1}{8}$	$=\frac{1}{32}$
2	$=\frac{1}{4}$	$=\frac{1}{16}$
4	$=\frac{1}{2}$	$=\frac{1}{8}$
	1	$=\frac{1}{4}$
	2	$=\frac{1}{2}$

Note.—If 100 lb. make 1 cwt., then 1, 2, 3, &c. pounds will be the same parts of 1 cwt. that 1, 2, 3, &c. cents are of $1.

CASE I.

When the given quantity is of one denomination only, and the price of one is also of one denomination.

RULE.

Multiply together the price of *one*, and the given number of articles, or yards, &c., and the product will be the answer in the same denomination with the price.

Or, if the price of 1 yard, &c. be an aliquot part of 1 of some higher denomination of that kind of money, then you may consider the given number of yards, &c. as so many units of the said higher denomination of money, and take such proportional part of this number as the price is of 1 of the said denomination, and you will have the value required.

EXAMPLES.

1. What will 734 pounds of cheese come to, at 4d. a pound?

Operation by the first method.

734 lb.
4 d.

2936 d.=12 *l.* 4s. 8 d. Ans.

Operation by the second method.

3)734 s. value at 1 s. a lb.

Ans. 244s. 8d.=12*l.* 4s. 8d. value at 4d. a lb.

In working by the second method, I consider the given quantity, viz. 734 lb., as so many shillings, and this sum is the value of the said quantity at 1 s. a lb.: Then, because 4d. is $\frac{1}{3}$ of 1s., I take $\frac{1}{3}$ of 734s., and have 244s. 8d. for the answer; which being reduced to pounds, is 12*l.* 4s. 8d.

The operation may be performed otherwise, thus:—

4d.=$\frac{1}{60}$*l.* 6,0)73,4*l.* value at 1*l.* a lb.

Ans. 12*l.* 4s. 8d. value at 4d. a lb.

Here I consider 734 lb. as 734*l.* of money; and, because 4d. is $\frac{1}{60}$ of L 1, I take $\frac{1}{60}$ of 734*l.*, and have 12*l.* 4s. 8d. for the answer.

Note.—If after the division of the given quantity by the aliquot part of L 1, &c. there is a remainder, its value may be found in the lower denominations, as in Compound Division. Thus, in the foregoing example, after dividing 734s. by 3, 2 shillings remain, which I reduce to pence, and then divide by 3, and the quotient is 8d., which I annex to the shillings of the answer.

2. What is the value of 75lb. at 6d. a lb.?

6d.$=\frac{1}{2}$s.$=\frac{1}{40}$*l*.: Therefore, 75s.$\div$2=37s. 6d.=1*l*. 17s. 6d. the Ans.

Or, L75$\div$40=1*l*. 17s. 6d. the ans. as before.

3. What is the value 92lb. of butter, at 12$\frac{1}{2}$ cents a lb.?

12$\frac{1}{2}$c.=\$$\frac{1}{8}$: Hence \$92$\div$8=\$11.50, Ans.

More Questions for exercise.

Find the value, or amount, of		L. s. d.
4528 quills, at 1q., or $\frac{1}{4}$d. each.	Ans.	4 .. 14 .. 4
642 eggs, at $\frac{3}{4}$d. each	——	2 .. 0 .. 1$\frac{1}{2}$
827 oranges, at 3d. each.	——	10 .. 6 .. 9
1274 pounds of cheese, at 8d. a lb.	——	42 .. 9 .. 4
257 bushels of rye, at 5s. a bush.	——	64 .. 5 .. 0
		\$
658 melons, at 3$\frac{1}{3}$ cents each.	——	21.933+
180 pounds of sugar, at 8$\frac{1}{3}$c.	——	15.
249 pounds of butter, at 16$\frac{2}{3}$c.	——	41.50
24 quires of paper, at 25c.	——	6.
75 yards of cloth, at 33$\frac{1}{3}$c.	——	25.

CASE II.

When the given quantity is of one denomination only, and the price of one is of different denominations.

RULE I.

1. When the price is an aliquot part of 1 of some higher denomination of money; then the like part of the quantity will be the answer in the said higher denomination, as in Case I.

2. When the price is not an aliquot part of 1 of any higher denomination of that kind of money; if it can conveniently be divided into two or more parts which shall be aliquot parts, either of 1 of some higher denomination, or of each other; then take the like parts of the given quantity, and add them together for the answer. But, if the price cannot conveniently be divided into such aliquot parts; or if the highest denomination of the price be the highest denomination of that kind of money; then multiply the given number of yards, or pounds, &c., by the price of one yard, &c., and take parts for the lower denominations of the

price, as directed above, and the sum of the product and the quotients, or parts, will be the answer.

RULE II.

Multiply the price of one yard, or pound, &c., by the given number of yards, or pounds, &c., by the rule for Compound Multiplication, and the product will be the answer.

EXAMPLES.

1. Required the value of 40 bushels of oats, at 2s. 6d. a bushel.

Operation by Rule I.

I find, by the table of aliquot parts, that 2s. 6d. are= L$\frac{1}{8}$: Hence, L40÷8=L5, the Ans.

Operation by Rule II.

Here I multiply the value of one bushel by 40, the number of bushels, which gives the value of the 40 bushels.

	s. d.
	2..6
	40
L.	
Ans.	5..0..0

2. What would 96 bushels of rye cost, at 4s. 8d. a bushel?

4s.=L$\frac{1}{5}$, and 8d.=L$\frac{1}{30}$: Therefore,

L.	L. s.		s. d.	
96÷ 5=	19..4	amount at	4..0	a bushel.
96÷30=	3..4	do. at	8	do.
Ans.	L22..8	do. at	4..8	do.

3. What do 252 yards of cloth amount to, at 8s. 4d. a yard?

Here I divide the price, viz. 8s. 4d., into these two parts, 5s. and 3s. 4d.; the former being $\frac{1}{4}$ and the latter $\frac{1}{6}$ of a pound. Then,

L.	L.		s. d.	
252÷4=	63	amount at	5..0	a yard.
252÷6=	42	do. at	3..4	do.
Ans.	L105	do. at	8..4	do.

4. Required the amount or value of 124 tons of hay, at 4*l.* 2s. 6$\frac{1}{2}$d. per ton.

$2s.=\frac{1}{10}$)124 L. amount at 1 *l.* per ton.
4 L.

		L. s. d.	
	496 L. amount at	4..0..0	per ton.
6d.=$\frac{1}{4}$)	12..8s. do. at	2..0	do.
$\frac{1}{2}$d.=$\frac{1}{12}$)	3..2 do. at	6	do.
	5..2d. do. at	$\frac{1}{2}$	do.
Ans.	L511..15..2 do. at	4..2..6$\frac{1}{2}$	do.

Explanation of Ex. 4th.—I first multiply 124, the quantity, into 4 *l.*, the highest denomination of the price, and the product, 496*l.*, is the amount at 4 *l.* per ton. Then, because 2 s.=$\frac{1}{10}$ of 1 *l.*, I take $\frac{1}{10}$ of the amount at 1 *l.* per ton, (viz. $\frac{1}{10}$ of 124 *l.*) for the amount at 2s. per ton. Then, because 6d.=$\frac{1}{4}$ of 2s., I take $\frac{1}{4}$ of the amount at 2s. (viz. $\frac{1}{4}$ of 12 *l.* 8s.) for the amount at 6d.; and, because $\frac{1}{2}$d.=$\frac{1}{12}$ of 6d., I take $\frac{1}{12}$ of the amount at 6 d. for the amount at $\frac{1}{2}$ d. Lastly, I add together the several amounts, or values, thus found, and their sum, 511 *l.* 15 s. 2d., is the answer.

5. What is the value of 246 bushels of wheat, at 1 D. 12$\frac{1}{2}$ c. a bushel?

c. $ $
12$\frac{1}{2}$=$\frac{1}{8}$)246 value at $1 a bushel.
30.75 do. at 12$\frac{1}{2}$ c. do.

Ans. $276.75 do. at 1.12\frac{1}{2}$ do.

6. What is the value of 84 yards of broadcloth, at 5 D. 33$\frac{1}{3}$ c. a yard?

c. $ $
33$\frac{1}{3}$=$\frac{1}{3}$)84 value at $1.
5 D.

420 value at $5.
28 do. at 33$\frac{1}{3}$c.

Ans. $448 do. at 5.33\frac{1}{3}$

More Questions for exercise.

Required the value of	L. s. d.		L. s. d.
162 bush. of wheat, at	0..6..8.	Ans.	54.. 0..0.
425 do. of rye, at	3..4.	——	70..16..8.
288 do. of oats, at	1..8.	——	24.. 0..0.
78 lb. of butter, at	1..6.	——	5..17..0.

		L. s. d.		L. s. d.
	yards of cloth, at	7 .. 4.	Ans.	23 .. 9 .. 4.
156	do. of broadcloth, at	2 .. 6 .. 8.	——	364 .. 0 .. 0.
37	tons of hay, at	5 .. 5 .. 0.	——	194 .. 5 .. 0.
		D. c.		$
265	bushels, at	1 .. 12½.	Ans.	298.125
87	cwt., at	4 .. 16⅔.	——	362.50
68	tons, at	10 .. 25.	——	697.
158	acres, at	20 .. 33⅓.	——	3212.666+

CASE III.

When the given quantity is of several denominations.

RULE.

Multiply the price of 1 lb., or 1 yd. &c., by the given number of pounds, &c., and then take parts for the lower denominations of the given quantity, from the price of I pound, &c., as in the preceding Cases; and the sum of the product and quotients will be the answer.

Or, work by the Rule of Three, which will usually be the better way.

EXAMPLES.

1. If 1 lb. of butter be worth 14 cents, what is the value of 7 lb. 13 oz.?

oz.	$		lb. oz.
8=½)	.14	value of 1 lb.	
	7	no. of lb.	
	——		
	.98	value of	7 .. 0
4=½)	.07	do. of	8
1=¼)	.035	do. of	4
	.008+	do. of	1
	——		——
Ans.	$1.093+	do. of	7 .. 13

Here, I first multiply the value or price of 1 lb. by 7; and the product is the value of 7 lb. Then I take parts for the 13 oz. as follows: Because 8 oz.=½ of 1 lb. I take ½ of the value of 1 lb., (viz. ½ of $ .14,) for the value of 8 oz.: and then, because 4 oz. =½ of 8 oz., I take ½ of the value of 8 oz. for the value of 4 oz.; and then I take ¼ of the value of 4 oz. for the value of 1 oz. Lastly, I add together these values of the several parts of the given quantity, and the amount, $1.093, is the whole value required.

2. If 1 yard of cloth be worth 1 D. 46 c., what is the value of 1 yd. 2 qr.?

```
qr. yd. $
 2=½)1.46  value of 1 yd.
      .73  do. of 2 qr.
     ----
Ans. $2.19  do. of 1 yd. 2 qr.
```

3. If 1 bushel of wheat cost 10 s. 8 d., what cost 1 peck and 2 quarts?

```
pk.  s.  d.
 1=¼)10 .. 8  value of 1 bush.
qt.  ------
 2=¼) 2 .. 8  do. of 1 peck.
      0 .. 8  do. of 1 qt.
     ------
Ans.  3 .. 4  do. of 1 pk. 2 qt.
```

4. What do 5 lb. 8 oz. of tea amount to, at 6 s. 4 d. a lb.? Ans. 1 *l.* 14 s. 10 d.

5. Required the amount of 20 lb. 5 oz. of cheese, at 6 d. a lb. Ans. 10 s. $1\frac{3}{4}$ d.+

More Questions for exercise.

Required the value of

		$
lb. oz.		
6 .. 9, at $1.25 a lb.	Ans.	8.203 +
10 .. 6, at 12 cents a lb.	——	1.245
48 .. 11, at 6 cents a lb.	——	2.92 +
yd. qr. na.		
4 .. 2 .. 2, at 64 cents a yd.	——	2.96
3 .. 1, at 40 cents a yd.	——	.325
1 .. 3 .. 0, at $4 50 a yd.	——	7.875

Note.—The rules for the three foregoing cases are applicable to any kind of money, and they will serve for solving all questions in Practice. There are some other contractions which may be used in finding the value of goods, when the price of *one* is shillings and pence; but, as the use of these denominations will probably be soon laid aside in this country, it is thought unnecessary to give rules for them.

THEORETICAL QUESTIONS.

1. What is Proportion? 2. How are numbers compared to each other? 3. How many numbers must there be to form a comparison? 4. When one number is compared to another, by what names are the first and last numbers dis-

tinguished? 5. How is the geometrical ratio of two numbers usually denoted? 6. How is the equality of two such ratios usually denoted? and what is such a series of four proportional terms called? 7. What are the first and last terms of an analogy called? and what the other two terms? 8. Is the product of the extremes equal to the product of the means? 9. How may either of the extremes be found when the other extreme and the two mean terms are given? 10. What is Simple Proportion? 11. What does the Single Rule of Three teach? 12. Why is it called the Rule of Three? 13. How many of the terms, or numbers, given in any question belonging to this rule, are contained in a supposition; and how many in a demand? 14. What are the terms contained in the supposition called? and what is the other term called? 15. Which two of the given terms are of the same kind? and which term is of the same kind with the answer or fourth term? 16. What is understood by *stating* a question in Simple Proportion? 17. What is the rule for stating such questions? [Here the learner should repeat the first part of the General Rule.] 18. What is to be done when the first and second terms are not both simple numbers of the same name? 19. How is the fourth term, or answer, then found? 20. How are operations in Simple Proportion proved? 21. What is the first method of contraction in Simple Proportion? 22. What is the second contraction? 23. ——— the third? 24. What is Practice? 25. What is meant by one number being an *aliquot part* of another? 26. What is the first case in Practice? Repeat the Rule. 27. What is the second case?—the Rule? 28. What is the third case?—the Rule?

COMPOUND PROPORTION, OR

THE DOUBLE RULE OF THREE,

Embraces such questions as require two or more statements in Simple Proportion. It is called the *Double Rule of Three*, because most of the questions belonging to it may be solved by two statements by the Single Rule of Three.

In questions belonging to this rule, there is always given an odd number of terms, viz. five, seven, or nine, &c. The terms are distinguished, as in Simple Proportion, into terms of *supposition*, and terms of *demand;* the number of the former always exceeding that of the latter, by one. Each term of supposition, except one, has a corresponding term of the same kind given in the demand; and the odd term of supposition, which has no corresponding term of demand given, is of the same kind with the answer or term required.

All questions in Compound Proportion may be solved by the following

RULE.*

1. State the given question as follows: Write that term of supposition which is of the same kind or quality with the answer sought, for the third term of the proportion. Take one of the other terms of supposition and one of the demanding terms of the same kind, and place one of them for a first term, and the other for a second, according to the directions given in the Single Rule of Three. Do the same with another term of supposition, and its corresponding term of demand; and so on with all the remaining terms, (if any,) of each kind, writing the terms in the first and second places in columns, under each other.

2. When any of the terms are compound numbers, reduce them to simple numbers; and always reduce the first and second terms of each line to the same denomination when they are of different denominations. Then, multiply together all the terms in the second and third places for a dividend, and those in the first place, or column, for a divisor. Lastly, divide the dividend by the divisor, and the quotient will be the answer, in the same denomination as the third term, or as that which the third was reduced to.

Note.—According to the foregoing Rule, when the third term is a compound number, it is to be reduced to a simple number. It may be proper, however, to inform the learner, that it is not absolutely necessary, and sometimes not so convenient, to perform this reduction: for the third

* The principles on which the operations in Compound and in Simple Proportion depend, are the same. In solving any question by the Double Rule of Three, all the dividends that would be used in solving the same question by two or more statements by the Single Rule of Three, are collected into one dividend; and all the divisors, into one divisor; by which means the answer is found by one statement, or set of operations.

term may be multiplied by the product of all the terms in the second place, and the result then divided by the continued product of the terms in the first place, as in Compound Multiplication and Division, and the quotient will be the answer. When this method is used, if there are decimal figures in the multiplier, (i. e. the product of the terms in the first place,) and divisor, (i. e. the product of the terms in the second place,) or in either of them, it will be proper to equalize the numbers of decimal places in them, and then reject the decimal points, as directed in the Note immediately after exam. 40th, in Simple Proportion.

Proof.—By the Single Rule of Three; or by varying the order of the question.

EXAMPLES.

1. If 6 men can build a wall 96 rods long in 18 days, when the day is 11 hours long; how many men must be employed to build 64 rods of the wall in 2 days, when the day is 12 hours long?

STATEMENT.

As 96 rods : 64 rods }
2 days : 18 days } :: 6 men : the ans.
12 hours : 11 hours }

Operation.

$64\times18\times11\times 6=76032$, the dividend.
$96\times 2\times12= 2304$, the divisor.

Then, $76032\div2304=33$ men, Ans.

Explanation.—In the foregoing question, the *supposition* is, *that* 6 *men can build* 96 *rods of wall in* 18 *days, when the day is* 11 *hours long;* and the *demand* is, *to find how many men must be employed to build* 64 *rods of the wall in* 2 *days, when the day is* 12 *hours long*; there being four numbers, or terms, in the supposition, and three in the demand.* The 96 rods of wall, mentioned in the sup-

* The number of terms in the question may be reduced to *five*, thus: 18 days of 11 hours each = 198 hours, and 2 days of 12 hours each = 24 hours. Then the question is as follows: If 6 men can build 96 rods of wall in 198 hours, how many men must be employed to build 64 rods of the wall in 24 hours?

Stated thus; As 96 rods : 64 rods }
24 hours : 198 hours } :: 6 men : 33 men, Ans.

position, is a term which is of the same kind with the 64 rods in the demand; the 18 days in the supposition and the 2 days in the demand, are terms of the same kind; and, the 11 hours in the supposition and the 12 hours in the demand, are corresponding or like terms. The 6 men is the term of supposition which has no corresponding term of demand given, and which is of the same kind with the answer required.—I state the question thus: I first set down 6, the given number of men, for the third term, because this is the term of supposition which is of the same kind with the answer. I then dispose of the other given terms in the first and second places, as follows: I begin with the 96 rods in the supposition, and the corresponding term of demand, 64 rods; and because it would not require as many men to build 64 rods of wall as it would to build 96 rods, in the same time, I write the less number of rods for the second term, and the greater for the first. Then, because it would require more men to do the required work in 2 days than it would to do it in 18 days, I write the greater number of days in the second place and the less in the first. Lastly, because it would not require as many men to perform the work when the day is 12 hours long as it would when the day is only 11 hours long, I write the less number of hours in the second place and the greater in the first.—Having thus stated the question, I proceed to resolve the statement; and, as all the terms are simple numbers, and the first and second terms of each line are of the same denomination, there are no reductions to be performed; so I multiply together all the terms in the second and third places, for a dividend, and those in the first column, or place, for a divisor: I then perform the division, and the quotient is the answer.

Any question in Compound Proportion may be solved, by two or more statements, by the Single Rule of Three. The foregoing question may be solved by three statements* by the Single Rule of Three, viz. as follows:

1st, As 96 rods : 64 rods :: 6 men : 4 men, the number of men that must be employed to build 64 rods of wall in 18 days, when the day is 11 hours long.

* If the number of terms in the question be reduced to *five*, as in the note at the bottom of page 165, the answer may be found by two statements by the Single Rule of Three.

2dly, As 2 days : 18 days :: 4 men : 36 men, the number that must be employed to build 64 rods of wall in 2 days of 11 hours each.

Lastly, As 12 hours : 11 hours :: 36 men : 33 men, the answer, or number of men that must be employed to build 64 rods of wall in 2 days of 12 hours each.

It will be well for the learner to solve some of the following questions by the Single Rule of Three.

2. If 5 men in 8 days eat 27 lb. 12 oz. of bread, how much bread will 4 men eat in 15 weeks?

Stated thus :—

As 5 men : 4 men }
8 days : 15 weeks } :: 27 lb. 12 oz. : the ans.

After performing the proper reductions, according to the Rule, the statement is as follows:

As 5 men : 4 men }
8 days : 105 days } :: 444 oz. : the ans.

Then, $\frac{4\times105\times444}{5\times8}=\frac{186480}{40}$=4662 oz.=291 lb. 6 oz. Ans.

3. If 9 bushels of oats will serve 7 horses 10 days, how many bushels, at the same rate, will serve 20 horses 3 weeks? Ans. 54 bush.

4. If a family of 19 persons expend $235 in 8 months, how much, at the same rate, will a family of 12 persons expend in 5 months? Ans. $92.763+

5. If 6 oxen, in 5 days, plough 11 acres, how many oxen would plough 44 acres in 12 days?

Days, 12 : 5 }
Acres, 11 : 44 } :: 6 oxen : 10 oxen, Ans.

6. If 3 masons, working 7 hours a day, build a wall in 6 days; how many hours a day must 4 masons work, to build it in 5 days?

Masons, 4 : 3 }
Days, 5 : 6 } :: 7 hours : 6 h. 18 min. Ans.

7. If 36 yards of cloth, 7 quarters wide, cost $504, what cost 120 yards, of the same quality, but only 5 quarters wide? Ans. $1200.

8. If the carriage of 13 cwt. 65 miles cost $9, how many

cwt. may be carried 40 miles, at the same rate, for \$15?

Ans. $35\frac{5}{14}$ cwt.

9. If \$100, in one year, gain \$6 interest, how much will \$75.28 gain in 9 months?

$$\left.\begin{array}{l} \$\ 100 : 75.28 \\ \text{Mo.}\ \ 12 : 9 \end{array}\right\} :: \text{Int. } \$6 : \text{Int. } \$3.3876, \text{ Ans.}$$

10. If \$100 will gain \$6 interest in 12 months, in what time will \$75.28 gain \$3.3876? Ans. 9 months.

11. If 3000 copies of a history of the United-States, each containing 11 sheets, require 66 reams of paper, how much paper will 5000 copies require, if the work be extended to $12\frac{1}{2}$ sheets? Ans. 125 reams.

12. If 6 men can build a wall 20 feet long, 6 feet high, and 4 feet thick, in 16 days; in what time will 24 men build a wall 200 feet long, 8 feet high and 6 feet thick?

$$\left.\begin{array}{lll} \text{Men,} & & 24 : 6 \\ \text{Length,} & \text{ft.} & 20 : 200 \\ \text{Height,} & \text{ft.} & 6 : 8 \\ \text{Thickness,} & \text{ft.} & 4 : 6 \end{array}\right\} :: \text{da. } 16 : \text{days } 80, \text{ Ans.}$$

13. Suppose 30 men perform a piece of work in 20 days; how many men will accomplish another piece of work 4 times as large, in a fourth part of the time?

$\frac{1}{4}$ of 20 days=5 days.

$$\left.\begin{array}{l} \text{Days, } 5 : 20 \\ \qquad\ \ 1 : 4 \end{array}\right\} :: 30 \text{ men} : 480 \text{ men, Ans.}$$

METHOD OF CONTRACTION.

The work may often be very much abbreviated as follows:

1. State the question as usual, and perform such reductions as may be necessary, in order to prepare the terms for multiplying and dividing. Then draw a horizontal line, and place all the numbers which are to be multiplied together for a dividend, (viz. all the terms in the second and third places,) in a row, above this line, with the sign of multiplication between every two of them; and place all the terms which are to be multiplied together for a divisor, below the line, in the same manner.

2. If any factor, or number, above the line, be the same as a factor below the line, cancel or strike out both the num-

bers; and, if any number above, and another below the line, can be divided by a common divisor, then divide them by it, and set down the quotients instead of the numbers so divided. Proceed in this manner, till it appears that there is no number greater than 1 that will divide a number above and another below the line. Then multiply together the remaining numbers above the line for a dividend, and those remaining below the line for a divisor, and the quotient will be the answer.*

Note.—This method of contraction may often be used in solving questions in Simple Proportion, and in other similar operations.

EXAMPLES.

1. If 2 masters, who have each 3 apprentices, earn \$50 in 6 days, by working 12 hours each day, how much will 4 masters, who have each 6 apprentices, earn in 25 days, by working 8 hours each day?

Statement.

$$\left.\begin{array}{lrcr} \text{Masters,} & 2 & : & 4 \\ \text{Apprent.} & 3 & : & 6 \\ \text{Days,} & 6 & : & 25 \\ \text{Hours,} & 12 & : & 8 \end{array}\right\} :: \overset{\$}{50} : \text{the ans.}$$

Operation by contraction.

$$\frac{\overset{2}{\cancel{4}}\times\cancel{6}\times25\times\overset{2}{\cancel{8}}\times50}{\underset{1}{\cancel{2}}\times3\times\cancel{6}\times\underset{3}{\cancel{12}}}=\frac{2\times25\times2\times50}{1\times3\times3}=\frac{5000}{9}=\$555\tfrac{5}{9},\ \text{Ans.}$$

Explanation of the contraction.—Having placed the factors which compose the dividend above a horizontal line, and those which compose the divisor below the line, I first cancel or dash the 6 in the upper row and the 6 in the under row; these being all the upper and lower numbers which are alike. I next divide the 4 in the upper row and the 2 in the under row by 2, and set down the quotients, dashing the 4 and the 2. I then divide the 8 in the upper row and the 12 in the under row by 4, and set down the quotients, dashing the 8 and the 12. Then I find that there is

* The truth and reason of this method of contraction are evident from what has been said respecting the 2d contraction in Simple Proportion.

no number greater than 1, which will measure or divide a number remaining above and another below the line: so, I multiply together the numbers remaining above the line, viz. 2, 25, 2 and 50, for a dividend, and those remaining below the line, viz. 1, 3 and 3, for a divisor; and the quotient is the answer.

Note.—The numbers which are marked with a little dash at the right and left, are *supposed* to be canceled or erased. The student had better make a mark with his pen or pencil quite across each number that is to be canceled, or omitted, which cannot be done in a printed book. In dividing the upper and lower numbers, when the quotient is 1, it will not be necessary to set it down. I have set down such quotients in the examples which I have given, merely for the purpose of illustrating the operations.

2. If 14 men consume 38 dollars worth of wheat in 10 months, when the price of wheat is $2 a bushel, how many men will consume $114 worth in 6 months, when the price is $1.75 a bushel?

Statement.

$$\left.\begin{array}{ll}\$\ 38 & :114 \\ \$\ 1.75 & :\ 2 \\ \text{Mo. } 6 & :\ 10\end{array}\right\} :: 14 \text{ men} : \text{the ans.}$$

Operation.

$$\frac{\overset{3}{,114'}\times\overset{1}{,2'}\times\overset{2}{,10'}\times\overset{2}{,14'}}{\underset{1}{,38'}\times\underset{\underset{.05}{,.35'}}{,1.75'}\times\underset{,3'}{,6'}}=\frac{1\times2\times2}{1\times.05}=80 \text{ men, Ans.}$$

Here, I divide the numbers 114 and 38 by 38, and the numbers 2 and 6 by 2, setting down the quotients, and canceling the dividends. There being then a 3 above and a 3 below the line, I cancel them. I next divide the numbers 10 and 1.75 by 5, and the quotients are 2 and .35, which I set down, and cancel the dividends. I then divide 14 and .35 by 7, and cancel them. Then I find that the remaining numbers cannot be reduced any lower: so I multiply together those which remain above the line, for a dividend, and those below the line, for a divisor, and the quotient is the answer.

3. If 8 horses eat 6 tons of hay in 60 days, how many tons will 16 horses eat in 90 days?

Statement.

$$\left.\begin{array}{l}\text{Horses, } 8 : 16 \\ \text{Days, } 60 : 90\end{array}\right\} :: 6 \text{ tons} : \text{the ans.}$$

Operation.

$$\frac{\overset{2}{\cancel{16}}\times\overset{3}{\cancel{90}}\times 6}{\underset{1}{\cancel{8}}\times\underset{2}{\cancel{60}}}=\frac{3\times 6}{1}=\frac{18}{1}=18 \text{ tons, Ans.}$$

If the student should wish to work out more questions by the method of contraction, he may solve, in this way, some of the questions which precede the rule for the contraction, or some of the questions in Simple Proportion.

THEORETICAL QUESTIONS.

1. What is Compound Proportion? 2. Do questions in Compound Proportion always contain an odd number of terms? 3. By what names are the terms distinguished? 4. Does each term in the supposition have a corresponding term of the same kind in the demand? 5. Is the term of supposition which has no corresponding term of demand given, of the same kind with the answer? 6. What is the rule for stating questions in Compound Proportion? 7. How is the answer then found? 8. What method of contraction may be used in Compound Proportion? Repeat the rule. 9. Can this method of contraction be used in other similar operations?

Note.—I have inserted questions on almost all the rules contained in the preceding part of this work, because those rules are the principal rules of Arithmetic, and ought to be well understood by learners before they proceed to the subsequent rules. Questions of this nature are of less importance in the remaining parts of Arithmetic, and I shall not swell the size of this work by inserting them. If any instructor, however, should think it would be useful to put such questions to his pupils on every subsequent rule, a little attention to each rule will enable him to put the proper questions to learners.

TARE AND TRET,

Are allowances which are made by merchants and tradesmen in selling goods, &c. by weight.

Tare is an allowance made to the buyer, for the weight of the box, cask, bag, &c. which contains the goods bought; and is either at so much per box, &c., at so much per cwt., or so much in the whole gross weight.

Tret is an allowance of 4 lb. in every 104 lb.,* which is sometimes made for waste, dust, &c.

Gross weight, is the whole weight of any sort of goods, together with the box, cask, or bag, &c. which contains them.

Suttle, is what remains after part of the allowance has been deducted from the gross.

Neat weight is what remains after all allowances have been made.

Note.—In the examples in the following Cases in Tare and Tret, 1 cwt.=112 lb.

CASE I.

When the tare is so much in the whole gross weight.

RULE.†

Subtract the tare from the gross, and the remainder will be the neat weight.

EXAMPLES.

1. In 558 lb. 8 oz. gross, tare 75 lb. 12 oz., how much neat weight?

	lb.	oz.	
	558 ..	8	gross weight.
	75 ..	12	tare, deducted.
Ans.	482 ..	12	neat weight.

2. What is the neat weight of 4 casks of indigo, each weighing 518 lb. gross, tare in the whole 140 lb.?

	518	lb.
	4	no. of casks.
	2072	whole gross wt.
	140	tare.
Ans.	1932	lb. neat wt.

* This is the tret allowed in London.

† The reason of this rule is exceedingly obvious. The rules for the following Cases in Tare and Tret are only particular applications of the rules of Proportion and Practice, and are easily understood.

3. What is the neat weight of 6 hogsheads of sugar, each weighing 9cwt. 2qr. 10lb. gross, tare in the whole 1cwt. 1qr. 2lb.? Ans. 56cwt. 1 qr. 2lb.

4. Find the neat weight of 3 tubs of butter, the gross weight and tare of which are as follow:

		lb. oz.		lb. oz.		
No. 1,	gross wt.	25 .. 8,	tare	8 .. 4,		
2,		18 .. 7,		5 .. 6,		lb. oz.
3,		14 .. 4,		4 .. 0.	Ans.	40 .. 9

CASE II.

When the tare is so much per box, or bag, &c.

RULE.

Multiply the tare per box, or bag, &c. by the number of boxes, &c., and the product will be the whole tare; which subtract from the whole gross weight, and the remainder will be the neat weight.

Or, if the weight of the contents of every box, or bag, &c. be the same, you may find the neat weight of the contents of one of the boxes, &c. by Case I., which multiply by the number of boxes, and the product will be the neat weight of the whole.

EXAMPLES.

1. In 8 bags of pepper, each weighing 85lb. 4oz. gross, tare per bag 2lb. 15oz., how much neat?

Operation by the first method.

lb. oz.		lb. oz.	
85 .. 4		2 .. 15	tare per bag
8	no. of bags.	8	no. of bags.
682 .. 0	whole gross wt.	23 .. 8	whole tare.
23 .. 8	whole tare, deducted.		
658 .. 8	neat wt. Ans.		

Operation by the second method.

lb.	oz.	
85 ..	4	gross weight of each bag.
2 ..	15	tare of do.
82 ..	5	neat weight of do.
	8	no. of bags.
Ans. 658 ..	8	whole neat weight.

2. Required the neat weight of 7 tierces of rice, which together weigh 30cwt. 2qr. 7lb. gross, tare per tierce 34lb.

34×7=238lb.=2cwt. 0qr. 14lb.=whole tare.

	cwt.	qr.	lb.	
Then, from	30 ..	2 ..	7	the whole gross wt.
subtract	2 ..	0 ..	14	the tare.
Ans.	28 ..	1 ..	21	neat weight.

3. In 12 firkins of butter, each weighing 56lb. gross, tare per firkin 11lb., how much neat? Ans. 540lb.

CASE III.

When the tare is so much per cwt.

RULE.

Take such proportional part, or parts, of the whole gross weight as the tare is of 1cwt., for the *whole* tare; which subtract from the gross, and the remainder will be the neat weight.

Or, multiply the pounds gross by the tare per cwt., and divide the product by the number of pounds in 1cwt., and the quotient will be the tare, in pounds; which subtract from the gross, and the remainder will be the neat weight.

EXAMPLES.

1. In 34cwt. 2qr. 8lb. gross, tare 14lb. per cwt., how much neat?

Operation by the first method.

lb. cwt.	Cwt.	qr.	lb.	
14 = $\frac{1}{8}$)	34 ..	2 ..	8	gross.
	4 ..	1 ..	8	tare.
Ans.	30 ..	1 ..	0	neat.

Here, because 14lb.=$\frac{1}{8}$ of a cwt., I take $\frac{1}{8}$ of the gross weight, for the whole tare; which I subtract from the gross, and the remainder is the neat weight.

Operation by the second method.

34 cwt. 2 qr. 8 lb.=3872 lb. gross weight.
14 lb. tare per cwt.

15488
3872

No. of lb. in 1 cwt.=112)54208(484 lb. whole tare.

Then, from 3872 lb. gross,
subtract 484 lb. tare.

Ans. 3388 lb.=30 cwt. 1 qr. neat weight.

2. What is the neat weight of 4 hogsheads of tobacco, each weighing 8 cwt. 2 qr. 12 lb. gross, tare 21 lb. per cwt.?

	Cwt. qr. lb.	
	8 .. 2 .. 12	gross wt. of 1 hhd.
	4	no. of hogsheads.
lb. $14=\frac{1}{8}$)	34 .. 1 .. 20	gross wt. of the 4 hhd.
$7=\frac{1}{2}$)	4 .. 1 .. 6	tare, at 14 lb. per cwt.
	2 .. 0 .. 17	do. at 7 do.
	6 .. 1 .. 23	do. at 21 do., deducted.
Ans.	27 .. 3 .. 25	neat weight.

3. In 228 lb. 12 oz. gross, tare 28 lb. per cwt., how much neat weight? Ans. 171 lb. 9 oz.

4. What is the neat weight of 7 barrels of potash, each weighing 201 lb. gross, tare 10 lb. per cwt.?

Ans. 1281 lb. 6 oz.

CASE IV.

When tret is allowed with tare.

RULE.

1. Find the tare as before, which subtract from the gross, and call the remainder *suttle*.

2. Divide the suttle by 26,* and the quotient will be the

* The reason of dividing by 26, is because 4 lb. is $\frac{1}{26}$ of 104 lb.: But if the tret should be at any other rate than 4 lb. per 104 lb., then other parts must be taken, according to the rate proposed.

tret, which subtract from the suttle, and the remainder will be the neat weight.

EXAMPLES.

1. In 688lb. gross, tare 64lb., tret 4lb. per 104lb., how much neat weight?

688lb. gross.
 64 tare, deducted.
——
26)624 suttle.
 24 tret, deducted.
——
Ans. 600 lb. neat wt.

2. In 1000lb. gross, tare 38lb., tret 4lb. per 104lb., how much neat weight? Ans. 925lb.

3. Required the neat weight of 4 tierces of rice, each weighing 546lb. gross, tare 78lb. per tierce, and tret as usual? Ans. 1800lb.

4. Required the *value* of the neat weight of 4 casks of tobacco, at 12 cents a lb.; their weights being 121 lb., 136lb., 184lb., and 105lb. gross, tare 16lb. per cwt. and tret 4lb. per 104lb. Ans. $54.

Note.—Another allowance, called *Cloff*, is sometimes made, of 2lb. in every 3cwt., (or $\frac{2}{3}$ of a lb, per cwt.,) which may be found by Case III.

INTEREST,

Is an allowance made by the Borrower to the Lender, for the use of money lent; and is usually computed at some rate per cent. per annum.*

The sum lent is called the *principal.*

The aggregate sum of the principal and interest is called the *amount.*

The *rate per cent.* is the interest (or money allowed for the use) of one hundred dollars, or pounds, &c. for any time specified, but usually for a year.†

* *Per cent* means *by the hundred*; and *per annum*, *by the year*. *Cent.* is a contraction of *centum*, the Latin for *hundred*.

† *Lawful* or *legal* interest, is that which is permitted by the laws of the State. It is different under different governments. In England the

Interest is of two kinds, *Simple* and *Compound.*

SIMPLE INTEREST,

Is that which is allowed on the Principal only.

PROBLEM I.

To find the interest of a given sum, for any given time, at a given rate per cent. per annum.

RULE I.*

1. Multiply the given principal by the rate per cent.; divide the product by 100, (or, in Decimals and Federal money, point off in the said product two more decimal places than usual in Multiplication of Decimals,) and the quotient, (or result,) will be the interest of the given sum for one year.

2. *For any number of years;* multiply the interest of the given sum for one year by the given number of years: And *for parts of a year*, work by the rules of Practice; that is, take such part, or parts, of the interest for one year as the time is of one year; (allowing 12 months to make 1 year, and 30 days 1 month;†) then add together the interest for the several parts of a year, and also, the interest for the whole years, if any, in the given time, and the sum will be the whole interest sought.

Or, work by the Rule of Three thus: As 1 year : is to the given time :: so is the interest of the given sum for one year : to the interest required.

---

established rate is 5 per cent. per annum; in the States of New England it is 6; and in some of the other States it is 7 per cent. The Courts of the United-States allow interest according to the practice of the State where the suit is commenced. The rules of the Courts in the States of Connecticut and New-York, for computing *legal* interest, will be given immediately before "Compound Interest."

* The first part of this rule is a contraction of a process in the Rule of Three; for, by the rule of Three, the interest of any sum for a year is found thus; as 100D. : is to the given sum :: so is the rate per cent. per annum : to the interest of the given sum for a year Therefore, the operation by the above rule is the same as by the Rule of Three, only the question is not stated in the usual form. The reason of the remaining part of the rule is obvious.

† The average number of days in each month is nearly 30 1-2; but if 30 days be called a month, it will be sufficiently exact for ordinary calculations.

Note.—When the *amount* is required, add the interest to the principal.

PROOF.—By the Single Rule of Three, or by Compound Proportion.

EXAMPLES.

1. Required the interest of 425D. 50c. for 1 year, at 6 per cent. per annum.

```
     $425.50 Principal.
           6 Rate per cent.
     -------
Ans. $25.5300 Interest.
```

Here, I multiply the principal by the rate per cent., and point off in the product two more decimal places than usual in Multiplication of Decimals, (which is the same as dividing by 100,) and I have $25.53, the interest sought.

2. Required the amount of $125 for 3 years, at 7 per cent. per annum.

```
 $125
    7
 ----
 8.75  Int. for 1 yr.
    3  No. of years.
 ----
26.25  Int. for 3 yr.
125.   Principal.
------
$151.25 Amount, Ans.
```

Here, I find the interest for 1 year, as in Ex. 1st. I then multiply the interest for 1 year by 3, the given number of years, and the product is the interest for 3 years; to which I add the given principal, and the sum is the amount sought.

3. Required the interest of 425 D. 17 c. for 5 years, 2 months and 20 days, at 6 per cent. per annum.

```
                $425.17
                      6
               --------
2 mo.=1/6  |    25.5102 Interest for 1 year.
           |          5
           |   --------       yr. mo. da.
           |   127.5510 Int. for 5 .. 0 .. 0
15 da.=1/4 |     4.251+ Do. for      2 .. 0
5 da.=1/3  |     1.062+ Do. for           15
           |      .354+ Do. for            5
               --------       -----------
      Ans. $133.218+ Do. for  5 .. 2 .. 20
```

Here, I first compute the interest of the given sum for the given number of years, as in Ex. 2d. I then find the interest for the parts of a year thus: 2 months being $\frac{1}{6}$ of a year, I take $\frac{1}{6}$ of the interest for 1 year, for the interest for 2 months: then, because 15 days=$\frac{1}{4}$ of 2 months, I take $\frac{1}{4}$ of the interest for 2 months, for the interest for 15 days; and, because 5 days=$\frac{1}{3}$ of 15 days, I take $\frac{1}{3}$ of the interest for 15 days, for the interest for 5 days. Lastly, I add together the several interests for the years, months, and days, in the given time, and the sum is the interest required.

N. B. In the last example, in dividing by the aliquot parts of a year, &c. I rejected all the fractional parts of a mill. The like omissions are also made in the following examples.

4. Required the interest of \$578 for 3 months, at 7 per cent. per annum.

Here, \$578×7=\$40.46, the interest for 1 year. Then, (because 3 months=$\frac{1}{4}$ of a year,) \$ 40.46÷4=\$10.115, the Ans.

5. Required the interest of \$42 for 1 year, at 5$\frac{1}{2}$ per cent.

Here, \$42×5$\frac{1}{2}$=\$2.31, Ans.

Requ red the interest of

\$ i		\$
784.65 for 1 year, at 6 per cent.	Ans.	47.079
240.20 for 4 years, at 7 per cent.	Ans.	67.256
180.48 for 8 years, at 5 per cent.	Ans.	72.192
64.40 for 2 years, at 4$\frac{3}{4}$ per cent.	Ans.	6.118
72.20 for 1 yr. and 6 mo., at 7 per cent.	Ans.	7.581
26.70 for 7 yr. and 8 mo., at 4 per cent.	Ans.	8.188
120 for 4 yr. 1 mo. 20 da., at 6 per cent.	Ans.	29.80
600 for 1 yr. 7 mo. 18 da., at 7 per cent.	Ans.	68.60

Required the amount of

\$		\$
75.45 for 1 year, at 6 per cent.	Ans.	79.977
400 for 4 years, at 5 per cent.	Ans.	480.
145 for 2 yr. and 3 mo., at 6 per cent.	Ans.	164.575
48 for 2 mo. and 15 da., at 5 per cent.	Ans.	48.50

18. Required the interest of 573*l*. 13s. 9$\frac{1}{2}$d for 2 years and 6 months, at 6 per cent.

```
                    L.   s.  d. q.
                   573 .. 13 .. 9 .. 2
                                     6
               --------------------------
              100)3442 ..  2 .. 9 .. 0
               --------------------------
6 months=½ yr.)34 ..  8 .. 5 .. 0+ Interest for 1 year.
                                 2 No. of years.
               --------------------------
                    68 .. 16 .. 10 .. 0 Interest for 2 years.
                    17 ..  4 ..  2 .. 2    Do.    for 6 months.
               --------------------------
            Ans. L86 ..  1 .. 0 .. 2+  Do.    for 2 yr. and 6 mo.
```

19. Required the interest of 25*l.* for 4 years, 2 months, and 15 days, at 6 per cent. Ans. 6*l.* 6s. 3d.

RULE II.* *For Federal Money.*

1. Reduce the given time to months; reducing the days, if any, to the decimal of a month.

2. Multiply together the given principal and the months (or months and decimal parts) in the given time; point off in the product two more decimal places than usual in Multiplication of Decimals, and you will have the interest of the given sum at 12 per cent.

3. When the given rate per cent. is an aliquot part of 12, take the like proportional part of the interest at 12 per cent. for the in terest required. When the rate per cent. is not an aliquot part of 12; then multiply the interest at 12 per cent. by the given rate, and divide the product by 12, and the quotient will be the interest required.

Note 1.—To reduce to the decimal of a month any number of days less than 30; divide the days by 30, and the quotient will be the decimal required: Or you may omit the cipher in the divisor, and divide by 3; taking care when the number of days is less than 3, to prefix a cipher to the quotient. The quotient should always be found to two decimal places when the true decimal consists of so many

* The interest of 100 dollars, at 12 per cent. per annum, is just 1 dollar per month; whence the reason of this rule is evident.—This method of computing interest is usually more concise and convenient than any other, especially when the principal is federal money.

places. When the given number of days is 30, call the days a month, or .99 of a month.

Note 2.—When the given principal is pounds, shillings and pence, the interest may be computed by the foregoing rule, if the parts of a pound be first reduced to the decimal of a pound. After having found the interest, or the amount, in pounds and decimal parts, you may then find the value of the decimal of a pound, as in Case II. Reduction of Decimals.

EXAMPLES.

1. Required the interest of $40 for 4 years, 2 months, and 17 days, at 6 per cent. per annum.

4 years and 2 months are=50 months.
30)17. Days.

50.56+ Months,=4 yr. 2 mo. 17 da.
40 D. Principal.

12÷6=2)20.2240 Interest at 12 per cent.

Ans. $10.112 Do, at 6 per cent.

Here, because 6, the given rate per cent., is $\frac{1}{2}$ of 12, I take $\frac{1}{2}$ of the interest at 12 per cent. for the interest at 6 per cent.

2. Required the interest of $450.27 for 24 days, at 7 per cent. per annum.

Here 24÷30=.8 of a month.

$450.27
.8

3.60216
7

12)25.214

$2.101+ Ans.

Here the rate per cent, is not an aliquot part of 12, and therefore I multiply the interest at 12 per cent. by the given rate, and divide the product by 12; which gives the interest required. In multiplying by the given rate per cent. I reject those figures of the multiplicand which are to the right hand of the place of mills; and the like omission may be made in every similar operation.

Required the interest of

$	yr. mo. da.		$
200	for 5 .. 7 .. 12,	at 6 per cent.	Ans. 67.40
48	for 2 .. 8 .. 18,	at 4 per cent.	Ans. 5.216

$	yr. mo. da.			$
278.75	for 8..21,	at 3 per cent.	Ans.	6.062+
150	for 1..4..3,	at 5 per cent.	Ans.	10.0625
650*l*.	for 1 yr. and 8 mo.,	at 6 per cent.	Ans.	L. 65.
75*l*. 15s.	for 6mo. and 24da.,	at 4 per cent.	Ans.	L. 1.717

PROBLEM II.

To find the interest of a given sum for a given number of days, at a given rate per cent. per annum.

RULE I.

Multiply together the principal, the number of days, and the rate per cent.; divide the product by 36500, and the quotient will be the interest required.*

EXAMPLES.

1. Required the interest of $730 for 50 days, at 6 per cent. per annum.

Here, $730×50×6=$219000, and $219000÷36500=$6, the Ans.

2. Required the interest of $481.75 for 25 days, at 7 per cent. per annum. Ans. $2.309+

3. Required the interest of $372.50 for 308 days, at $4\frac{1}{2}$ per cent. per annum. Ans. $14.144+

Note.—When the rate is 5 per cent. per annum; divide the product of the principal and the number of days by 7300, and the quotient will be the interest required.†

4. Required the interest of $250 for 63 days, at 5 per cent. per annum.

Here $250×63=$15750, and $15750÷7300=$2.157+, the Ans.

5. Required the interest of $850, from August 15th, 1829, till February 15th, 1830, at 5 per cent. per annum. Ans. $21.424+

---

* The reason for dividing by 36500, is, that there are 365 days in a common year, and dividing by 36500 is the same as dividing by 365 and by 100.

† This is a contraction of the foregoing rule: Dividing by 7300 gives the same result as multiplying by 5 and dividing by 36500.

The number of days in any part of a year may be found by the following table.

A Table, showing the number of Days from any day in any given month to the same day of any other month.

From any day of,	To the same day of,											
	1.Mo. Jan.	2.Mo. Feb.	3.Mo. Mar.	4.Mo. April.	5.Mo. May.	6.Mo. June.	7.Mo. July.	8.Mo. Aug.	9.Mo. Sept.	10.M. Oct.	11.M. Nov.	12.M. Dec.
	Days.	Days.	Days.	Days.	Days.	Days.	Days.	Days.	Days.	Days.	Days.	Days.
1st Mo. Jan.	365	31	59	90	120	151	181	212	243	273	304	334
2d Mo. Feb.	334	365	28	59	89	120	150	181	212	242	273	303
3d Mo. March,	306	337	365	31	61	92	122	153	184	214	245	275
4th Mo. April,	275	306	334	365	30	61	91	122	153	183	214	244
5th Mo. May,	245	276	304	335	365	31	61	92	123	153	184	214
6th Mo. June,	214	245	273	304	334	365	30	61	92	122	153	183
7th Mo. July,	184	215	243	274	304	335	365	31	62	92	123	153
8th Mo. Aug.	153	184	212	243	273	304	334	365	31	61	92	122
9th Mo. Sept.	122	153	181	212	242	273	303	334	365	30	61	91
10th Mo. Oct.	92	123	151	182	212	243	273	304	335	365	31	61
11th Mo. Nov.	61	92	120	151	181	212	242	273	304	334	365	30
12th Mo. Dec.	31	62	90	121	151	182	212	243	274	304	335	365

Illustration of the foregoing table.—To find the number of days from August 15th, 1829, to February 15th, 1830, I find *August* in the left hand column, and *Feb.* at the top of the table; and against the former month, and under the latter, I find 184, the number of days required. In like manner the number of days from any day of any given month to the *same day* of any other month may be found.

If the given days of the months be *different*, the difference must be added to, or subtracted from the tabular number, as the case shall require. Thus, to find the number of days from March 4th, to July 20th, (of the same year,) I find, by the table, that the number of days from the 4th of March to the 4th of July is 122; to which I add 16, (the difference between 4 and 20,) and the sum is 138 days, the time required. Again, to find the number of days from May 21st to Dec. 11th, I look in the table, and find that from May 21st to Dec. 21st there are 214 days; from which number I subtract 10, (the difference between 21 and 11,) and the remainder is 204, the number of days required.

In leap years, if the last day of February be included in the given time, the tabular number must be increased 1. Thus, from the 1st day of Feb. to the 1st day of May, in leap year, there are 89+1=90 days.

The following method of computing interest for days is very accurate, and when the principal is Federal money, and the rate per cent. is an aliquot part of 36, it is very convenient.

RULE II.—*For Federal Money.*

1. Multiply together the given principal and the number of days, and point off in the product three more decimal places than usual. Subtract from this product one-seventieth part of itself, and the remainder will be the interest at 36 per cent.

2. When the given rate per cent. is an aliquot part of 36, the like proportional part of the interest at 36 per cent. will be the interest required. When the rate per cent. is not an aliquot part of 36; multiply the interest at 36 per cent. by the given rate; divide the product by 36, (or divide

twice by 6,) and the quotient will be the interest required.*

Note 1.—To deduct $\frac{1}{70}$ of the product of the principal and days, divide the said product by 7, set the quotient one place farther to the right hand than usual in short division, and subtract it from the dividend.

Note 2.—The 2d Note under Rule II. Prob. I. is also applicable to this Rule.

EXAMPLES.

1. What is the interest of 84D. 28c. for 40 days, at 6 per cent. per annum?

```
          $84.28
              40
         -------
       7)3.37120
           .048
         -------
36÷6=6)3.323
         -------
   Ans. $.553+
```

Here, because the rate per cent. is $\frac{1}{6}$ of 36, I take $\frac{1}{6}$ of the interest at 36 per cent., for the interest required. In performing the operation I reject all the figures beyond the place of mills. Fractional parts of a mill may always be rejected in computing interest by this rule.

2. What is the interest of $100 for 147 days, at 6 per cent.? Ans. $2.415

3. Required the interest of $78 for 64 days, at 4 per cent. Ans. 54c. 6m.+

4. Required the interest of 675D. 50c. from the 4th of October, 1829, till the 1st of February, 1830, at 7 per cent. Ans. $15.5365

5. Required the interest of $150 from April 1st till June 20th, of the same year, at 5 per cent. Ans. $1.642+

Note 3.—When the rate per cent. is not an aliquot part of 36, it will sometimes be convenient to first compute the interest at some other rate, which is an aliquot part of 36, and then increase or diminish the interest thus found, by some part or parts of itself which will give the true interest required. Thus, if the interest at 6 per cent. be increased by $\frac{1}{6}$ of itself, the sum will be the interest at 7 per cent.; or, if the interest at 6 per cent. be diminished by $\frac{1}{6}$ of itself, the

* The interest of $1, at 36 per cent. per annum, (reckoning 365$\frac{1}{4}$ days to the year,) is very nearly $\frac{69}{70}$ of a mill per day; whence the reason of this rule is obvious.

remainder will be the interest at 5 per cent.: Also, $\frac{1}{2}$ of the interest at 9 per cent. is the interest at $4\frac{1}{2}$ per cent., &c. It will be well for the learner to solve questions 4th and 5th in this way.

Note 4.—When the rate is 7 per cent. the interest may be computed as follows: From twice the given principal subtract $\frac{1}{12}$ part of the said principal; multiply the remainder by the number of days; point off in the product four more decimal places than usual in multiplying decimals, and you will have the interest required.*

6. What is the interest of $25.86 for 80 days, at 7 per cent. per annum?

```
12)25.86D.
       2
   ------
   51.72
   —2.155
   ------
   49.565
       80
   ------
Ans. $.3965200
```

7. Find, by the method laid down in the last Note, the interest of $72, for 172 days, at 7 per cent. Ans. $2.3736

PROBLEM III.

To calculate interest on Accounts Current.†

RULE.

When you wish to close an account current, and intend to charge interest on every particular entry, the shortest method of computing the interest, is the following:

1. Find the number of days from the date of the first charge to the close of the account, or time of settlement. Proceed in the same manner with all the accounts, both on the Dr. and Cr. sides.

2. Multiply each sum on the Dr. side by its correspond-

* This is a very easy and accurate method of computing interest at 7 per cent. The method is a contraction of the 2d general rule for computing interest for days.

† An *account current* contains a statement of the mercantile transactions of one person with another, when immediate payments are not made.

ent number of days, and add all these products into one sum total. In like manner, multiply each sum on the Cr. side by its correspondent number of days, and find the sum of the products.

3. Find the difference between the two sums of products, (by subtracting the less from the greater,) which difference multiply by the rate of interest per cent.; divide the product by 36500, and the quotient will be the interest required.—The interest, thus found, must be added to the amount of the debts when the sum of the products on the Dr. side exceeds the sum of the products on the Cr. side; but when the latter sum exceeds the former, then the interest belongs on the Cr. side.

Note.—When goods are sold on a stipulated credit, i. e. for three, six, or nine months, or whatever time may be agreed on; that time must be taken off, in computing interest on the value of such goods, which is readily done by beginning to count the time of each charge so much after its date.

EXAMPLES.

Ex. 1. James Smith, Baltimore, in Account Current with David Wanzer, New-York.

Dr. Cr.

1829.		$	1829.		$
Feb. 11	To balance by ac't. furnish'd,	186.50	March 24	By Flour,	167
			April 6	By Cash,	348
26	To am't. of Sug.	214.75	Sept. 26	By Bill on James Hicks,	250
June 20	To Goods,	515.25			

To close this account on the 10th day of Dec. 1829; who is indebted, and how much, allowing interest at the rate of 7 per cent. per annum?

Operation.

Dr. side. From		Days.	Cr. side. From		Days.
Feb. 11	to Dec. 10	=302	March 24	to Dec. 10	=261
—— 26	to do.	=287	April 6	to do.	=248
June 20	to do.	=173	Sept. 26	to do.	= 75

$ Days. $	$ Days. $
186.50×302=56323.00	167×261=43587
214.75×287=61633.25	348×248=86304
515.25×173=89138.25	250× 75=18750
916.50 207094.50	765 148641

148641.00 subtracted.

58453.50
7 rate per cent.

36500)409174.50(11.21+ $ interest due to D. Wanzer.

	$	
Then, to	916.50	the amount of the debts,
add	11.21	the interest due.
From this,	927.71	
subtract	765.00	the amount of Cr.
Ans.	$162.71	Due to D. Wanzer.

Ex. 2. J. Fox, New-Orleans, in Account Current with A. Davis, New-York.

Dr.				Cr.	
1829.		$	1829.		$
May 19	To Goods,	512.75	June 11	By Cash,	200.50
Aug. 23	To Flour,	274.00	Nov. 8	By Sugar,	580.00
Oct. 4	To Goods,	200.00	Dec. 1	By Cotton,	340.50
Nov. 18	To Linen,	300.00			

How much remained due on this account on the 26th of June, 1830; computing interest at 6 per cent. per annum?

Ans. $187.067+

PROBLEM IV.

To compute interest on Notes, Bonds, &c., having partial payments endorsed.

There has hitherto been much diversity of opinion relative to the computation of *lawful* interest on notes, bonds, &c., on which partial payments have been made; and several different rules have been adopted by the Courts in the different States. Those which follow are most in use. The difference of these rules depends on the principle assumed in respect to the *time when interest becomes due.*

RULE I.*

The following rule is generally thought to allow too little interest. It has, however, been adopted in some of the States.

1. Find the amount of the given principal for the whole time.

2. Compute the interest on each payment, from the time when it was made up to the time of settlement, and add all the payments and their interests into one sum. Subtract this sum from the amount of the principal, and the remainder will be the sum due at the time of settlement.

RULE II.

The following rule has been established by the Courts of law in the State of New-York. The method is also frequently practised in some of the other States; and it is, perhaps, as equitable and convenient as any now in use.

Compute the interest on the given principal, from the time when the interest commenced to the first time when a payment was made, which, either alone, or together with the preceding payments, if any, exceeds the interest then due: add that interest to the principal, and from the amount subtract the payment, or the sum of the payments made

* According to this rule, interest is not considered due until the obligation is paid; and hence interest is allowed on all the endorsements, from the time they were severally made to the time of final settlement. Although the rule is evidently founded on an erroneous principle, it was formerly in general use, and is still used by many persons in this country. It may answer for very short periods of time; but when the time is long it will often give very erroneous results; for it may so happen, that in a course of years, the obligation may be canceled, and even the holder of the note brought into debt to the giver, by the payment of the interest only, without any part of the principal being paid. Thus, if any sum of money whatever should lie at interest 28 years, at 6 per cent. per annum, and the interest should be paid annually, and endorsed on the obligation; at the end of said term of time the amount of the several endorsements, (found according to the above rule.) would exceed the amount of the principal; and consequently the lender of the money would fall in debt to the borrower, without receiving any thing more than the annual interest of the sum lent.

within the time for which the interest was computed, and the remainder will be a new principal; on which compute interest and subtract the payments as upon the first principal, and proceed in this manner to the time of final settlement.

Note.—According to the preceding rule, no interest is ever to be computed on endorsments; for the payments are to be applied to keep down the interest on the principal, so that neither the *interest* nor *payments* shall ever *draw interest.*

RULE III.

The following rule was established in the State of Connecticut, in the year 1784, by the Superior Court of the said State.

1. Compute the interest to the time of the first payment, if that be one year or more from the time when the interest commenced; add the said interest to the principal, and from the amount subtract the payment. Take the remainder for a new principal, and compute the interest on it up to the time of the next payment, if that be one year or more from the time of the former payment; and proceed in like manner from one payment to another, till all the payments are absorbed; provided the time between one payment and another be one year or more.

2. But, if any payment be made before one year's interest hath accrued, then compute the interest on the principal sum due on the obligation for one year;* add it to the principal; and compute the interest on each payment made within that time, from the time when it was made up to the time to which the interest was computed on the principal: subtract the amount of these payments and their interests from the amount of the principal and its interest, and the remainder will be a new principal; with which proceed as before.

3. If any payments be made of a less sum than the interest arisen at the time of such payments, no interest is to be computed on them, but only on the principal sum, for any period.

---

* If a year does not extend beyond the time of final settlement; but if it does, then compute the interest on the principal sum up to the time of settlement.

EXAMPLE 1st,

To be calculated according to the preceding rules.

For value received I promised to pay A. B., or order, five hundred dollars, on demand, with interest.

Danbury, Jan, 1, 1827. C. D.

On this note were the following endorsements:

March 1, 1828,	received	40	dollars.
July 1, 1828,	"	12	do.
Sept. 1, 1829,	"	8	do.
Nov. 1, 1829,	"	350	do.

What was the balance due on the 1st of July, 1830, allowing interest at the rate of 6 per cent. per annum?

By Rule I.

$	
500	Principal.
105	Interest on do. to July 1, 1830,=3½ years.
605	Amount.
40	First payment, made March 1, 1828.
5.60	Interest on do. to July 1, 1830,=2yr. 4mo.
12.	Second payment, made July 1, 1828.
1.44	Interest on do. to July 1, 1830,=2 years.
8.	Third payment, made Sept. 1, 1829.
0.40	Interest on do. to July 1, 1830,=10 months.
350.	Fourth payment, made Nov. 1, 1829.
14.	Interest on do. to July 1, 1830,=8 months.
431.44	Amount of all the payments, deducted.
173.56	Balance due July 1, 1830, *Answer.*

By Rule II.

$	
500	Given principal, January 1, 1827.
35	Interest on do. to 1st payment,=14mo.
535	Amount.
40	1st payment deducted.

[*Carried over.*

495 New principal, due March 1, 1828.
9.90 Interest on do. to 2d payment,=4mo.

504.90 Amount.
12. 2d payment deducted.

492.90 New principal, due July 1, 1828.
39.432 Interest on do. to 4th payment,=16mo.

532.332 Amount.
358. Sum of the 3d and 4th payments deducted.

174.332 New principal, due Nov. 1, 1829.
6.973 Interest on do. up to July 1, 1830,=8mo.

181.305+ Amount, due July 1, 1830, *Ans.*

Note.—Each of the payments, except the 3d, exceeds the interest due on the note at the time of payment. The 3d payment, ($8,) being less than the interest then due, is added to the 4th; and the interest is computed on the principal sum from the time of the 2d payment to that of the 4th.

By Rule III.

$
500 Given principal.
35 Int. on do. to 1st payment,=14mo.

535 Amount.
40 1st payment deducted.

495 New principal, due March 1, 1828.
29.70 Int. on do. for 1 year, viz. up to March 1, 1829.

524.70 Amount.

12. 2d payment, made July 1, 1828.
0.48 Int. on do. up to March 1, 1829,=8mo.

12.48 Amount of 2d payment and int. deducted.

[*Carried up.*

512.22 New principal, due March 1, 1829.
 30.733 Int. on do. for 1yr., viz. up to March 1, 1830.

542.953 Amount.

 8. 3d payment, which does not bear interest.
350. 4th payment, made Nov. 1, 1829.
 7. Int. on do. up to March 1, 1830,=4mo.

365. Amount of 3d and 4th payments and int. deducted.

177.953 New principal, due March 1, 1830.
 3.559 Int. on do. up to July 1, 1830,=4mo.

181.512 Amount, due July 1, 1830, *Ans.*

Note.—The interval of time between the 1st and 2d payments being less than a year, the interest is computed on the new principal of March 1. 1828, for 1 year, as directed in the 2d article of Rule III. The interest is also computed on the new principal of March 1, 1829, for 1 year, which extends beyond the time of the 4th payment. The 3d payment, ($8,) being less than the interest then due, (which is $15.366,) no interest is computed on it. *See the 3d article of Rule* III.

2. A note was given, August 10th, 1824, for $1500, and afterwards endorsed as follows:

Sept. 10th, 1824,	received	8	dollars.
Oct. 25th, 1825,	"	500	do.
Nov. 15th, 1826,	"	40	do.
Oct. 25th, 1827,	"	800	do.

What was due on the 10th day of January, 1828, computing interest according to Rule II. at the rate of 7 per cent. per annum? Ans. $441.868+

3. On January 1, 1827, Samuel Trusty owed me $800, on which I was to receive interest at 6 per cent. On July 1, 1827, he paid me $100; on March 1, 1828, $15; on Sept. 1, 1828, $85; and on Jan. 1, 1830, $500. What was due October 1, 1830, by the three preceding rules?

Ans. By Rule I. $225.05; by Rule II. $238.943+, and by Rule III. $238.766+

COMPOUND INTEREST,

Is that which arises from the interest being added to the principal, and (continuing in the hands of the borrower) becoming part of the principal, at the end of each stated time of payment.

RULE.

1. Find the *amount* of the given principal, by Simple Interest, for the first year, and this amount will be the principal for the second year: Then find the amount of this principal for one year, and it will be the principal for the third year; and so on for any number of years.

2. When the given time does not consist wholly of entire years; find the amount of the principal for the whole years, if any, as directed above, and then find the amount for the parts of a year.

3. From the last amount subtract the given principal, and the remainder will be the compound interest for the whole time.

Note.—The interest is here supposed to be payable *annually*. When it is payable more or less frequently, then calculate the interest up to the time when it is due, and add it to the principal, to form a new principal, and so proceed.

EXAMPLES.

1. What is the compound interest of $400, for 3 years, at 6 per cent. per annum?

Operation.

$400 given principal.
6
———
24.00 interest for 1 year.
400.
———
424. amount for 1 year.
6
———
25.44
424.
———

[*Carried up.*

```
 449.44     amount for 2 years.
      6
 -------
 26.9664
449.44
 -------
476.406     amount for 3 years.
400.        given principal deducted.
 -------
$76.406+    compound interest for 3 years, Ans.
```

It is not usually necessary to carry the work beyond mills. When the fractional parts of a mill are rejected, if the figure next to the right hand of the place of mills exceeds 4, it will be proper to add 1 to the mills.

2. What is the compound interest of $100, for 2 years, at 7 per cent. per annum? Ans. $14.49

3. What is the compound interest of $500, for 4 years, at 6 per cent.? Ans. $131.238+

4. Required the amount of $78.40, for 2 years, at 5 per cent. per annum, compound interest. Ans. $86.436

5. What will $450 amount to in 3 years, at 7 per cent. per annum? Ans. $551.269+

6. What will $1000 amount to in 2 years and 8 months, at 6 per cent. per annum? Ans. $1168.544

To solve question 6th, find the amount for 2 years, and then find the amount of that sum for 8 months.

7. What is the compound interest of $1000, for 4 months, at 2 per cent. per month? Ans. $82.432+

To solve question 7th, find the amount of the principal for the first month, and then find the amount of that sum for the second month, and so on.

Note.—When the principal is pounds, shillings, &c., it will generally be best to reduce the parts of a pound to the decimal of a pound, and then find the amount, or interest, as in federal money.

8. Required the amount of 70*l.* 10s. 6d. for 2 years, at 6 per cent. per annum, compound interest.

Ans. L 79.24189=79*l.* 4s. 10d.+

REBATE OR DISCOUNT,

BY SIMPLE INTEREST.

Discount is an allowance made for the payment of any sum of money before it becomes due, and is the difference between that sum, due some time hence, and its present worth.

The *present worth* of any sum or debt, due some time hence, is such a sum, as if put to interest, would, in that time, and at the rate per cent. for which the discount is to be made, amount to the sum or debt then due.*

RULE.

Find, by Simple Interest, the amount of $100, or L100, &c., for the given time, at the proposed rate. Then say,

As the said amount of $100, or L100, &c. : is to the given sum or debt :: so is $100, or L100, &c. : to the present worth of the sum or debt.—Subtract the present worth, thus found, from the given sum, and the remainder will be the discount.

Or, as the amount of $100, or L100 : is to the given sum :: so is the interest of $100 or L100 : to the discount; which being subtracted from the given sum will give the present worth.

Proof.—Find the amount of the present worth of the debt, at the given rate and time, and, if the work be right, that amount will be equal to the debt.

EXAMPLES.

1. What must be discounted for the ready payment of a

* The interest of any sum, for any given time, is greater than the discount on the same sum, at the same rate, and for the same time; the discount being the interest of the present worth of the given sum or debt; not of the debt itself It is the common practice, however, to deduct the interest, instead of the discount; the parties not attending to the real difference between discount and interest. Thus, when 100 dollars are discounted for a year, in this way, at 6 per cent., 6 dollars are deducted, and the person to whom the debt is due receives only 94 dollars. If he were to lend the 94D on interest, for a year, at 6 per cent, the amount would be only 99D 64c, which is 36 cents less than the debt. This would, in fact, be discounting at the rate of about 6.4 per cent, instead of 6 per cent—In Bank Discount, the interest is always considered as the discount.

debt of $212, due a year hence, at 6 per cent. per annum?

The interest of $100 for a year, at 6 per cent., is $6, and the amount is $106. Therefore,

As $106 : $212 :: $100 : $200, the present worth. Then, $212—200=$12, the discount, Ans.

Or, as $106 : $212 :: $6 : $12, the discount, Ans.

This answer may be proved by finding the amount of $200, (the present worth,) for a year, at 6 per cent.; which will be found to be $212, equal to the debt.

2. How much must be discounted for the ready payment of a debt of $456, due 2 years hence, at 7 per cent. per annum? Ans. $56.

3. What is the present worth of $600, due 4 years hence, at 5 per cent.? Ans. $500.

4. What is the difference between the simple interest of $1000 for 10 years, at 6 per cent. per annum, and the discount of the same sum for the same time and rate?

Ans. $225.

Note 1.—The present worth of any sum, for any given number of months, may be found as follows: Multiply the number of months by the rate per cent.,and add the product to 1200: then, as this sum : is to 1200 :: so is the debt : to its present worth.

5. Bought goods, amounting to $645.75, on 8 months credit; how much ready money must I pay, discount at 6 per cent. per annum?

8×6=48, and 1200+48=1248. Then, as 1248 : 1200 :: $645.75 : $620.913+ Ans.*

6. What sum of ready money must be paid for a bill of 202D. 70c., due 2 months and 21 days (or 2.7 months) hence, discount at 6 per cent. per annum? Ans. $200.

Note 2.—When the rate per cent. is 6, the discount may be found thus: To 200 add the given number of months: then, as this sum : is to the given number of months :: so is the debt : to the discount.

---

* The reason of the foregoing rule may be shown from this example, thus: As 12 months : 8 months :: $6, the interest of $100 for 12 months : $$\frac{48}{12}$, the interest of $100 for 8 months. Then, as 100\frac{48}{12}$: $100; or by reduction of both to 12ths, as 1248 : 1200 :: $645.75 : $620.913+ the present worth required.

7. What is the discount of $250, due 3 years and 4 months (or 40 months) hence, at 6 per cent.?

As 240 : 40 :: $250 : $41.666+ Ans.

Note 3.—The present worth of any sum, for any given number of days, may be found as follows: Multiply the given number of days by the rate per cent., and add the product to 36500: then, as this sum : is to 36500 :: so is the debt : to its present worth.*

8. Required the present worth of a debt of $730, due 134 days hence, at 6 per cent.

134×6=804, and 36500+804=37304. Then, as 37304 : 36500 :: $730 : $714.266+ Ans.

Note 4.—To find the discount on a given sum for 60 days, at 6 per cent., as practised in the banks; divide the given sum by 100, (or, in federal money, remove the decimal point two places to the left,) and the quotient, (or result,) will be the discount required.

For any other number of days than 60, multiply the discount for 60 days by the given number of days, and divide by 60.

9. Required the bank discount on $730, for 60 days, at 6 per cent. Ans. $7.30

10. Required the bank discount on $1000, for 48 days, at 6 per cent.

The discount for 60 days is $10, which being multiplied by 48, and the product divided by 60, gives $8, Ans.

Although the foregoing method of discounting is practised at many of the banks in this country, it is very erroneous; a year being reckoned only 360 days, and the interest being allowed instead of the true discount. Some persons compute simple interest by the same rule; but it always gives the interest about one-seventieth too great; and, therefore, the method is too erroneous for use when the given time exceeds a month.

---

* This rule depends upon the same principle as that given in Note 1st, and the reason of it may be illustrated in a similar manner, from example 8th, as follows: As 365 days : 134 days :: $6 : $\$\frac{804}{365}$, the interest of $100 for 134 days. Then, as $\$100\frac{804}{365}$: $100; or by reduction to 365ths, as 37304 : 36500 :: $730 : $714.266+ the present worth required.

Note 5.—When several sums are to be paid at different times, find the discount, or present worth, of each particular payment, separately, and when so found, add them into one sum.

11. If a legacy is left me of $2000, of which $500 are payable in 6 months, $800 in one year, and the rest at the end of 3 years; how much ready money ought I to receive for said legacy, allowing 6 per cent. discount?

Ans. $1833.372+

EQUATION OF PAYMENTS,

Is the finding a time to pay, at once, several debts, due at different times, so that no loss shall be sustained by either party.

RULE.

Multiply each debt by the time that must elapse before it will become due; then divide the sum of the products, thus obtained, by the sum of all the debts, and the quotient will be the time required.*

Note.—When the products are taken, all the times, and likewise all the debts must be in the same denomination. The answer will be in the denomination which the periods of time mentioned in the question are reduced to.

EXAMPLES.

1. If one person owes to another $300 payable in 4 months, $400 payable in 6 months, and $500 payable in a year; in what time might the whole debt be paid without loss to either party?

Operation.

$	mo.	products.
300 ×	4 =	1200
400 ×	6 =	2400
500 ×	12 =	6000
1200		1200)9600(8 months, Ans.

* This is the method commonly employed; but it is not exactly equitable, because the interest is allowed instead of the discount, on each

2. If a person owes $100 payable at present, and $500 payable in 4 years; at what time may both debts be justly paid at a single payment?

$ yr.
100×0= 0
500×4=2000

600)2000(3⅓ years=3yr. 4mo. Ans.

3. What is the equated time for the payment of three debts; the first of $70.25, payable in 1 month; the second of $140.50, due at 4 months; and the third of $210,75, due at 5 months? Ans. 4 months.

4. A owes B $800, of which $200 are payable in 1 year, $280 in 2 years, $120 in 3 years, and the rest in 4 years; but by agreement A is to pay the whole at one time: The equated time is required.

Ans. 2⅖ years; or 2 years, 4 months, 24 days.

5. What is the equated time for the payment of two debts, the one of $100, due at the end of 40 days, and the other of $400, due in 80 days? Ans. 72 days.

COMMISSION, INSURANCE, &c.

Commission is the sum which a merchant charges for buying or selling goods for another.

Brokerage is a smaller allowance of the same nature, paid usually for negotiating bills, or transacting other money concerns.

Insurance, or *Assurance*, is a contract by which one party, on being paid a certain sum or premium by another, on account of property that is exposed to risk, engages in case of loss, to pay the owner of the property the sum insured on it.

---

payment which is made before it becomes due. The method, however, is much more convenient than any other, and it is sufficiently exact for use in all ordinary cases. A more accurate rule will be given in "Equation of Payments by Decimals."

PROBLEM I.

To compute the commission, brokerage, insurance, or any other allowance, on a given sum, at a given rate per cent.

RULE.

Multiply the sum by the rate per cent., and divide the product by 100, (or if the given sum be federal money, multiply it by the rate per cent. and point off two more decimal places than usual in the product,) and the result will be the answer.

EXAMPLES.

1. Find the commission on $250.25, at 2 per cent.

$250.25
2
———
$5.0050 Ans.

Here, I point off in the product of the given sum and rate per cent. two more decimal places than usual in multiplication of federal money, which is the same as dividing by 100.

2. Find the commission on $850, at $2\frac{1}{2}$ per cent. Ans. $21.25

3. Find the brokerage on $2400, at $\frac{3}{4}$ per cent. Ans. $18.

4. Required the premium of insurance on $512.50, at 4 per cent. Ans. 20.50

5. A man's house, estimated at $3500, was insured against fire, for $1\frac{3}{4}$ per cent. a year: what insurance did he annually pay? Ans. $61.25

PROBLEM II.

To find how much must be insured on property worth a given sum, so that in case of loss, both the value of the property and the premium of insurance may be repaid.

RULE.

Subtract the rate per cent. from $100; then, as the remainder : is to $100 :: so is the value of the property : to the sum to be insured.

EXAMPLES.

1. How much must be insured at 8 per cent. on an ad-

venture of $6440, so that in case of loss, not only the value of the adventure, but also the premium of insurance may be paid?

Here, as $92 (=100—8) : $100 :: $6440 : $7000, Ans.

The truth of this operation is proved by finding the premium on $7000 at 8 per cent. This is found to be $560. Hence, in case of the adventure being lost, the owner will receive not only $6440, the value of the adventure, but also $560, the premium; and thus he will sustain no loss whatever.

2, What must be the sum insured, at $5\frac{1}{2}$ per cent., on goods worth $4725, so that in case of loss, the owner may be repaid both the value of the goods and the premium of insurance? Ans. $5000.

BARTER,

Is the exchanging of one commodity for another; value for value, according to rates or prices agreed upon by the parties concerned.

RULE.

Find, (by the Rule of Three, or by Practice, &c.) the value of that commodity whose quantity is given; then find what quantity of the other, at the proposed rate, can be purchased for the same money, and it will be the answer.

EXAMPLES.

1. What quantity of flax, at $12\frac{1}{2}$ cents per lb., must be given in barter for 5 yards of cloth, at 45 cents per yard?

$.45×5=$2.25, value of the 5yds. of cloth.

Then, as $.125 : $2.25 :: 1lb. : 18lb. Ans.

2. How many pounds of butter, at 20 cents per lb., must be given in barter for 14 yards of cloth, at 25 cents per yard? Ans. $17\frac{1}{2}$lb.

3. How much wheat, at $1.25 a bushel, ought to be given for 50 bushels of rye, at 70 cents a bushel?

Ans. 28 bush.

4. How many bushels of salt, at $62\frac{1}{2}$ cents a bushel,

must be given for 20 bushels of oats, at 25 cents a bushel?

Ans. 8 bush.

5. How many bushels of corn, at 5s. a bush., ought to be given in barter for 84 bushels of wheat, at 7s. 6d. a bush.?

Ans. 126 bush.

6. Sold goods to the value of $214, and received in payment 120 bushels of corn, at 60 cts. a bushel—the remainder is to be paid in wheat, at $1.25 a bushel: how much wheat will pay the balance? Ans. 113⅗ bush.

7. A gives B 250 yards of cloth, at 24 cents per yard, for 300lb. of loaf sugar: what was the sugar worth per lb.?

Ans. 20 cents.

8. Two merchants barter: A has 20cwt. of cheese, at 1*l.* 1s. 6d. per cwt.; B has 8 pieces of cloth, at 3*l.* 14s. per piece—I desire to know who must receive the difference, and how much? Ans. B must receive of A 8*l.* 2s.

Note.—Sometimes, in bartering, one commodity is rated above the ready money price; then, to find the bartering price of the other, say,

As the ready money price of one commodity : is to that of the other :: so is the bartering price of the former : to that of the latter.

Then, find the quantity required, according to either the bartering or ready money price.

9. A has ribands at 24 cents per yard, ready money; but in barter he will have 27 cents. B has broadcloths at 3D. 84c. per yard ready money: at what rate must B value his cloth per yard, to be equivalent to A's bartering price, and how many yards of riband, at 27 cents per yard, must then be given by A for 200 yards of B's broadcloth?

As $.24 : $3.84 :: $.27 : $4.32, B's bartering price.

$4.32×200=$864, value of the 200yds. of cloth.

Then, as $.27 : $864 :: 1 yd. : 3200 yds. of riband.

Ans. B must value his broadcloth at $4.32 per yard, and he must receive 3200 yards of riband for the 200 yards of cloth.

10. A has 200 yards of linen cloth, worth 25 cents a yard, ready money, which he barters with B at 31 cents a yard, taking sugar of him at 10 cents a pound, which is worth but 8 cents a lb. ready money: who gets the best bargain?

620lb. of sugar, at barter price, will pay for the 200yd. of linen, at the bartering price.

Value of the 200yd. of linen at cash price = \$50.00
Do. of the 620lb. of sugar at do. = 49.60

Difference, \$ .40

Ans. B gets the best bargain, by 40 cents.

LOSS AND GAIN,

Is a method of computing the profit or loss on the purchase and sale of goods. The rules for making calculations of this kind, are only particular applications of the Rule of Three.

CASE I.

When the buying and selling prices are given, to find what is gained or lost by selling.

RULE.

First, find the value of the commodity at the price it cost; then find its value at the price sold at; and the difference between these will be the gain or loss.

Or, as 1 yard, or 1lb. &c. : is to the given quantity :: so is the gain or loss on 1yd. or 1lb. &c. : to the whole gain or loss.

EXAMPLES.

1. A merchant sold 100 yards of cloth at 1D. 50c. a yard, which cost him 1D. 25c. a yard: how much did he gain by the sale?

\$1.50×100=\$150 Selling price.
1.25×100=\$125 Prime cost.

Ans. \$ 25 Whole gain.

Or, by the second method thus: \$1.50−1.25=\$.25, the gain per yard. Then, as 1yd. : 100yd. :: \$.25 : \$25, Ans.

2. Bought 25 yards of broadcloth at 5 D. a yard, and sold the same at 5 D. 75 cts. a yard: how much did I gain by the bargain? Ans. \$18.75

3. Bought a piece of baize containing 42 yards, for 11D. 81c., and sold it at 31 cents a yard: what was the gain or loss on the whole piece? Ans. $1.21 gain.

4. Bought 11cwt. of sugar at 3*l.* 8d. per cwt., but could not sell the same for more than 2*l.* 16s. per cwt.: how much did I lose on the whole? Ans. 2*l.* 11s. 4d.

5. Bought a pipe of wine at $1.75 per gallon; paid the freight $3.46; paid for carting the same $2.52; and by accident 46 gallons leaked out: at what rate must I sell the remainder per gallon to gain on the whole $6?

Ans. $2.906.

Note.—When goods are bought or sold on credit, you must calculate, (by Discount,) the present worth of their price, in order to find the true gain or loss.

6. Bought 204 yards of broadcloth, at $2.25 a yard, and sold the whole for $510, on 4 months credit: what did I gain or lose, allowing discount at 6 per cent. a year?

	$		$		$		$	
As	102	:	510	::	100	:	500	Present worth.
				$2.25×204=			459	Prime cost.
							$ 41	Gain, Ans.

7. Bought 412 bushels of rye for $206, and sold the same at 60 cents a bushel, on 6 months credit: what did I gain, allowing discount at 6 per cent. per annum? Ans. $34.

CASE II.

To find what is gained or lost per cent.

RULE.

As the prime cost : is to $100 or L100 :: so is the gain or loss on the cost : to the gain or loss per cent.

EXAMPLES.

1. If I buy cloth at 88 cents per yard, and sell it at 1D. 10c. per yard: what do I gain per cent., or in laying out $100?

Sold at	$1.10	per yard.
Cost,	.88	do.
Gain,	.22	do.

Then, as $.88 : $100 :: $.22 : $25 Ans.

2. If tea be bought for $87\frac{1}{2}$ cents per lb. and sold at 1D. $12\frac{1}{2}$c. per lb., what is the gain per cent.?

Ans. $28.571+

3. Bought a pipe of wine for $150, and sold the same at $1.25 per gallon: did I gain or lose by the sale, and how much per cent.? Ans. Gained 5 per cent.

4. If I buy cloth at 6s. 8d. per yard, and sell the same at 7s. 4d. per yard, what do I gain per cent.? Ans. L 10.

5. Bought Wheewell's Mechanics and Dynamics for $7.25, Laplace's System of the World for $5.75, Bonnycastle's Algebra for $6.375, and Simpson's Fluxions for $5.75: sold all those books for $30; how much is the gain per cent? Ans. $19.402+

6. If I buy cloth at $4.16 per yard, on 8 months credit, and sell it at $3.90 per yard, ready money, what do I lose per cent., allowing 6 per cent. discount on the purchase price? Ans. $2.50

CASE III.

To know how a commodity must be sold to gain or lose a certain rate per cent.

RULE.

As $100 : is to $100 with the gain per cent. added, or loss per cent. subtracted :: so is the purchase price : to the selling price.

EXAMPLES.

1. How must tea, which cost 92 cents per lb., be sold per lb. to gain 25 per cent.?

$100 : $125 (=100+25) :: $.92 : $1.15 Ans.

2. How must pork, which cost $4\frac{1}{2}$ cents per lb., be sold to gain 20 per cent.? Ans. At 5c. 4m. per lb.

3. Bought cloth at 50 cents per yard, which not proving as good as I expected, I am willing to lose 10 per cent.: at what price per yard must I sell it? Ans. 45 cents.

4. Bought goods to the amount of $875, and by selling the same gained 25 per cent.: what did I get for the goods?

Ans. $1093.75

CASE IV.

When the gain or loss per cent. and the selling price are given, to find the first cost.

RULE.

As $100 increased by the gain per cent. or diminished by the loss per cent. : is to $100 :: so is the selling price : to the prime cost.

Note.—Cases III. and IV. prove each other.

EXAMPLES.

1. If 25 per cent. be gained by selling tea at $1.15 per lb., what was the prime cost per lb.?

As $125 : $100 :: $1.15 : $.92 Ans.

2. If 20 per cent. be gained by selling pork at 5c. 4m. per lb., what was the prime cost per lb.? Ans. 4½ cents.

3. If 12 per cent. be lost by selling 120 yards of broadcloth for $422.40, what was the prime cost per yard?

Ans. $4.

CASE V.

If by goods sold at a given rate there is so much gained or lost per cent., to find what would be gained or lost per cent. if sold at another rate.

RULE.

As the first price : is to the second :: so is $100 increased by the gain per cent. or diminished by the loss per cent. : to a fourth number. If this fourth number exceeds $100, the excess is the required gain per cent.; but if it be less than 100, that deficiency is the loss per cent.

EXAMPLES.

1. Sold a quantity of wheat at $1.50 per bushel, and thereby gained 25 per cent.: what should I have gained or lost per cent. if I had sold the wheat at $1.08 per bushel?

$1.50 : $1.08 :: 125 : 90. Then 100—90=10.

Ans. I should have lost 10 per cent.

2. If I sell sugar at $8 per cwt., and thereby gain 12 per cent.; what should I gain per cent. by selling it at $9 per cwt.? Ans. 26 per cent.

3. If by selling coffee at 24 cents per lb. I gain 20 per cent., what should I gain or lose per cent. by selling it at 20 cents per lb.? Ans. *Nothing*.

FELLOWSHIP,

Is a rule, by which any sum or quantity may be divided into any number of parts, which shall be in any given proportion to one another.

By this rule are adjusted the gains or loss or charges of partners in company; or the effects of bankrupts, or legacies in case of a deficiency of assets or effects, &c.

Fellowship is either *Single* or *Double*. It is Single, or Simple, when the shares or portions are to be proportional each to one single given number only; as when the stocks of partners are all employed for the same time, or when they are considered without regard to time: And Double, or Compound, when each portion is to be proportional to two or more numbers; as when the stocks of partners are employed for different times.

SINGLE, OR SIMPLE FELLOWSHIP.

GENERAL RULE.

Add together the numbers which denote the proportion of the shares. Then say,

As the sum of the said proportional numbers : is to each particular proportional number :: so is the whole sum to be parted or divided : to the share or part corresponding to each respective proportional number.

Or, As the sum of the stocks of all the partners : is to each partner's particular stock :: so is the whole gain or loss : to each partner's share of the said gain or loss.

Or, if the given sums be *federal money*, you may divide the whole gain or loss by the whole stock, as in Division of Decimals, and the quotient will be the gain or loss on the dollar; which being multiplied into each partner's share of the stock, will give the required shares of the partners.

To prove the work: Add all the shares or parts together, and the sum will be equal to the whole stock, or number to be shared, if the work be right.

EXAMPLES.

1. To divide the number 240 into three such parts, as shall be in proportion to each other as the three numbers 1, 2 and 3.

Here 1+2+3=6, the sum of the proportional numbers. Then,

6 : 1 :: 240 : 40 the 1st part.
6 : 2 :: 240 : 80 the 2d part. } Ans.
6 : 3 :: 240 : 120 the 3d part.

Sum of all, 240 Proof.

2. Two merchants, A and B, enter into partnership: A's stock is $2500, and B's $1500. Required the share of each in a gain of $200.

Here $2500+1500=$4000, the sum of the stocks. Then,

$ $ $ $
4000 : 2500 :: 200 : 125 A's share. } Ans.
4000 : 1500 :: 200 : 75 B's share.

200 Proof.

Or, otherwise thus: $200÷4000=$.05, the gain on each dollar of the whole stock. Then,

$ $
2500×.05=125 A's share. } Ans.
1500×.05= 75 B's share.

3. Three partners, A, B, and C, purchased goods to the amount of $800; of which A paid $120, B $200, and C $480; and by selling the goods they gained $160. What was each partner's share of the gain?

Ans. A's $24, B's $40, and C's $96.

4. Three merchants, A, B, and C, freighted a ship from Madeira for Liverpool, with 216 tuns of wine; of which A owned 96, B 72, and C 48 tuns; the mariners meeting with a storm at sea were constrained for the safety of their lives to cast 45 tuns thereof overboard: How many of the 45 tuns did each merchant lose, according to his rate of the adventure? Ans. A 20, B 15, and C 10 tuns.

5. A bankrupt owes to A $900, to B $850, to C $640, to D $150, to E $750, and to F $310; and his whole estate amounts to only $1800. If his estate be delivered up to these creditors, how much will he pay in the dollar, and what sum will each creditor receive?

Ans. 50 cents in the dollar; and A will receive $450, B $425, C $320, D $75, E $375, and F $155.

6. A and B purchased a house jointly for $4000, and

afterwards let it for the yearly rent of $650: what share of the yearly rent must each receive, the one having contributed $1850, and the other $2150?

Ans. $300.625, and $349.375

7. If a tax of $1200 be laid on a town, and the inventory or grand list of the town is $500000; what is A's tax, whose list amounts to $4500? Ans. $10.80

8. A and B, trading together, gained $100. A put in $640; B put in so much that he received $60 of the gain: I demand B's stock?

As $40 : $60 :: $640 : $960, B's stock, Ans.

9. Pewter is composed of 112 parts of tin, 15 of lead, and 6 of brass: How much of each ingredient is requisite to make 266lb. of pewter?

Ans. 224lb. of tin, 30lb of lead, and 12lb. of brass.

10. 76 parts of nitre, 14 of charcoal, and 10 of sulphur, compose gunpowder: How much of each of these ingredients will be requisite to form 50lb. of gunpowder?

Ans. 38lb. of nitre, 7lb. of charcoal, and 5lb. of sulphur.

11. Proof spirits are composed of 48 parts of alcohol, or pure spirit, and 52 parts of water: how much of each of these are contained in 25 gallons of proof spirits?

Ans. 12 gallons of alcohol, and 13 gallons of water.

CONTRACTION.

The work may often be much abbreviated by dividing the particular stocks, or proportional numbers, by any number that will measure or divide them, and then using the quotients instead of the numbers so divided.

EXAMPLES.

1. Two merchants, A and B, trading together, gained $812—A's stock was $2400, and B's $3200: What was each man's share of the gain?

Divisor.	Stocks.	Quotients.
800	$2400	3
	3200	4
	5600	7 Sum.

7 : 3 :: $812 : $348 A's share.
7 : 4 :: $812 : $464 B's share. } Ans.

2. Three men, trading together, gained $450—A's stock was $900, B's $1800, and C's $4500: What was each man's share of the gain?

Ans. A's $56.25, B's $112.50, and C's $281.25.

DOUBLE, OR COMPOUND FELLOWSHIP,

Is concerned in cases in which the stocks of partners are employed or continued for different times.

RULE.

Multiply each partner's stock by the time of its continuance; then divide the quantity, as in Single Fellowship, into shares, in proportion to these products, by saying,

As the total sum of all the said products :
Is to each partner's particular product ::
So is the whole gain or loss :
To each partner's share of the gain or loss.*

Note 1.—The several stocks, and also the times, must be reduced to the same denomination, when they are of different denominations.

Note 2.—The operation may frequently be contracted by dividing either the several stocks, or products, by a common divisor, and then using the quotients instead of the numbers so divided, as in Single Fellowship.

PROOF.—The method of proof is the same as in Single Fellowship.

EXAMPLES.

1. Two merchants traded together—A put in $1200 for 6 months, and B $400 for one year; and they gained $280: what was each partner's share of the gain?

* When the times are equal, the shares of the gain or loss are evidently as the stocks, as in Single Fellowship; and when the stocks are equal, the shares are as the times; wherefore, when neither are equal, the shares must be as their products.

Operation.

$ mo.
1200× 6=7200
400×12=4800

12000

12000 : { 7200 / 4800 } :: $280 : { $168 A's share. / $112 B's share. } Ans.

Or, by contraction, as follows:

Divisor.	Stocks.	Quotients.	mo.
400	$1200	3	3× 6=18
	400	1	1×12=12
			30

30 : { 18 / 12 } :: $280 : { $168 A's share. / $112 B's share. } Ans.

2. Two merchants traded in company; A put in $215 for 6 months, and B $390 for 9 months; but by misfortune they lose $200: how must they share the loss?

Ans. A's loss is $53.75, and B's $146.25

3. Three persons received $665 interest: A had put in $4000 for 12 months, B $3000 for 15 months, and C $5000 for 8 months. How much is each person's part of the interest? Ans. A's part is $240, B's $225, and C's $200.

4. A, B and C, hold a pasture in common, for which they pay 45D. 60c. a year. A put in 8 oxen for 6 weeks, B 12 oxen for 8 weeks, and C 12 oxen for 12 weeks: How much must each pay of the yearly rent?

Ans. A $7.60, B $15.20, and C $22.80

5. Two merchants enter into partnership for 18 months. A at first put into stock $500, and at the end of 8 months he put in $100 more. B at first put in $800, and 4 months after that took out $200. At the end of the 18 months they find they have gained $500: what is each man's share of the gain?

500×18=9000
100×10=1000

A's prod. 10000
B's prod. 11600

Sum, 21600

800×18=14400
200×14= 2800

11600

$21600 : \left\{ \begin{matrix} 10000 \\ 11600 \end{matrix} \right\} :: \$500 : \left\{ \begin{matrix} \$231.48+ \text{ A's share.} \\ \$268.51+ \text{ B's share.} \end{matrix} \right\}$ Ans.

6. A and B companied—A put in $1000 on the 1st of January; but B could not put in any till the 1st of May: what did he then put in, to have an equal share with A at the end of the year?

As 8mo. : 12mo. :: $1000 : $1500 Ans.

EXCHANGE,

Is the method of finding what sum of the money of one country is equivalent to any given sum of the money of another.

By the *par of exchange* between two countries, is meant the intrinsic value of the money of one country compared with that of the other, and estimated by the weight and fineness of the coins. If the exchange be made at the intrinsic value of the money of the different countries, it is said to be *at par;* but if the money of one country be estimated at more or less than its intrinsic value, the exchange is said to be *above par*, or *below par*.

Owing to changes in the course of trade, and to the de-demand for money, &c., the relative value of the money of two countries is liable to frequent changes. Hence the *course of exchange*, or the *current price of exchange*, must vary with these circumstances, and be sometimes *above*, and sometimes *below par*. Tables of the course of exchange are published daily in the great commercial cities.

There are two kinds of money; *real* and *nominal*, or *imaginary*. All gold, silver, and copper coins, are called *real money*: and the *imaginary money* is a denomination used to express money, of which there is no real species current, precisely of the same value; as a *livre* in France, and a *pound* and *penny* in the United-States. In some countries they keep their accounts and calculate their payments in imaginary money.

Of the Currencies of Great Britain, and the United-States, &c.

Accounts are kept in England, Ireland, the British Provinces in America, and by many persons in the United-States, in *pounds, shillings, pence*, and *farthings;* 1 pound being=20 shillings, 1 shilling=12 pence, and 1 penny=4 farthings. The real values of these denominations, however, are different in different countries and states: Thus, in England, L1=$\$4\frac{4}{9}$=$4.444+, and consequently 1s.=$\$\frac{2}{9}$=$.222+, and 4s. 6d. or 54 pence=$1; in Ireland L1=$\$4\frac{4}{39}$=$4.102+, and 4s. 10½d. or 58.5 pence=$1; in Nova-Scotia, the Canadas, and the adjacent British Provinces, L1=$4, and 5s.=$1; in the New-England States, and in the States of Virginia, Kentucky, and Tennessee, L1=$\$3\frac{1}{3}$=$3.333+, and 6s.=$1; in New-York and North-Carolina, L1=$\$2\frac{1}{2}$, and 8s.=$1; in New-Jersey, Pennsylvania, Delaware and Maryland, L1=$\$2\frac{2}{3}$=$2.666+, and 7s. 6d. or 90 pence=$1; in South-Carolina and Georgia, L1=$\$4\frac{2}{7}$=$4.285+, and 4s. 8d. or 56 pence=$1.

Note.—English money is usually denominated *sterling money.* For the sake of brevity, I shall call the currency of Nova-Scotia, the Canadas, &c., *Nova-Scotia currency;* that of New-England, Virginia, &c., *New-England currency;* that of New-York and North-Carolina, *New-York currency;* that of New-Jersey, Pennsylvania, &c., *Pennsylvania currency;* and that of South-Carolina and Georgia, *South-Carolina currency.*

PROBLEM I.

To reduce pounds, shillings, pence, and farthings, of the various currencies, to Federal money.

RULE I.

1. Where a dollar is an exact number of shillings, as in the currencies of New-England, New-York, and Nova-Scotia; reduce the given sum to shillings and decimal parts; then divide it by the number of shillings in a dollar, and the quotient will be the answer in dollars and decimal parts.

2. Where a dollar is not an exact number of shillings; reduce the given sum to pence; then divide it by the number of pence in a dollar, and the quotient will be the answer.

RULE II.

If there are parts of a pound in the given sum, reduce them to the decimal of a pound, and annex this decimal to the number of pounds, if any; then multiply this sum by the number of dollars in L1 of the given currency, and the product will be the answer.

Note.—To reduce parts of a pound to the decimal of a pound, or parts of a shilling to the decimal of a shilling, &c., work by the rule for Case III. Reduction of Decimals. The decimal part of each denomination ought to be found to three places of decimals, when the true decimal contains so many places.

EXAMPLES.

1. Reduce 24*l.* 11s. 8½d. Nova-Scotia currency, to federal money.

Operation by Rule I.

```
L.  s.      4 | 2q.
24 .. 11   12 | 8.5d.
 20           | ------
----       5)491.708s.
491s.       ---------
      Ans. $ 98.341+
```

Here I reduce the given sum to shillings and decimal parts, and have 491.708s., which I divide by 5, the number of shillings in a dollar, and the quotient is the answer.

Operation by Rule II.

```
 4 | 2q.
   | ----
12 | 8.5d.
   | -------
20 | 11.708s.
   | -------
   24.585l.
         4
   -------
$ 98.340+ Ans.
```

Here I first reduce the parts of a pound, viz. 11s. 8½d., to the decimal of a pound; then I multiply the whole sum by 4, the number of dollars in a pound, and the product is the answer.

2. Reduce 28*l.* 11s. 6d. New-England currency, to federal money. Ans. $95.25

3. Reduce 48*l.* 8s. New-York currency, to federal money. Ans. $121.

4. Reduce 36*l.* 11s. 8½d. Pennsylvania currency, to federal money.

36*l.* 11s. 8d. 2q.=8780.5d., and 8780.5÷90=$97.561+, the Ans.—Here, I first reduce the given sum to pence and decimal parts; then I divide it by 90, the number of pence in a dollar, and the quotient is the answer.

5. Change 54*l*. 16s. 9¾d. South-Carolina currency, into federal money.

Here, 54*l*. 16s. 9d. 3q.=13161.75d., and 13161.75÷56=$235.031+, the Ans.

Or, otherwise thus: Since 8×7=56, consequently 13161.75÷7=1880.25, and 1880.25÷8=$235.031+, the answer, as before.

6. Change 48*l*. 14s. 8¼d. sterling, or English money, into federal money.

Here, 48*l*. 14s. 8d. 1q.=11696.25d., and 11696.25÷54=$216.597+, the Ans.

Or, since 9×6=54, you may divide the pence by 6, and the quotient by 9, and the last quotient will be the answer.

7. Reduce 15s. 6¼d. sterling, to federal money.

Ans. $3.449+

8. Change 278*l*. 15s. 9d. Irish money, into federal money.

Here, 278*l*. 15s. 9d.=66909d., and 66909÷58.5=$1143.743+, the Ans.

9. Change 100*l*. Irish money, into federal money.

Ans. $410.256+

PROBLEM II.

To reduce Federal money to pounds, shillings, &c.

RULE.

Multiply the dollars and decimal parts, either by the number of shillings, or by the number of pence in a dollar, and the product will be the answer in the same denomination with the multiplier.

Or, you may divide the given sum by the number of dollars in L1 of the currency to which the said sum is to be reduced, and the quotient will be the answer in pounds and decimal parts.

Note.—When the answer contains a decimal fraction, the value of the fraction may be found as in Case II. Reduction of Decimals.

EXAMPLES.

1. Reduce 46D. 75c. to pounds, &c. Nova-Scotia currency.

Operation.

$46.75
5s.=$1.

233.75s.
12

9.00d. value of .75s.

2,0)23,3s.

L11 .. 13s.

Ans. 11*l*. 13s. 9d.

Or, otherwise thus:—

L1=$4)$46.75

L11.6875=11*l*. 13s. 9d. Ans.

Explanation.—In working by the first method, I multiply the given sum by 5, the number of shillings in a dollar, and the product is the answer in shillings and decimal parts; which being reduced to pounds, &c. is 11*l*. 13s. 9d. In working by the second method, I divide the given sum by 4, the number of dollars in L1, and the quotient is L11.6875; the decimal parts of which I reduce to shillings and pence, and then I have 11*l*. 13s. 9d. for the answer, as before.

2. Reduce 365D. 20c. to New-England currency.
Here $365.20×6=2191.20s.=109*l*. 11s. 2¼d.+ Ans.
3. Reduce $100 to New-England currency. Ans. 30*l*.
4. Reduce $100 to New-York currency. Ans. 40*l*.
5. Reduce $100 to Pennsylvania currency.
Here $100×90=9000d.=37*l*. 10s. Ans.
6. Reduce $100 to South-Carolina currency.
Ans. 23*l*. 6s. 8d.
7. Reduce $100 to sterling money. Ans. 22*l*. 10s.
8. Change $75.085 into sterling money.
Ans. 16*l*. 17s. 10½d.+
9. Reduce $100 to pounds, &c. Irish money.
Here 100×58.5=5850d.=24*l*. 7s. 6d. Ans.
10. Change $75.085 into Irish money.
Ans. 18*l*. 6s. ¼d.+

PROBLEM III.

To reduce pounds, shillings, pence, and farthings, from one currency to another.

RULE.–Work by the Single Rule of Three thus: As the value of a dollar in the currency given: is to the value of a dollar in the currency required :: so is the sum given : to its value required.

Or, work by the particular rules or theorems in the following table.

To reduce pounds, shillings, &c. from the currency of	To the currency of						
	N. Eng.	N. York.	Penn.	S. Car.	Nov. Sco.	England.	Ireland.
N. England,		Add $\frac{1}{3}$.	Add $\frac{1}{4}$.	× by $\frac{7}{9}$.	Sub. $\frac{1}{6}$.	Sub. $\frac{1}{4}$.	× by $\frac{13}{16}$.
New-York,	Sub. $\frac{1}{4}$.		Sub. $\frac{1}{16}$.	× by $\frac{7}{12}$.	× by $\frac{5}{8}$.	× by $\frac{9}{16}$.	× by $\frac{39}{64}$.
Pennsylva.	Sub. $\frac{1}{5}$.	Add $\frac{1}{15}$.		× by $\frac{28}{45}$.	Sub. $\frac{1}{3}$.	× by $\frac{3}{5}$.	× by $\frac{13}{20}$.
S. Carolina,	× by $\frac{9}{7}$.	× by $\frac{12}{7}$.	× by $\frac{45}{28}$.		Add $\frac{1}{14}$.	Sub. $\frac{1}{28}$.	× by $\frac{117}{112}$.
Nova-Scotia,	Add $\frac{1}{5}$.	× by $\frac{8}{5}$.	Add $\frac{1}{2}$.	Sub. $\frac{1}{15}$.		Sub. $\frac{1}{10}$.	Sub. $\frac{1}{40}$.
England,	Add $\frac{1}{3}$.	× by $\frac{16}{9}$.	× by $\frac{5}{3}$.	Add $\frac{1}{27}$.	Add $\frac{1}{9}$.		Add $\frac{1}{12}$.
Ireland,	× by $\frac{16}{13}$.	× by $\frac{64}{39}$.	× by $\frac{20}{13}$.	× by $\frac{112}{117}$.	Add $\frac{1}{39}$.	Sub. $\frac{1}{13}$.	

Note 1.—To find any rule in the foregoing table; look for the given currency in the left hand column, and then look along the top for the currency required, and directly below this, in a line with the given currency, you will find the rule. Thus, to find the rule for reducing New-York currency to Penn. currency, find *New-York* in the left hand column, and *Penn.* at the top of the table, and in a line with the former and under the latter, is the rule, viz. *subtract* $\frac{1}{16}$.—When the rule directs to *add*, or *subtract*, then add to the given sum, or subtract from it, such part of itself as the rule directs, and the result will be the answer. Where the rule directs to multiply by a fraction, multiply the given sum by the numerator of the fraction, and divide the product by the denominator, and the quotient will be the answer.

The particular rules in the table are founded on the par of exchange, and they are nothing more than common contractions of the operations by the Rule of Three.

EXAMPLES.

1. Change 48*l*. 15s. 6d. New-England currency, into New-York currency.

Operation by the Rule of Three.

```
   s.  s.   L.   s.  d.
As 6 : 8 :: 48 .. 15 .. 6 : the 4th term.
                        8
            -------------
         6)390 ..  4 .. 0
            -------------
   Ans. L 65 ..  0 .. 8
```

Operation by the tabular rule.

```
     L.   s.  d.
   3)48 .. 15 .. 6
   +16 ..  5 .. 2
   ---------------
Ans. L65 ..  0 .. 8
```

Here I add to the given sum $\frac{1}{3}$ of itself, and the amount is the answer.

2. Change 10*l*. 14s. 7½d. New-York currency, into New-England currency.

Operation by the Rule of Three.

As 8s. : 6s. :: 10*l*. 14s. 7d. 2q. : 8*l*. 0s. 11d. 2½q. Ans.

Operation by the tabular rule.

L.	s.	d.	q.
4)10 ..	14 ..	7 ..	2
—2 ..	13 ..	7 ..	3+

Here I subtract from the given sum ¼ of itself, and the remainder is the answer.

Ans. L 8 .. 0 .. 11 .. 3—

3. Reduce 252*l.* 18s. 4d. Pennsylvania currency, to English or Sterling money.

Performed by the Rule of Three as follows:—

As 90d. : 54d. :: 252*l.* 18s. 4d. : 151*l.* 15s. Ans.

Performed by the tabular rule as follows:—

L.	s.	d.
252 ..	18 ..	4
		3
5)758 ..	15 ..	0

Here I multiply by ⅗; that is, I multiply the given sum by 3 and divide the product by 5.

Ans. L151 .. 15 .. 0

In the preceding examples, the operations by the tabular rules are contractions of the operations by the Rule of Three. Thus, in example 1st, the first term of the proportion or analogy, increased by ⅓ of itself, is equal to the second term; (that is, 6 increased by ⅓ of itself, equals 8;) and hence, (by the 3d contraction in Simple Proportion,) the third term, viz. the given sum, increased by ⅓ of itself, is equal to the fourth term, or answer. In the 2d example, the first term of the proportion diminished by ¼ of itself, is equal to the second term; and therefore the third term, or given sum, diminished by ¼ of itself, is the answer. In the 3d example, by dividing the first and second terms of the proportion by 18, the quotients are 5 and 3; and hence, (by the 2d contraction in Simple Proportion,) the fourth term, or answer, is found by multiplying the third term, or given sum, by 3, and then dividing the product by 5.—The other rules in the table are found in a similar manner. *See the Note immediately after the 3d Contraction in Simple Proportion.*

4. Reduce 100*l.* New-York currency, to each of the other currencies mentioned in the foregoing Table.

Ans. 75*l.* New-England currency, 93*l.* 15s. Pennsylvania currency, 58*l.* 6s. 8d. South-Carolina currency,

62*l*. 10s. Nova-Scotia currency, 56*l*.5s. Sterling, 60*l*. 18s. 9d. Irish.

5. Reduce 100*l*. New-England curreney, to each of the other currencies before mentioned.

Ans. 133*l* 6s. 8d. New-York currency, 125*l*. Penn. cur., 77*l*. 15s. 6$\frac{2}{3}$d. S. Carolina cur., 83*l*. 6s. 8d. Nova-Scotia cur., 75*l*. Sterling, 81*l*. 5s. Irish.

Note 2.—Sometimes the rate of exchange is stated at a certain sum per cent.; that is, 100*l*. in one country are worth so much more than 100*l*. in the other. When this is the case, add the rate per cent. to 100*l*.; then consider in which currency 100*l*. are worth the most: If in the currency required, make the amount of 100*l*. and the rate per cent. the first term of a proportion, and 100*l*. the second term; but if the rate per cent. be in favor of the given currency, then make the said amount the second term, and 100*l*. the first; and, in each case, make the given sum the third term. Then resolve the statement as usual, and the result, or fourth term, will be the answer; as in the two following examples.

6. Halifax is indebted to London, 1000*l*. Nova-Scotia currency: how much is that in Sterling money, if the exchange be at 11 per cent. in favor of London?

Here, 100*l*. in the required currency are=100+11=111*l*. in the given currency: Therefore,

As 111 : 100 :: 1000*l*. : 900*l*. 18s.+ Ans.

7. Liverpool is indebted to Boston 750*l*. sterling: how much is that in New-England currency, if the exchange be at 33$\frac{1}{3}$ per cent. in favor of Liverpool?

As 100 : 133$\frac{1}{3}$ :: 750*l*. : 1000*l*. Ans.

PROBLEM IV.

To reduce the Coins, Currencies, Weights, and Measures, of any country to those of any other.

Rule.—Work by the Rule of Three, or by Practice, &c.

Note.—In order to solve the following questions, it will be necessary for the learner to consult the 19th, 20th, and 21st, sets of Tables in Reduction.

EXAMPLES OF EXCHANGING COINS AND CURRENCIES.

Of French Money.

Note.—*Francs* and *centimes* are written in the same manner as *dollars* and *cents* are; the francs being considered *integers*, and the centimes *decimal parts*. So, 250 francs and 28 centimes, are written thus, 250.28fr. A cipher must be prefixed to the centimes when their number is less than 10: thus, 7 francs and 5 centimes are equal to 7.05 francs.

Ex. 1. Reduce 1578 francs, 24 centimes, to federal money.

1 franc is equal to $18\frac{3}{4}$ cents, =$.1875: Therefore, as 1 franc : 1578.24 francs :: $.1875 : $295.92, the answer.

2. Change $98.64 into francs.

As $.1875 : $98.64 :: 1fr. : 526.08 francs, =526 francs, 8 centimes, Ans.

3. Change 100 francs, 20 centimes, into federal money.

Ans. $18.7875

4. Change $112.725 into French money.

Ans. 601fr. 20cen.

Of the Netherlands.

5. Reduce 40 guilders, 18 stivers, 1 grote, 7 phennings, to federal money.

1 guilder=40 cents=$.40: Therefore,

As 1 guild. : 40 guild. 18 stiv. 1 gr. 7 phen. :: $.40 : the 4th term: Or, by reducing the first and second terms to phennings, to make them both simple quantities of the same denomination,—As 320 phen. : 13103 phen. :: $.40 : $16.378+ Ans.

6. Change $127.41 into Dutch guilders.

As $.40 : $127.41 :: 1 guild. : 318.525 guilders, =318 guilders, 10 stivers, 1 grote, Ans.

Of Spain.

7. Reduce 50 piastres, 5 rials, 17 marvadies of plate, to federal money.

1 rial of plate=10 cents=$.10 : therefore,

As 1 rial : 50 pias. 5 ri. 17 marv. :: $.10 : the 4th term: Or, by reducing the first and second terms to marvadies,—As 34 marv. : 13787 marv. :: $.10 : $40.55, the answer.

Or, otherwise thus: Since 1 rial of plate is equal to 1 dime, it is evident that if the given sum be reduced to rials and decimal parts, it will then be the answer in dimes. Now, 50 piastres, 5 rials, 17 marvadies of plate, are=405.5 rials, or dimes; and 405.5 dimes are=\$40.55, the answer, as before.

8. Reduce \$48.155 to rials of plate.

\$48.155=481.55 dimes, or rials of plate,=481 rials, $18\frac{7}{10}$ marvadies, Ans.

9. Reduce 1000 rials vellon to federal money.

1 rial vellon=5 cents=\$.05: therefore, 1000×.05=\$50, Ans.

Of Portugal.

Note 1.—Since 1000 reas of Portugal are equal to 1 milrea, it is evident that any number of reas less than 1000 may be considered as decimal parts of a milrea, and expressed like other decimal fractions. So, 24 milreas and 587 reas are=24.587 milreas. A cipher must be prefixed to the number of reas when it consists of two figures, and two ciphers when it consists of only one figure. Thus, 8 milreas and 45 reas=8.045 milreas; and 2 milreas, 7 reas,= 2.007 milreas.

10. Reduce 45 milreas, 78 reas, to federal money.

45 milreas, 78 reas,=45.078 milreas; and 1 milrea=\$1.24: hence, 45.078×1.24=\$55.89672, Ans.

Note 2.—To reduce any given sum of federal money to Portuguese money, at par, divide the dollars, (or dollars and decimal parts,) by 1.24, carrying on the quotient as far as three decimal places when there is a remainder: then the integral part of the quotient will be the milreas, and the decimal figures the reas.

11. Reduce \$19.53 to milreas.

Here 19.53÷1.24=15.750 milreas,=15 milreas, 750 reas, Ans.

12. Change \$8.712 into Portuguese money.

Ans. 7.025+ milreas, or 7 milreas and 25 reas.

East-India Money.

13. Reduce 80 pagodas of India to federal money.

1 pagoda=\$1.84: hence, \$1.84×80=\$147.20, Ans.

14. Reduce 128 rupees of Bengal to federal money.
1 rupee=50 cents=\$.50, and 128×.50=\$64, Ans.

15. Change 47 tales of China into federal money.
1 tale of China=\$1.48, and \$1.48×47=\$69.56, Ans.

Money of the ancient Jews, or Israelites.

16. The patriarch Abraham gave 400 shekels of silver for the cave and field of Machpelah: how much federal money is equal to that sum?

1 shekel of silver=\$.507: therefore, \$.507×400=\$202.80, the answer.

17. The gold and silver used in building the tabernacle, (Exodus XXXVIII. 24, 25,) amounted to 29 talents 730 shekels of gold, and 100 talents 1775 shekels of silver: Required the value of the whole in federal money.

1 talent=3000 shekels: therefore, 29 talents 730 shekels of gold=87730 shekels, and 100 talents 1775 shekels of silver=301775 shekels. Then,

Shek.	\$	\$	
87730×	8.51$\frac{1}{9}$=	746679.77+	value the gold.
301775×	.507=	152999.92+	do. of the silver.

Ans. \$899679.69+

Note.—In the preceding rules and examples in Exchange, the course of exchange between the United-States and other countries is supposed to be *at par.* The course of exchange, however, is, (as I have before stated,) often either above or below par; and, in such cases, the value of the given sum, or bill of exchange, must be calculated accordingly. In procuring bills of exchange on Great Britain, a premium is usually paid, which varies from 4 to 16 per cent., according to the state of trade.—When the course of exchange is at a given rate per cent. above or below par, it will generally be the most convenient way, to first perform the exchange, or find the value of the given sum, *at par*, and then find the *true value* according to the 2d Note in Prob. III.; as in the following example.

18. A merchant in New-York owes a merchant in Liverpool 375*l.* 10s. 6d. sterling: how many dollars must the merchant in New-York pay for a bill to that amount, if the premium be 12$\frac{1}{2}$ per cent.?

First, as 4s. 6d. : 375*l.* 10s. 6d :: \$1 : \$1669, the value of the bill at par.

Then, as \$100 : \$112½ :: \$1669 : \$1877.625 Ans.

If the pupil be made fully acquainted with what precedes, respecting the exchanges of money, he will find very little difficulty, with the assistance of tables, in applying the same principles to similar cases, which the proposed limits of this work do not admit of my illustrating individually. To facilitate this, the following tables* are annexed, which, together with those given in Reduction, will be found to contain what is most useful and necessary on the subject.

HAMBURGH.—12 fennings=1 schilling; 16 schillings=1 mark; 3 marks=1 rix dollar=1 American dollar.

There are two kinds of money in Hamburgh, called *banco*, or bank money, and *currency*, or current money. In the former of these, all bills of exchange are valued and paid. It is of purer metal than the currency, and usually bears a premium of about 20 per cent. This premium on bank money, or difference between bank and current money, is called the *Agio*.

LEGHORN.—12 denari di pezza=1 soldo di pezza; 20 soldi di pezza=1 pezza of 8 reals. Also, 12 denari di lira=1 soldo di lira; 20 soldi di lira=1 lira; 5¾ lira, moneta-buona,=1 pezza of 8 reals.

Par of exchange with Leghorn, 1 pezza of 8 reals=86½ cents, nearly.

GENOA.—The same table as for Leghorn serves for Genoa; besides, 4 lire and 12 soldi=1 scudio di cambio, or crown of exchange; 10 lire and 14 soldi=1 scudio d'oro marche, or gold crown.—Pezza, or dollar of exchange,=85 cents, nearly.

NAPLES.—10 grains=1 carlin; 10 carlins=1 ducat regno; 1 ducat regno=75½ cents, nearly.

VENICE.—12 denari=1 soldo; 20 soldi=1 lira; 6 lire and 4 soldi=1 ducat current, or of account; 8 lire=1 ducat effective; 1 lira peccola, new coin,=7 cents, 9 mills, nearly.

PETERSBURGH.—100 copecs=1 ruble; 1 ruble=\$1.

VIENNA.—4 fennings=1 creutzer; 60 creutzers=1 flo-

* These tables I copy from *Ryan's Arithmetic.*

rin; 90 creutzers, or $1\frac{1}{2}$ florins=1 rix dollar of account; 1 florin=$66\frac{1}{2}$ cents, nearly.

STOCKHOLM.—12 fennings, or oers,=1 skilling; 48 skillings=1 rix dollar=1 American dollar.

COPENHAGEN.—12 fenings=1 skilling; 16 skillings=1 mark; 6 marks=1 rix dollar=1 American dollar.

The following exercises, which the pupil will find to be easily resolved, will serve to illustrate some of the foregoing tables.

19. Reduce 5127 marks, 5 schillings, Hamburgh money, to federal money. Ans. $1709.104+

20. Reduce 467 pezzi, 12 soldi, 6 denari, of Leghorn, to federal money; exchange at $86\frac{1}{2}$ cents per pezza.
Ans. $404.495+

21. Reduce 1200 ducats regno, 8 carlins, 9 grains, Naples currency, to federal money; exchange at $75\frac{1}{2}$ cents per ducat regno. Ans. $906.671+

EXAMPLES OF EXCHANGING WEIGHTS AND MEASURES.

1. The great bell of Moscow weighs 12500.6 Russian poods: how many pounds is that, 2 poods being equal to 71 lb. Avoirdupois?

Poods. poods. lb. lb.
As 2 : 12500.6 :: 71 : 443771.3 Ans.

2. The head of Goliath's spear weighed 600 shekels of iron: how many pounds is that, Avoirdupois weight?

1 shekel=$\frac{1}{2}$ oz.: therefore, 600÷2=300oz.=18lb. 12oz. Ans.

3. If 4 ells of Holland be equal to 3 American yards, how many ells of Holland are equal to 75 American yards?
Ans. 100.

4. How many English, or American miles, are equal to 55 Irish miles? Ans. 70.

5. The house of the forest of Lebanon, built by King Solomon, was 100 cubits long: What was its length in American feet?

1 cubit=21 inches, and 21×100=2100in.=175ft. Ans.

6. The magnificent temple built by Solomon, was 60 cubits long, 20 wide, and 30 high: What were its dimensions in feet?

Ans. Length 105ft., width 35ft., height $52\frac{1}{2}$ft.

7. The brazen sea, in Solomon's temple, contained 3000 baths: reduce these to wine gallons. Ans. $22687\frac{1}{2}$ gal.

CONJOINED PROPORTION,

Is when the coins, weights, or measures, of several countries, are compared in the same question; or, it is joining several proportions together, and by the relation which several antecedents have to their consequents, the proportion between the first antecedent and the last consequent is discovered, as well as the proportion between the others in their several respects.

Questions in Conjoined Proportion may be solved by the Single Rule of Three, or by Compound Proportion; but the following particular rules are generally preferable.

CASE I.—*When it is required to find how many of the first sort of coin, weight, or measure, mentioned in the question, are equal to a given quantity of the last.*

RULE.—Reduce all the quantities which are of the same kind to the same denomination, when they are of different denominations. Then place the numbers alternately, in two columns, the antecedents at the left hand, and the consequents at the right, and let the last number stand on the left hand. Lastly, multiply together all the numbers in the first column, for a dividend, and those in the second column, for a divisor, and the quotient will be the answer.

Note 1.—Compound quantities must be reduced to simple quantities, and each consequent and the next following antecedent must be of the same denomination.

Note 2.—The operation may often be much abbreviated by the method of contraction used in Compound Proportion; and the work may be proved by Compound Proportion, or by as many statings in Simple Proportion as the nature of the question may require.

Ex. 1. If 20lb. English make 19lb. Flemish, and 19lb. Flemish 25lb. at Turin; how many pounds English are equal to 50lb. at Turin?

Antecedents.	*Consequents.*
20 lb. Eng.	=19 lb. Flemish
19 lb. Flem.	=25 lb. at Turin
50 lb. at Turin	=the term sought.

Then, 20×19×50=19000, the dividend; and 19×25=475, the divisor. Then, 19000÷475=40lb. Ans.

Or, by contraction, thus:—

$$\frac{20\times{,}19'\times{,}50'}{{,}19'\times{,}25'}\overset{2}{}=\frac{40}{1}=40\text{lb. Ans.}$$

2. If 11 Irish miles be equal to 14 English miles, 31 English miles to 30 Scotch miles, and 120 Scotch miles to 31 German miles; how many Irish miles are equal to 28 German miles? Ans. 88.

3. If 1 bushel of wheat be equal in value to 2 bushels of rye, 1 bushel of rye to 3 bushels of oats, and 5 bushels of oats to 2 bushels of barley; how many bushels of wheat are equal in value to 20 bushels of barley?

Ans. $8\frac{1}{3}$ bush.

CASE II.—*When it is required to find how many of the last sort of coin, weight, or measure, mentioned in the question, are equal to a given quantity of the first.*

RULE.—Place the numbers as in Case I., only let the last number stand on the right hand: Then multiply together all the numbers in the first column, for a divisor, and those in the second column, for a dividend, and the quotient will be the answer.

Observe the same Notes here as in Case I.

Ex. 1. If 20lb. English make 19lb. Flemish, and 38lb. Flemish 50lb. at Turin; how many pounds at Turin are equal to 40lb. English?

lb.	lb.
20 Eng.	=19 Flem.
38 Flem.	=50 at Tu.
	40 Eng.

Then, by contraction;—

$$\frac{{,}19'\times 50\times{,}\overset{2'}{40'}}{{,}20'\times{,}\underset{2'}{38'}}=50\text{ lb. Ans.}$$

2. If 1 English crown be equal to 11 American dimes, 10 American dimes to 1 Spanish dollar, and 124 Spanish dollars to 100 milreas of Portugal; how many milreas of Portugal are equal to 62 English crowns? Ans. 55.

3. If 10lb. of sugar be equal in value to 1lb. of tea, 2lb. of tea to 22lb. of raisins, and 5lb. of raisins to 40 cents; what is the value of 5lb. of sugar? Ans. 44 cents.

ARBITRATION OF EXCHANGES.

By this term is understood how to choose or determine the best way of remitting money from abroad with advantage; which is performed by Conjoined Proportion, as in the following

EXAMPLES.

1. Suppose a merchant in Boston has effects at Amsterdam to the amount of $4500, which he can remit by way of Lisbon at 840 reas per dollar, and thence to Boston at 1D. 40c. per milrea, (or 1000 reas:) Or, by way of Nantz, at 6 livres per dollar, and thence to Boston at 20 cents per livre: It is required to arbitrate these exchanges, that is, to choose that which is most advantageous.

Operation.

First; by Conjoined Proportion, Case II.

If 1D. at Amsterdam=840 reas at Lisbon,
1000 reas at Lisbon =1.40 D. at Boston,
how many D. at Boston=4500 D. at Amsterdam?

$$\frac{840\times1.40\times4500}{1\times1000}=\$5292, \text{ by way of Lisbon.}$$

Secondly; If 1D. at Amst. =6 livres at Nantz,
1 livre at Nantz=.20 D. at Boston,
how many D. at Boston=4500 D. at Amst.?

$$\frac{6\times.20\times4500}{1\times1}=\$5400, \text{ by way of Nantz.}$$

Then, $5400—5292=$108, advantage by way of Nantz, Ans.

2. A merchant in Liverpool can draw directly for 1000 piastres in Leghorn, at 50d. sterling per piastre; but he chooses to remit the sum to Cadiz, at 19 piastres for 7000 marvadies; thence to Amsterdam, at 680 marvadies for 189d. Flemish; and thence to Liverpool, at 9d. Flemish

for 5d. sterling: What is gained by this circular remittance, and what is the value of a piastre to him?

Operation.

First; 1000×50=50000 pence, the value of the 1000 piastres by the direct remittance.

2dly; If 19 pias. at Leghorn=7000 marv. at Cadiz,
680 marv. at Cadiz = 189 d. Flemish,
9 d. Flemish = 5 d. sterling,
how many d. sterling=1000 pias. at Leghorn?

$\frac{7000\times189\times5\times1000}{19\times680\times9}$=56888d. 2q.+, the value of the 1000 piastres by the circular remittance; which being divided by 1000, gives 56d. 3q.+ for the value of 1 piastre. Then 56888½d.—50000d.=6888½d.=28*l*. 14s. ½d., the sum gained.

Ans. The sum gained by the circular remittance is 28*l*. 14s. ½d.+ sterling, and the value of a piastre is 56¾d.+ sterling.

VULGAR FRACTIONS.

Having briefly introduced Vulgar Fractions immediately after Compound Division, and given some general definitions, and a few problems therein, which were necessary to prepare the student for learning decimals, &c., I refer the learner to those general definitions in pages 110 and 111.

Vulgar Fractions are either *proper*, *improper*, *single*, *compound*, or *mixed*.

A *proper* fraction has the numerator less than the denominator; as ½, or ¾.

An *improper* fraction has the numerator either equal to, or greater than the denominator; as $\frac{4}{4}$, $\frac{8}{7}$, $\frac{25}{11}$, &c.

A *single* or *simple* fraction, is a single expression, denoting any number of parts of the integer; as ½, or $\frac{7}{5}$.

A *compound* fraction, is a fraction of a fraction; or several single fractions connected, with the word *of* between them; as ½ of $\frac{4}{5}$ of $\frac{7}{8}$, or ¾ of $\frac{2}{6}$ of 8, &c.

A *mixed number*, is composed of a whole number and a fraction; as 4½, or 12¾, &c.

A whole number may be expressed like a fraction, by writing 1 below it, as a denominator: So 4 is equal to $\frac{4}{1}$, and 25 is equal to $\frac{25}{1}$.

A fraction denotes division; and its value is equal to the quotient obtained by dividing the numerator by the denominator: So $\frac{12}{4}$ is equal to 3, and $\frac{20}{5}$ is equal to 4.—Hence then, if the numerator be less than the denominator, the value of the fraction is less than 1: But if the numerator be greater than the denominator, the fraction is greater than 1: And if the numerator be the same as the denominator, the fraction is just equal to 1.

The *common measure* of two or more numbers, is some number which will divide each of them without a remainder; and the greatest number that will do this, is called the *greatest common measure:* Thus, 2 is a common measure of 8 and 12, but their greatest common measure is 4.

A number which can be measured by two or more numbers, is called their *common multiple;* and, if it be the least number which can be so measured, it is called the *least common multiple:* Thus, 24 is a common multiple of 2, 3 and 4, but their least common multiple is 12.

A *prime number*, is one which can be measured only by itself or a unit; as 3, 7, 11, &c.

PROBLEM I.

To find the greatest common measure of two or more given numbers.

RULE.

1. If there be two numbers only, divide the greater by the less; then divide this divisor by the remainder; and so on, dividing always the last divisor by the last remainder, till nothing remains; and the last divisor of all will be the greatest common measure sought.

2. When there are more than two numbers, find the greatest common measure of two of them, as before; then do the same for that common measure and another of the given numbers; and so on, through all the numbers to the last; then will the greatest common measure last found be the answer.

Note.—If it happen that the common measure, found

by the above rule, is 1; then the given numbers are incommensurable, or in their lowest terms.

EXAMPLES.

1. What is the greatest common measure of 459, 972, and 273? Ans. 3.

Operation.

```
             459)972(2
                 918
                 ---
                  54)459(8
                     432
                     ---
Last divisor,*        27)54(2
                         54

                  27)273(10
                     27
                     ---
Last divisor,         3)27(9
                        27
```

Here, I find that 27 is the greatest common measure of 459 and 972.—I next find the greatest common measure of 27 and 273; as follows:

Here, I find that 3 is the greatest common measure of 27 and 273; and this, by the rule, is the greatest common measure of the three given numbers.

Required the greatest common measure

Of 180 and 204. Ans. 12.
Of 246 and 372. Ans. 6.
Of 48, 56 and 74. Ans. 2.
Of 522, 918 and 1998. Ans. 18.

PROBLEM II.

To find the least common multiple of two or more given numbers.

RULE.†

Arrange the given numbers in succession, and find by

---

* The truth of the foregoing rule may be shown from this example thus: Since 27 measures 54, it also measures $(54\times 8)+27$, or 459, the second dividend, being one of the given numbers; and since 27 measures 459, it also measures $(459\times 2)+54$, or 972, another of the given numbers. So 27 is the common measure of 459 and 972. It is also the *greatest* common measure; for suppose there be a greater; then, since the greater measures 459 and 972, it also measures the remainder 54; and since it measures 54 and 459, it also measures the remainder 27; that is, the greater measures the less, which is absurd; therefore 27 is the greatest common measure.—In the same manner the demonstration may be applied to one or more additional numbers.

† The reason of this rule may be shown from the first example: Now, $2\times 2\times 3\times 9\times 7$, or 756, the product of the first divisor and the num-

inspection a number which will measure as many of them as possible. By this number divide all the given numbers which it measures, and write the quotients and the undivided numbers in a line beneath. Proceed in the same manner with the numbers in this line; and thus continue the process, till no number greater than 1 will measure any two of the numbers last found. Then multiply all the numbers in the last line and all the divisors used in the operation, continually together, and the result will be the least common multiple required.

Note.—If no two of the given numbers have any common measure greater than unity, then the continual product of the numbers will be their least common multiple.

EXAMPLES.

1. What is the least common multiple of 4, 6, 9, and 14?

Operation.

2)4, 6, 9, 14

3)2, 3, 9, 7

2, 1, 3, 7

Then, $2\times3\times2\times1\times3\times7=252$, the Ans.

Here I survey the given numbers, and find that 2 will measure or divide three of them, viz. 4, 6, and 14: I therefore divide them by 2, and set down their quotients, 2, 3 and 7, under them respectively, and 9, the number not divided, in the same line. Then, I perceive that 3 will measure two of the numbers in the second line, viz. 3 and 9; so I divide them by 3, and set down the quotients, and also 2 and 7, the numbers not divided, in a line beneath. There being no two of the numbers in the third line which have any common measure greater than 1, the numbers are in their lowest terms: So, I multiply all those numbers and the two divisors, continually together, and their product, 252, is the answer.

---

bers in the second line, is evidently a multiple of each of the given numbers; 4, the first number, being contained in it $3\times9\times7$, or 189 times; 6, the second number, $2\times9\times7$, or 126 times, &c. Again, $2\times3\times2\times1\times3\times7$, or 252, the product of the first and second divisors and the numbers in the third line, is also a common multiple of the given numbers; 4 being contained in it $3\times1\times3\times7$, or 63 times; 6 the second number, $2\times1\times3\times7$, or 42 times, &c.; and it is evidently the *least* common multiple of those numbers.

What is the least common multiple

Of 6 and 8?	Ans. 24.
Of 3, 5, 8 and 10?	Ans. 120.
Of 2, 5 and 7?	Ans. 70.
Of 1, 2, 3, 4, 5, 6, 7, 8 and 9?	Ans. 2520.

REDUCTION OF VULGAR FRACTIONS,

Is the bringing of them out of one form, or denomination, into another, in order to prepare them for the operations of Addition, Subtraction, &c.

CASE I.

To abbreviate or reduce fractions to their lowest terms.

RULE.

Find (by Prob. I.) the greatest common measure of the terms of the given fraction. Then divide both the terms by their greatest common measure, and the quotients will make the fraction required.

Or, you may work by the rule given in the Introduction to Vulgar Fractions.

EXAMPLES.

1. Reduce $\frac{56}{119}$ to its lowest terms.

56)119(2
 112
 ——
 7)56(8
 56

Then, 7)$\frac{56}{119}=\frac{8}{17}$, Ans.

Here, I first find the greatest common measure of 56 and 119, which is 7. Then I divide both terms of the given fraction by 7, and it gives $\frac{8}{17}$ for the answer.

2. Reduce the following fractions to their lowest terms; viz. $\frac{114}{285}$, $\frac{208}{684}$, $\frac{825}{960}$, and $\frac{5184}{6912}$.

Answers, $\frac{2}{5}$, $\frac{52}{171}$, $\frac{55}{64}$, and $\frac{3}{4}$.

CASE II.

To reduce an improper fraction to its equivalent whole or mixed number.

RULE.*

Divide the numerator by the denominator, and the quotient will be the whole or mixed number sought.

EXAMPLES.

1. Reduce $\frac{42}{5}$ to its equivalent mixed number.

Here $42 \div 5 = 8\frac{2}{5}$ Ans.

2. Reduce the following improper fractions to their equivalent numbers; viz. $\frac{147}{10}$, $\frac{26}{25}$, and $\frac{24}{4}$.

Answers, $14\frac{7}{10}$, $1\frac{1}{25}$, and 6.

CASE III.

To reduce a mixed number to an improper fraction of the same value.

RULE.

Multiply the whole number by the denominator of the fraction, and add the numerator to the product; then set that sum above the denominator, for the fraction required.

Note.—This Case is the converse of Case II., and consequently these two Cases prove each other.

EXAMPLES.

1. Reduce $8\frac{2}{5}$ to an improper fraction.

Here, $8 \times 5 = 40$, and $40 + 2 = 42$, the numerator: Hence, $\frac{42}{5}$ is the improper fraction required.

2. Reduce the following mixed numbers to improper fractions; viz. $14\frac{7}{10}$, $127\frac{4}{17}$, and $1\frac{1}{25}$.

Answers, $\frac{147}{10}$, $\frac{2163}{17}$, and $\frac{26}{25}$.

CASE IV.

To reduce a whole number to an equivalent improper fraction, having a given denominator.

RULE.

Multiply the whole number by the given denominator; then set the product over the said denominator, and you will have the fraction required.†

---

* The reason of this rule, and of that for the next Case, is evident from the nature of fractions.

† This is no more than first multiplying a quantity by some number, and then dividing the result by the same number, which it is evident does not alter its value.

EXAMPLES.

1. Reduce 5 to a fraction, whose denominator shall be 8.

Here, $5\times8=40$, and $\frac{40}{8}$ the Ans.

2. Reduce 7 to a fraction whose denominator shall be 4.

Ans. $\frac{28}{4}$.

3. Reduce 1 to a fraction whose denominator shall be 5.

Ans. $\frac{5}{5}$.

CASE V.

To reduce a compound fraction to a simple one of equal value.

RULE.*

1. Reduce all whole and mixed numbers to their equivalent improper fractions.

2. Multiply together all the numerators for a new numerator, and all the denominators for a new denominator, and they will form the simple fraction sought; which reduce to its lowest terms, if necessary.

Note.—The process may often be very much abbreviated by the method of contraction used in Compound Proportion.

EXAMPLES.

1. Reduce $\frac{1}{2}$ of $\frac{2}{3}$ of $\frac{3}{4}$, to a simple or single fraction.

Here, $\frac{1\times2\times3}{2\times3\times4}=\frac{6}{24}=\frac{1}{4}$ Ans.

Or, by contraction, as follows: $\frac{1\times\not{2}\times\not{3}}{\not{2}\times\not{3}\times4}=\frac{1}{4}$ Ans.

2. Reduce $\frac{4}{5}$ of $\frac{7}{10}$ of $8\frac{1}{2}$ to a simple fraction.

$8\frac{1}{2}=\frac{17}{2}$, and $\frac{4\times7\times17}{5\times10\times2}=\frac{476}{100}=4\frac{19}{25}$ Ans.

---

* The truth of this rule may be shown as follows: Let the compound fraction be $\frac{2}{3}$ of $\frac{5}{7}$. Now $\frac{1}{3}$ of $\frac{5}{7}$ is $\frac{5}{7}\div3$, which is $\frac{5}{21}$; consequently $\frac{2}{3}$ of $\frac{5}{7}$ will be $\frac{5}{21}\times2$, or $\frac{10}{21}$; that is, the numerators are multiplied together, and also the denominators, as in the Rule. When the compound fraction consists of more than two single ones; having first reduced two of them as above, then the resulting fraction and a third will be the same as a compound fraction of two parts; and so on to the last of all.

Reduce the following compound fractions to simple ones, viz.

$\frac{2}{7}$ of $\frac{7}{8}$ of $\frac{1}{2}$.	Ans. $\frac{14}{112}=\frac{1}{8}$.
$\frac{1}{2}$ of $12\frac{1}{7}$.	Ans. $\frac{85}{14}=6\frac{1}{14}$.
$\frac{2}{3}$ of $\frac{3}{4}$ of $\frac{4}{7}$ of 7.	Ans. $\frac{168}{84}=2$.

CASE VI.

To reduce fractions having different denominators, to equivalent fractions having a common denominator.

RULE I.

Multiply each numerator into all the denominators except its own, for a new numerator; and all the denominators continually together, for a common denominator: this written under the several new numerators will give the fractions required.*

Note 1.—When any of the given quantities are compound fractions, or whole or mixed numbers; first reduce them to the form of simple fractions, and then proceed according to the Rule.

EXAMPLES.

1. Reduce $\frac{1}{2}$, $\frac{3}{5}$, and $\frac{4}{7}$, to equivalent fractions having a common denominator.

$1\times5\times7=35$	the new numerator	for $\frac{1}{2}$.
$3\times2\times7=42$	ditto	for $\frac{3}{5}$.
$4\times2\times5=40$	ditto	for $\frac{4}{7}$.
$2\times5\times7=70$	the common denominator.	

Therefore, the equivalent fractions are $\frac{35}{70}$, $\frac{42}{70}$, and $\frac{40}{70}$, Ans.

2. Reduce $\frac{1}{8}$, $\frac{4}{5}$, and $\frac{7}{12}$ to a common denominator.
Ans. $\frac{60}{480}$, $\frac{384}{480}$, and $\frac{280}{480}$.

3. Reduce $\frac{2}{5}$, $\frac{4}{7}$, $\frac{1}{8}$ and $\frac{1}{10}$ to a common denominator.
Ans. $\frac{1120}{2800}$, $\frac{1600}{2800}$, $\frac{350}{2800}$, $\frac{280}{2800}$.

4. Reduce $\frac{1}{4}$ of 2, and $\frac{1}{2}$ of $5\frac{1}{2}$, to simple fractions having a common denominator. Ans. $\frac{4}{8}$ and $\frac{22}{8}$.

---

* This is multiplying the numerator and the denominator of each fraction by the same number, (viz. by the product of the denominators of all the other fractions,) and hence the values of the fractions are not altered.

In the second rule, the common denominator is a multiple of all the given denominators, and consequently will divide by any of them: Therefore, proper parts may be taken for all the numerators, as required.

Note 2.–Sometimes two fractions may be readily reduced to a common denominator, by either multiplying or dividing both terms of one of the fractions by some number which will make the denominator the same as that of the other fraction; as in the following examples.

5. Reduce $\frac{1}{2}$ and $\frac{5}{8}$ to a common denominator.

Here, multiplying both terms of $\frac{1}{2}$ by 4, gives $\frac{4}{8}$; and $\frac{4}{8}$ and $\frac{5}{8}$ are the fractions required.

6. Reduce $\frac{2}{7}$ and $\frac{8}{14}$ to a common denominator.

Here, $2)\frac{8}{14}=\frac{4}{7}$; and $\frac{2}{7}$ and $\frac{4}{7}$ the Ans.

RULE II.

To reduce any given fractions to others of equal value, which shall have the least common denominator possible.

1. Find (by Prob. II. page 232) the least common multiple of all the denominators of the given fractions, and it will be the least common denominator required.

2. Divide the least common denominator by the denominator of each fraction, and multiply the quotient by the numerator, for a new numerator. Then, under each of the new numerators write the common denominator, and you will have the fractions required.

EXAMPLES.

1. Reduce $\frac{1}{7}$, $\frac{5}{14}$ and $\frac{11}{21}$ to equivalent fractions having the least common denominator possible.

7)7, 14, 21 — $7\times1\times2\times3=42$, least common denominator.
1, 2, 3

$(42\div7)\times1=6$ the new numerator for $\frac{1}{7}$.
$(42\div14)\times5=15$ ditto for $\frac{5}{14}$.
$(42\div21)\times11=22$ ditto for $\frac{11}{21}$.

Ans. $\frac{6}{42}$, $\frac{15}{42}$ and $\frac{22}{42}$.

2. Reduce $\frac{1}{2}$, $\frac{3}{4}$, and $\frac{5}{8}$, to equivalent fractions having the least common denominator possible.

Ans. $\frac{4}{8}$, $\frac{6}{8}$, $\frac{5}{8}$.

3. Reduce $\frac{1}{2}$, $\frac{2}{3}$, $\frac{5}{6}$, and $\frac{7}{8}$ to their least common denominator. Ans. $\frac{12}{24}$, $\frac{16}{24}$, $\frac{20}{24}$, $\frac{21}{24}$.

CASE VII.

To reduce a fraction from one denomination to another.

RULE.*

First, reduce the given fraction to such a compound one as will express the value of the given fraction, by comparing it with all the denominations between it and that denomination you would reduce it to; then reduce this compound fraction to a single one, by Case V., and it will be the answer.

EXAMPLES.

1. Reduce $\frac{4}{5}$ of a penny to the fraction of a pound.

$\frac{4}{5}$ of a penny is $=\frac{4}{5}$ of $\frac{1}{12}$ of $\frac{1}{20}$ of a pound, $=\frac{4}{1200}=\frac{1}{300}$ of a pound, Ans.

Or, since 240 pence make 1 pound, $\frac{4}{5}$ of a penny $=\frac{4}{5}$ of $\frac{1}{240}=\frac{4}{1200}=\frac{1}{300}$ of a pound, the answer, as before.

2. Reduce $\frac{1}{300}$ of a pound to the fraction of a penny.

$\frac{1}{300}$ of a pound is $=\frac{1}{300}$ of $\frac{20}{1}$ of $\frac{12}{1}=\frac{240}{300}=\frac{4}{5}$ of a penny, Ans.

Or, $\frac{1}{300}$ of a L. $=\frac{1}{300}$ of $\frac{240}{1}=\frac{240}{300}=\frac{4}{5}$d. Ans.

3. Reduce $\frac{2}{7}$ of a pwt. to the fraction of a lb. Troy.

$\frac{2}{7}$ of a pwt $=\frac{2}{7}$ of $\frac{1}{20}$ of $\frac{1}{12}$ of a lb. $=\frac{2}{1680}=\frac{1}{840}$ Ans.

Note.—The answers to the following questions are in their lowest terms.

4. Reduce $\frac{1}{280}$ of a lb. Avoirdupois, to the fraction of a dram. Ans, $\frac{32}{35}$.

5. Reduce $\frac{1}{2}$ of a rod to the fraction of a mile. Ans. $\frac{1}{640}$.

6. Reduce $\frac{1}{20}$ of a foot to the fraction of an inch. Ans. $\frac{3}{5}$.

7. What part of a square rod is $\frac{1}{240}$ of an acre? Ans. $\frac{2}{3}$.

8. Reduce $\frac{1}{2}$ of a gill to the fraction of a gallon. Ans. $\frac{1}{64}$.

9. What part of a bushel is $\frac{7}{8}$ of a quart? Ans. $\frac{7}{256}$.

10. Reduce $\frac{2}{5}$ of an English crown, at 5 shillings sterling, to the fraction of a guinea, at 21 shillings.

$\frac{2}{5}$ of a crown $=\frac{2}{5}$ of $\frac{5}{1}$ of $\frac{1}{21}$ of a guinea $=\frac{10}{105}=\frac{2}{21}$, Ans.

CASE VIII.

To find the value of a fraction in parts of the integer; as of money, weight, &c.

---

* It is evident from the first and second examples, that this rule is, in effect, the same as the rules for reducing whole numbers from one denomination to another.

RULE.*

1\. *If the integer be a simple quantity;* multiply the numerator of the given fraction, by the parts in the next inferior denomination, and divide the product by the denominator: If any thing remains, multiply it by the parts of the next inferior denomination, and divide again by the denominator; and so on, as far as may be necessary: Then, the several quotients, placed in order, will be the value of the fraction required.

2\. *When the integer is a compound quantity;* multiply it by the numerator of the given fraction, and divide the product by the denominator, by Compound Multiplication and Division, and the quotient will be the answer.

Note.—This and the following Case are the same with Problems II. and III. in the "Introduction to Vulgar Fractions."

EXAMPLES.

1\. What is the $\frac{4}{5}$ of 2*l*. 6s.?

```
     L.   s.
     2 .. 6  the integer.
          4
   ---------
  5)9 .. 4
   ---------
Ans. 1 .. 16 .. 9d. 2⅗q.
```

Here the integer is a compound quantity, and I multiply it by the numerator of the fraction, and divide the product by the denominator, as directed in the 2d article of the Rule.

2\. What is the value of $\frac{7}{16}$ of a dollar?

Ans. 43c. 7$\frac{1}{2}$m.

3\. How much is $\frac{1}{8}$ of a dollar, in the currency of Pennsylvania, $1 being equal to 7s. 6d.? Ans. 11$\frac{1}{4}$ pence.

4\. How much is $\frac{61}{128}$ of a lb. Avoirdupois?

Ans. 7oz., 10dr.

5\. How much is $\frac{1}{5}$ of a mile? Ans. 1fur. 24rd.

* The numerator of a fraction being considered as a remainder in Division, and the denominator as the divisor, this rule is of the same nature as Compound Division, or the valuation of remainders in the Rule of Three, before explained.

6. How much is $\frac{1}{2}$ of $\frac{3}{4}$ of a gallon of wine?

$\frac{1}{2}$ of $\frac{3}{4}=\frac{3}{8}$, and $\frac{3}{8}$ of a gallon=1qt. 1pt. Ans.

7. How much is $\frac{2}{3}$ of $\frac{1}{2}$ of 7 bushels, 4 quarts?

Ans. 2bush. 1pk. 4qt.

CASE IX.

To reduce any given quantity to the fraction of a greater denomination of the same kind.

RULE.

Reduce the given quantity to the lowest term mentioned, for a numerator; then reduce the integral part to the same term, for a denominator; which will be the fraction required.*

EXAMPLES.

1. Reduce 2 feet, 4 in. $2\frac{1}{4}$b.c. to the fraction of a yard.

Operation.

ft. in. b.c.	1 yard.
2 .. 4 .. $2\frac{1}{4}$	3
12	—
—	3 ft.
28 in.	12
3	—
—	36 in.
86 b.c.	3
4	—
—	108 b.c.
345 fourths of a b.c. Numerator.	4
	—
	432 fourths of a b.c. Denominator.

Ans. $\frac{345}{432}=\frac{115}{144}$.

Here I first reduce the given quantity to the lowest term in it, which is fourths of a barley corn, for a numerator; and then I reduce the integer, viz. 1 yard, to fourths of a barley corn, for a denominator.

2. Reduce 1*l.* 16s. 9d. $2\frac{2}{5}$q. to the fraction of 2*l.* 6s.

Ans. $\frac{8832}{11040}$,$=\frac{4}{5}$.

To solve the last question; reduce 1*l.* 16s. 9d. $2\frac{2}{5}$q. to

* This Case is the reverse of Case VIII., and the proof is evident from that.

fifths of a farthing, for a numerator; and reduce 2*l*. 6s. to fifths of a farthing, for a denominator.

The answers to the following questions are in their lowest terms.

3. Reduce 43 cents $7\frac{1}{2}$ mills to the fraction of a dollar. Ans. \$$\frac{7}{16}$.

4. Reduce 4oz. $9\frac{1}{7}$dr. to the fraction of a pound Avoirdupois. Ans. $\frac{2}{7}$lb.

5. What part of a mile is 1 furlong, 24 rods? Ans. $\frac{1}{5}$.

6. A man divided a farm, containing 481 acres of land, between his two sons: He gave 288 acres, 2 roods, 16sq. rods, to his eldest son, and the rest to the youngest son. What part of the farm did he give to each?

Ans. $\frac{3}{5}$ of the farm to the elder son, and $\frac{2}{5}$ to the younger.

ADDITION OF VULGAR FRACTIONS.

RULE.

Reduce compound fractions to single ones; whole and mixed numbers to improper fractions; fractions of different integers to those of the same denomination; and all of them to a common denominator: Then, the sum of the numerators written over the common denominator, will give the sum of the fractions, required.*

Note 1.—It will generally be the best way to reduce the fractions to their *least* common denominator, by the 2d Rule for the 6th Case in Reduction of Vulgar Fractions.

EXAMPLES.

1. Add $\frac{2}{5}$ and $\frac{4}{7}$ together.

* Fractions, before they are reduced to a common denominator, are entirely dissimilar, and therefore cannot be incorporated with one another; but when they are reduced to a common denominator, and made parts of the same thing, their sum, or difference, may then be as properly expressed by the sum or difference of the numerators, as the sum or difference of any two quantities whatever, by the sum or difference of their individuals; whence the reason of the rules for the Addition and Subtraction of fractions, is manifest.

$\frac{2}{5}$ and $\frac{4}{7}$, being reduced to a common denominator, are $\frac{14}{35}$ and $\frac{20}{35}$. Then, $14+20=34$, the sum of the numerators, which being written over the common denominator, gives $\frac{34}{35}$ for the answer.

2. Add $\frac{1}{2}$, $\frac{1}{3}$ of $\frac{7}{8}$, $5\frac{1}{4}$, and 6 together.

$\frac{1}{3}$ of $\frac{7}{8}=\frac{7}{24}$; $5\frac{1}{4}=\frac{21}{4}$; and $6=\frac{6}{1}$. Then, $\frac{1}{2}$, $\frac{7}{24}$, $\frac{21}{4}$, $\frac{6}{1}$, reduced to their least common denominator, become $\frac{12}{24}$, $\frac{7}{24}$, $\frac{126}{24}$, $\frac{144}{24}$; the sum of which is $\frac{289}{24}=12\frac{1}{24}$, Ans.

3. What is the sum of $\frac{1}{5}$ and $\frac{2}{5}$? Ans. $\frac{3}{5}$.

4. What is the sum of $\frac{1}{2}$, $\frac{2}{5}$ and $\frac{6}{7}$? Ans. $1\frac{53}{70}$.

5. What is the sum of $\frac{5}{8}$, $4\frac{1}{2}$, and $\frac{1}{4}$ of $\frac{7}{10}$? Ans. $5\frac{12}{40}$.

Note 2.—In adding mixed numbers that are not compounded with other fractions, you may first find the sum of the fractions, to which add the integers of the given mixed numbers.

6. What is the sum of $5\frac{1}{2}$, $8\frac{4}{7}$, and 14?

Here, $\frac{1}{2}+\frac{4}{7}=\frac{7}{14}+\frac{8}{14}=\frac{15}{14}=1\frac{1}{14}$.

Then, $1\frac{1}{14}+5+8+14=28\frac{1}{14}$, Ans.

7. Add $\frac{4}{5}$ and $8\frac{1}{2}$ together. Ans. $9\frac{3}{10}$.

8. Find the sum of $4\frac{2}{7}$, $18\frac{1}{2}$, $\frac{1}{2}$ of $\frac{4}{5}$, and 5. Ans. $28\frac{13}{70}$.

Note 3.—To add fractions of money, or weight, &c., reduce fractions of different integers to those of the same.

Or, you may find the value of each fraction, by Case VIII. in Reduction, and then add them in their proper terms.

9. Add $\frac{4}{5}$ of a foot to $\frac{1}{2}$ of a yard.

1st Method.

$\frac{4}{5}$ of a foot$=\frac{4}{5}$ of $\frac{1}{3}=\frac{4}{15}$ yd.
Then $\frac{4}{15}+\frac{1}{2}=\frac{23}{30}$ yd.$=$2ft. 3in. $1\frac{4}{5}$ b. c. Ans.

2d Method.

	ft.	in.	b. c.
$\frac{1}{2}$ yd.=	1	6	0
$\frac{4}{5}$ ft.=		9	$1\frac{4}{5}$
Ans.	2	3	$1\frac{4}{5}$

10. Add $\frac{2}{7}$ of a lb. Avoirdupois, to $\frac{1}{8}$ of an oz.

Ans. 4oz. $11\frac{1}{7}$dr.

11. Add $\frac{3}{4}$ of a mile to $\frac{7}{16}$ of a furlong.

Ans. 6fur. 28 rods.

12. Add together, $\frac{1}{4}$ of a week, $\frac{1}{3}$ of a day, and $\frac{1}{2}$ of an hour. Ans. 2da. 2h. 30min.

SUBTRACTION OF VULGAR FRACTIONS.

RULE.

Prepare the fractions as in Addition; then the difference of the numerators written above the common denominator, will give the difference of the fractions sought.

EXAMPLES.

1. From $\frac{7}{8}$ take $\frac{1}{4}$.

Here $\frac{7}{8}-\frac{1}{4}=\frac{7}{8}-\frac{2}{8}=\frac{7-2}{8}=\frac{5}{8}$ Ans.

2. From $\frac{1}{2}$ take $\frac{4}{5}$ of $\frac{1}{8}$.

$\frac{4}{5}$ of $\frac{1}{8}=\frac{4}{40}=\frac{1}{10}$; and $\frac{1}{2}-\frac{1}{10}=\frac{5}{10}-\frac{1}{10}=\frac{4}{10}=\frac{2}{5}$ Ans.

3. From $\frac{5}{9}$ take $\frac{3}{8}$. Ans. $\frac{13}{72}$.
4. From $\frac{3}{4}$ take $\frac{2}{7}$ of $\frac{5}{8}$. Ans. $\frac{4}{7}$.
5. From $\frac{2}{5}$ of 7 take $\frac{1}{2}$ of $\frac{7}{8}$. Ans. $2\frac{29}{80}$.
6. From $14\frac{2}{7}$ subtract $2\frac{1}{3}$. Ans. 11 [illegible].

Note 1.—One mixed number may be subtracted from another, in the following manner: Subtract the fractional part of the subtrahend from that of the minuend, and annex the remainder to the difference of the integers. When the lower fraction is greater than the upper one; add to the upper fraction a unit, (or its equivalent improper fraction,) from the sum subtract the lower fraction, and set down the remainder: then add 1 to the lower whole number, and subtract as usual.

Also, a fraction may be subtracted from a whole number, by taking the numerator of the fraction from its denominator, and placing the remainder over the denominator, and then taking 1 from the whole number.

(7)	(8)	(9)
From $7\frac{4}{5}$	From $8\frac{1}{2}$	From 5
take $4\frac{1}{5}$	take $5\frac{2}{3}$	take $\frac{2}{7}$
Ans. $3\frac{3}{5}$	Ans. $2\frac{5}{6}$	Ans. $4\frac{5}{7}$

In example 8th, I first reduce the fractions to a common denominator, and then they are $\frac{3}{6}$ and $\frac{4}{6}$. Then, because the lower fraction is greater than the upper one, I add a unit, or $\frac{6}{6}$, to the upper fraction, and the sum is $\frac{9}{6}$, from which I subtract the lower fraction, and set down the remainder. I then add 1 to the lower whole number, and subtract as usual.

Note 2.—When the fractions are of different denominations, you may, if you please, find the values of them,(by Case VIII. in Reduction,) and then find the difference of those values, as in Compound Subtraction.

10. From $\frac{3}{5}$ of an oz. Troy, take $\frac{7}{8}$ of a pwt.
Ans. 11pwt. 3gr.

11. From $\frac{4}{5}$ of a lb. Avoirdupois, take $\frac{1}{8}$ of an oz.
Ans. 12oz. $10\frac{4}{5}$dr.

12. From $\frac{2}{3}$ of a bushel, take $\frac{1}{2}$ of $\frac{7}{8}$ of a quart.
Ans. 2pk. $4\frac{43}{48}$qt.

MULTIPLICATION OF VULGAR FRACTIONS.

RULE.

Reduce whole and mixed numbers, (if any,) to improper fractions; then multiply all the numerators together for a numerator, and all the denominators together for a denominator, which will give the product required.*

EXAMPLES.

1. Required the product of $\frac{2}{5}$ and $\frac{7}{8}$.

Here, $\frac{2}{5}\times\frac{7}{8}=\frac{2\times7}{5\times8}=\frac{14}{40}=\frac{7}{20}$, the Ans.

2. What is the continued product of $\frac{1}{2}$ of 5, $\frac{1}{5}$ of $\frac{2}{7}$, and $3\frac{1}{2}$?

$\frac{1}{2}$ of $5=\frac{1}{2}$ of $\frac{5}{1}$, and $3\frac{1}{2}=\frac{7}{2}$.

Then, $\frac{1\times5\times1\times2\times7}{2\times1\times5\times7\times2}=\frac{70}{140}=\frac{1}{2}$ Ans.

Contracted thus: $\frac{1\times5'\times1'\times2'\times7'}{2\times1'\times5'\times7'\times2'}=\frac{1}{2}$ Ans.

3. Multiply $\frac{4}{15}$ by $\frac{5}{24}$. Ans. $\frac{1}{18}$.

4. Required the product of $\frac{3}{7}$, $\frac{4}{9}$, and $\frac{14}{15}$. Ans. $\frac{8}{45}$.

* Multiplication of any thing by a fraction, implies the taking some part or parts of the thing; it may therefore be truly expressed by a compound fraction; which is resolved by multiplying together the numerators and the denominators.

5. What is the product of $\frac{1}{3}$ of $\frac{3}{7}$, and $\frac{7}{8}$ of $\frac{5}{14}$?
Ans. $\frac{5}{112}$.

6. What is the continued product of $\frac{1}{2}$, $\frac{5}{7}$ of $\frac{7}{8}$, $8\frac{2}{5}$, and 4?
Ans. $10\frac{1}{2}$.

Note.—A single fraction is best multiplied by an integer, by dividing the denominator by the integer; but if it will not exactly divide, then multiply the numerator by the integer.

7. Required the product of $\frac{5}{8}$ and 4.

Here, $\frac{5}{8 \div 4}=\frac{5}{2}=2\frac{1}{2}$, the Ans.

8. Required the product of $\frac{2}{7}$ and 5.

Here, $\frac{2\times5}{7}=\frac{10}{7}=1\frac{3}{7}$, the Ans.

9. Multiply $\frac{2}{5}$ by 5. Ans. 2.

DIVISION OF VULGAR FRACTIONS.

RULE.*

Prepare the fractions as in Multiplication: then divide the numerator of the dividend by that of the divisor, and the denominator of the dividend by that of the divisor, if they will exactly divide; but if not, then invert the terms of the divisor, and multiply the dividend by it, as in Multiplication.

EXAMPLES.

1. Divide $\frac{15}{28}$ by $\frac{5}{7}$.

Here, (by the first method,) $\frac{15}{28}\div\frac{5}{7}=\frac{3}{4}$ Ans.

2. Divide $\frac{8}{9}$ by $\frac{1}{9}$ of $\frac{7}{8}$.

Performed by the second method, as follows:

Here $\frac{8}{9}\times\frac{9}{1}\times\frac{8}{7}=\frac{576}{63}=9\frac{1}{7}$ Ans.

Contracted thus: $\frac{8\times 9'\times 8}{9'\times 1\times 7}=\frac{64}{7}=9\frac{1}{7}$ Ans.

3. Divide $\frac{16}{25}$ by $\frac{4}{5}$. Ans. $\frac{4}{5}$.

* Division being the reverse of Multiplication, the reason of the Rule is evident.

4. Divide $\frac{2}{3}$ by $\frac{4}{17}$. Ans. $\frac{34}{12}=2\frac{5}{6}$.

5. Divide $5\frac{1}{8}$ by $7\frac{1}{2}$. Ans. $\frac{41}{60}$.

6. Divide $\frac{1}{2}$ of 15 by $\frac{1}{4}$ of $\frac{7}{8}$ of $5\frac{1}{2}$. Ans. $6\frac{18}{77}$.

Note.—A single fraction is best divided by an integer, by dividing the numerator by it; but if it will not exactly divide, then multiply the denominator by it.

7. Divide $\frac{4}{5}$ by 2. Here $\frac{4\div 2}{5}=\frac{2}{5}$ Ans.

8. Divide $\frac{7}{8}$ by 5. Here $\frac{7}{8\times 5}=\frac{7}{40}$ Ans.

SIMPLE PROPORTION, OR

THE SINGLE RULE OF THREE, IN VULGAR FRACTIONS.

RULE.

1. State the question as in the Single Rule of Three in whole numbers.

2. Prepare the fractions as in Multiplication of Vulgar Fractions, and reduce the first and second terms of the proportion to the same name, or denomination, when they are of different denominations.

3. Invert the first term of the proportion, as in Division of Vulgar Fractions; then multiply the three terms continually together, and the product will be the answer in the same name with the third term of the proportion.*

EXAMPLES.

1. If $2\frac{1}{6}$pwt. of standard silver be worth $\frac{1}{2}$ of $\frac{1}{4}$ of a dollar, what is the value of $\frac{1}{7}$ of an ounce?

pwt. oz. \$

Stated thus: as $2\frac{1}{6}$: $\frac{1}{7}$ :: $\frac{1}{2}$ of $\frac{1}{4}$: the Ans.

By preparing the fractions, as directed in the 2d article of the Rule, the question stands thus: as $\frac{13}{6}$pwt. : $\frac{20}{7}$pwt. :: \$$\frac{1}{8}$: the Ans.

Then, $\frac{6}{13}\times\frac{20}{7}\times\frac{1}{8}=\$\frac{120}{728}=16$c. $4\frac{76}{91}$m. Ans.

* This is only multiplying the 2d and 3d terms together, and dividing the product by the first, as in the Single Rule of Three in whole numbers.

2. If $\frac{1}{4}$ of a yard of cloth cost $\frac{7}{20}$ of a L. what cost $14\frac{1}{2}$ yards? Ans. 20*l*. 6s.

3. If $\frac{5}{7}$ oz. be worth $\frac{11}{12}$*l*. what is the value of 1oz? Ans. 1*l*. 5s. 8d.

4. If half a gill of rum cost $\frac{3}{100}$ of a dollar, what cost $31\frac{1}{2}$ gallons? Ans. \$60.48

5. A person having $\frac{2}{5}$ of a vessel, sells $\frac{1}{4}$ of his share for \$280: what is the whole vessel worth, at that rate? Ans. \$2800.

6. How many yards of shalloon that is $\frac{3}{4}$ of a yard wide, will line $4\frac{1}{2}$ yards of cloth which is $1\frac{1}{2}$yd. wide?

As $\frac{3}{4}$yd. width : $1\frac{1}{2}$ yd. width :: $4\frac{1}{2}$ yd. length : 9 yards length, Ans.

COMPOUND PROPORTION,

IN VULGAR FRACTIONS.

RULE.

1. State the question as in Compound Proportion in whole numbers.

2. Prepare the fractions as in Multiplication of Vulgar Fractions, and reduce the corresponding quantities in the first and second places to the same denomination, when they are of different denominations.

3. Invert the fractions which will compose the divisor; then multiply all the fractions together, and the product will be the answer in the same name with the third term of the proportion.

EXAMPLES.

1. If 9 persons use $1\frac{1}{8}$lb. of tea per month, how much will a family of 8 persons use in $\frac{3}{4}$ of a year, at that rate?

Stated thus: $\left.\begin{matrix}\text{Persons, } 9 : 8 \\ \text{Years, } \frac{1}{12} : \frac{3}{4}\end{matrix}\right\} :: \overset{\text{lb.}}{1\frac{1}{8}}$: the Ans.

By preparing the fractions, as directed in the Rule, the question stands thus:

As $\left.\begin{matrix}\frac{9}{1} : \frac{8}{1} \\ \frac{1}{12} : \frac{3}{4}\end{matrix}\right\} :: \overset{\text{lb.}}{\frac{9}{8}}$: the Ans.

Then, $\frac{1}{9}\times\frac{12}{1}\times\frac{8}{1}\times\frac{3}{4}\times\frac{9}{8}=\frac{2592}{288}=9$lb. Ans.

$$\text{Contracted thus: } \frac{\overset{3}{1'} \times 12' \times 8' \times 3 \times 9'}{9' \times 1' \times 1 \times 4' \times 8'} = 9\text{lb. Ans.}$$

2. If 2 men, in $\frac{1}{4}$ of a year, expend \$56$\frac{1}{4}$, how much money will defray the expences of 6 persons for 5$\frac{1}{3}$ months, at the same rate? Ans. \$300.

3. If a man earn \$4$\frac{1}{10}$ in 8$\frac{1}{2}$ days, how much would 20 men earn in 100 days, at that rate?

Ans. 964D. 70c. 5$\frac{15}{17}$m.

4. If \$100 dollars principal gain \$5$\frac{1}{2}$ interest in 12 months, how much will \$40$\frac{1}{2}$ gain in 2$\frac{1}{3}$ months?

Ans. 43c. 3$\frac{1}{3}$m.

5. If 2$\frac{1}{2}$ yards of cloth, $\frac{7}{8}$ of a yard wide, cost \$2$\frac{4}{5}$, what is the value of 3$\frac{1}{2}$ yards, $\frac{5}{8}$ of a yard wide, of the same quality? Ans. \$2.40

DUODECIMALS,*

Are fractions so called because they decrease by *twelves*; inches being twelfths of a foot, which is the integer; seconds twelfths of an inch; thirds twelfths of a second, and so on; as in the following table.

1 Foot (ft.)	= 12	Inches, marked *in.*, or	′
1 Inch	= 12	Seconds,	″
1 Second	= 12	Thirds,	‴
1 Third	= 12	Fourths,	⁗
&c.			

Duodecimals are chiefly useful to ascertain the superficial or solid contents of such things as are measured by feet, inches, &c.

Addition and Subtraction of Duodecimals are performed as in Compound Addition and Subtraction.

* This word is derived from the Latin word *duodecim*, which signifies *twelve*.

MULTIPLICATION OF DUODECIMALS.

RULE.*

1. Place the multiplier under the multiplicand, in such a manner that the feet of the multiplier shall stand under the lowest denomination of the multiplicand.

2. Multiply the multiplicand by each term or denomination of the multiplier, separately, as in Compound Multiplication, (always carrying one for every twelve, from each denomination to the next higher,) and place the right hand term of each product under that denomination of the multiplicand by which it is produced.

3. Add together the several partial products, as in Compound Addition, and their sum will be the total product required.

Note 1.—If there are no feet in the multiplier, supply their place with a cipher; and if any other denomination between the highest and lowest, either in the multiplier or multiplicand, be wanting, write a cipher in its place.

Note 2.—The marks which designate the denominations of duodecimals, are called *indices;* the index of feet being 0; that of inches 1, that is, one mark; that of the seconds 2 marks, &c. When any two of the denominations are multiplied together, the index of their product is equal to the sum of the indices of the two factors. Thus, if feet be multiplied by feet, the product will be feet; for the sum of the indices is $0+0=0$: If inches be multiplied by seconds, the product will be thirds; for $1+2=3$; &c. In multiplication of duodecimals, the terms of the product are of the same denominations as those terms of the multiplicand which stand over them.

EXAMPLES.

1. Multiply 8 feet, 6 inches, 9 seconds, by 6 feet, 5 inches, 7 seconds.

---

* The reason of this rule will be obvious by considering the denominations below the integer as fractional parts of the integer, and then multiplying as in Vulgar Fractions. Thus, feet multiplied by inches give inches; for 2 feet multiplied by 4 inches=2ft.$\times\frac{4}{12}$ft.$=\frac{8}{12}$ft.=8 inches: Inches multiplied by inches give seconds; for 2in.$\times$5in.$=\frac{2}{12}$ft.$\times\frac{5}{12}$ft.$=\frac{10}{144}$ft.$=\frac{10}{12}$in.=10 seconds; &c.

Operation.

Ft.	'	"		
8 ..	6 ..	9		
		6 ..	5 ..	7
	4 ..	11 ..	11 ..	3
3 ..	6 ..	9 ..	9	
51 ..	4 ..	6		
55 ..	4 ..	3 ..	8 ..	3

Here, I first place the multiplier so that 6, the feet, stand under 9, the seconds (or lowest denomination) of the multiplicand. I then begin with 7 and 9, the lowest denominations of the multiplier and multiplicand, and say 7×9 is 63; which being 5 times 12, and 3 over, I set down 3, and carry 5 to the next denomination. Then, 7×6 is 42, and 5 which I carried makes 47—I set down 11 and carry 3. Then, 7×8 is 56, and 3 which I carried makes 59—I set down 11 and carry 4; and as there is no other term to multiply, I set down the 4 in the next place to the left. In like manner I multiply the multiplicand by the inches, and then by the feet of the multiplier, and set down the right hand term of each product under the term I multiply by. Lastly, I add together the partial products thus found, and the answer is 55ft. 4in. 3" .. 8‴ .. 3⁗.

2. Multiply 14 feet and 2 seconds, by 5 inches, 6 seconds.

Here there are no feet in the multiplier, and I supply their place with a cipher: I also put a cipher in the place of inches in the multiplicand.

	Ft.	in.	"		
	14 ..	0 ..	2		
		0 ..	5 ..	6	
		7 ..	0 ..	1 ..	0
	5 ..	10 ..	0 ..	10	
Ans.	6 ..	5 ..	0 ..	11 ..	0

	Ft. ' "		Ft. ' "		Ft. ' " ‴
3. Multiply	7 .. 2 .. 8	by	5 .. 2 .. 6.	Ans.	37 .. 7 .. 4 .. 8
4. ———	12 .. 4 .. 0	by	2 .. 7 .. 0.	—	31 .. 10 .. 4 .. 0
5. ———	14 .. 6 .. 0	by	5 .. 8.	—	6 .. 10 .. 2 .. 0
6. ———	28 .. 2 .. 0	by	24 .. 6 .. 0.	—	690 .. 1 .. 0 .. 0

PRACTICAL QUESTIONS.

Note.—To find the area, or superficial content, of any *parallelogram*;* such as a board of equal width from end to end, or the floor of a square room, &c.; multiply the length by the breadth, and the product will be the answer.

To find the solid content of any *parallelopiped*;* such

* See the definitions of these terms in the "Mensuration of Superfices and Solids."

as a stick of timber hewn square, of equal bigness from end to end, or a load of wood, &c.; multiply continually together the length, breadth, and height, and the last product will be the answer.

1. What is the superficial content of a board which is 17 feet 7 inches long, and 1 foot 5 inches wide?

Ans.* 24sq. ft. 10′.. 11″.

2. How many square yards does the floor of a room contain, which is 24 feet 8 inches long, and 17 feet 6 inches wide? Ans. 47sq. yd. 8ft. 8′.

3. If a stick of hewn timber be 12ft. 10in. long, 1ft. 7in. wide, and 1ft. 9in. thick, how many solid or cubic feet does it contain? Ans.† 35cub. ft. 6′.. 8″.. 6‴.

4. How many cubic feet of wood in a load which is 8ft. 6in. long, 2ft. 3in. wide, and 3ft. 9in. high?

Ans. 71cub. ft. 8′.. 7″.. 6‴

INVOLUTION,

Is the raising of *powers* from any given number, as a *root.*

The product arising from the multiplication of any given number by itself, is called the *second power* or *square* of the number: if the second power be multiplied by the said given number, the product is the *third power*, or *cube*; if the third power be multiplied by the given number, the product is the *fourth power*, or *biquadrate*, &c.; the given number itself being called the *first power* or *root*.

Thus, 5 is the root, or 1st power of 5.
5×5=25 is the 2d power, or square of 5.
5×5×5=125 is the 3d power, or cube of 5, &c.

In this manner is calculated the following table of the first eight powers of the nine digits.

* The inches in this answer are 12ths of a square foot, and the seconds are 144ths of a square foot, or square inches.

† In this answer, the inches are 12ths, the seconds 144ths, and the thirds 1728ths of a cubic foot.

TABLE OF POWERS.

Root.	Square.	Cube.	4th pow.	5th pow.	6th pow.	7th pow.	8th pow.
1	1	1	1	1	1	1	1
2	4	8	16	32	64	128	256
3	9	27	81	243	729	2187	6561
4	16	64	256	1024	4096	16384	65536
5	25	125	625	3125	15625	78125	390625
6	36	216	1296	7776	46656	279936	1679616
7	49	343	2401	16807	117649	823543	5764801
8	64	512	4096	32768	262144	2097152	16777216
9	81	729	6561	59049	531441	4782969	43046721

The *index* or *exponent* of any power, is the number denoting the height or degree of that power; and it is 1 more than the number of multiplications used in producing the same. So, 1 is the index or exponent of the first power or root, 2 of the 2d power or square, 3 of the 3d power or cube, &c.

Powers, which are to be raised, are usually denoted by writing the index, in small figures, at the upper and right side of the given number: Thus, 5^2 denotes the 2d power of 5, and 12^3 denotes the 3d power of 12, &c.

If two or more powers be multiplied together, their product is that power whose index is equal to the sum of the indices of the factors, or powers multiplied together: Or, the multiplication of powers answers to the addition of their indices. Thus, the product of the 2d and 3d powers of any number, is the 5th power of that number; and, if the 2d power be multiplied by itself, the product is the 4th power, &c.

PROBLEM.

To involve or raise a given number to any proposed power.

RULE.

Multiply the given number, or first power, continually by itself, till the number of multiplications is 1 less than the index of the power to be found, and the last product will be the power required.

Note 1.—When the given number is to be involved to a high power, it will not be necessary to find all the lower powers, in order to obtain the one required: For, if a few of the lower or leading powers be found, as directed in the foregoing Rule, we may then find the answer by multiplying together some of these, viz. those powers whose indices, when added together, will be equal to the index of the power required.

Note 2.—It is evident, from the foregoing Rule, that when the given quantity, or first power, is greater than 1, its value will be *increased* by involution; but when it is less than 1, its value will be *diminished;* and when it is equal to 1, it will be neither increased nor diminished.

EXAMPLES.

1. Required the 5th power of 8.

Here, $8\times8\times8\times8\times8=32768$, the Ans.

Or, otherwise thus: $8\times8=64$, the 2d power of 8; then, $64\times64=4096$, the 4th power of 8, and $4096\times8=32768$, the 5th power of 8, as before.

The same result may also be obtained by multiplying together the 2d and 3d powers of 8.

2.	What is the 2d power of 45?	Ans. 2025.
3.	——— the 3d power of 12.5?	Ans. 1953.125
4.	——— the 4th power of 71?	Ans. 25411681.
5.	——— the 5th power of .25?	Ans. .0009765625
6.	——— the 6th power of .4?	Ans. .004096

Note 3.—Since *vulgar fractions* are multiplied together by taking the products of their numerators and of their denominators, we may involve a vulgar fraction to any power assigned, by raising each of its terms to the power required: Or, we may reduce the vulgar fraction to a decimal, and then involve it according to the foregoing Rule. When a mixed number is to be involved, we may either reduce the whole to an improper fraction, or reduce the vulgar fraction to a decimal, and then proceed as before.

7. What is the square of $\frac{4}{5}$? Here $\frac{4\times4}{5\times5}=\frac{16}{25}$, Ans.

Or, $\frac{4}{5}=.8$, and $.8\times.8=.64$, the Ans.

8. What is the 3d power or cube of $\frac{2}{7}$? Ans. $\frac{8}{343}$.

9. Required the third power of $2\frac{1}{2}$.

Here $2\frac{1}{2}=\frac{5}{2}$, and $\frac{5}{2}\times\frac{5}{2}\times\frac{5}{2}=\frac{125}{8}=15\frac{5}{8}$, Ans.

Or, $2\frac{1}{2}=2.5$, and $2.5\times2.5\times2.5=15.625$, Ans.

10. What is the 4th power of $\frac{1}{5}$? Ans. $\frac{1}{625}$, or .0016

EVOLUTION,

Is the reverse of Involution; being the method of extracting or finding the roots of any given powers or numbers.

The root of any given power, or number, is such a number as being multiplied into itself a certain number of times will produce that power. The *first* root of any given number, is the number itself; the *second* or *square* root, is that number whose second power or square equals the given number; the *third* or *cube* root, is that whose third power or cube equals the given number, &c. Thus, the square root of 4 is 2, because 2×2, or the square of 2, is 4; and the cube root of 27 is 3, because $3\times3\times3=27$.

Any power of a given number may be found *exactly;* but there are many numbers of which a proposed root cannot be exactly found; yet, by means of decimals, we may approximate, or approach towards the root, to any assigned degree of exactness.

Those roots which only approximate, are called *irrational* or *surd* roots; and those which can be found quite exact, are called *rational* roots. A number is called a *complete* power of any kind, when its root of the same kind is rational; and an *imperfect* power, when the root is irrational. Thus, 4 is a complete square, its square root being 2; and 27 is a complete cube, its cube root being 3: but 27 is not a complete square, because its square root is a surd, or an irrational number, which cannot be found *exactly.*

When the root of any given power or number consists of, or can be expressed by, a single digit, it may be found by trials, or by a table of powers. Roots which consist of two or more figures may be extracted by the rules for the following problems.

PROBLEM I.

To extract the Second, or Square Root, of any given number.

RULE*

1. Distinguish the given number into periods of two figures each, by setting a point or dot over the unit figure, and one over every second figure from the unit's place, to the left in whole numbers, and to the right in decimals.

2. Find, by trial, the greatest complete square number that can be had in the first or left hand period, and the root of this square will be the first figure of the required root; which figure place at the right hand of the given number, after the manner of a quotient in Division. Subtract the square, thus found, from the left hand period, and to the remainder bring down and annex the next period of figures, for a *dividual.*

3. Double the figure of the root already found, for a *defective divisor.*

4. Find how many times the defective divisor may be

* The principle on which the first article of this rule depends, is, *that the product of any two numbers can have, at most, but so many places of figures as are in both the factors, and at least but one less;* which may be thus demonstrated:—Take two numbers consisting of any number of places; but let them be the least possible of those places, viz unity with ciphers, as 100 and 10: then their product will be 1 with so many ciphers annexed as are in both the factors, and it will evidently consist of as many places lacking one as there are in both the factors: Thus, $100\times10=1000$, which product has one place less than 100 and 10 together. Again, take two numbers, consisting of any number of places, which shall be the greatest possible of those places, as 99 and 9 Now, 99×9 is less than 99×10; but 99×10, or 990, contains only so many places of figures as are in 99 and 9; whence it is evident that the product of any two numbers cannot have more places of figures than are in both the factors.—It evidently follows from this demonstration, that a square number cannot have more places of figures than double the places of the root, and at least but one less; and that a cube number cannot have more places than triple the places of the root, and at least but two less; and so on for higher powers. Consequently the square root of any number will consist of as many figures as there are periods of two figures in the given number; and the cube root of any number, of as many figures as there are periods of three figures in the number, or cube. &c; and hence the reason of pointing numbers into periods of two figures each for the square root, and periods of three figures for the cube root, &c., as directed in the rules for extracting roots, is evident.

The principle on which the remaining part of the rule for extracting

had in the dividual exclusive of its right hand figure; place the result in the quotient, for the second figure of the root, and annex the same figure to the defective divisor, calling the divisor thus increased, the *complete divisor*.

5. Multiply the complete divisor by the figure of the root last found, for a *subtrahend*, which subtract from the dividual, (as in common Division,) and to the remainder annex the next period of figures, for a new dividual.

6. Double the figures of the root already found, for a new defective divisor. Then find another figure of the root, as before; and continue the operation in like manner till you have brought down all the periods.

Note 1.—If, when the integral figures of the given power are distinguished into periods, the left hand period is deficient, (in the number of its figures,) it must nevertheless be considered a period. But, if there are decimals in the given square, and the last or right hand period is deficient, a cipher must be annexed to complete the period; and, in extracting the roots of higher powers, as many ciphers must be annexed as the power requires, to make up a full period.

Note 2.—As the defective divisors are less than the true

---

the square root depends is, *that the square of the sum of any two numbers is equal to the squares of the numbers with twice their product.* Thus, the number 45 is equal to $40+5$; and the square of 45 is equal to the squares of 40 and of 5, with twice the product of 40×5; that is, equal to $1600+2\times(40\times5)+25,=1600+400+25,=2025$. Now, in extracting the square root of 2025, according to the above Rule, we separate the number into these two parts, 2000 and 25. Thus, 2000 contains 1600, the square of 40, with the remainder 400: the first part of the root is, therefore, 40, (or 4 tens) and the remainder of the square is $400+25$, or 425. Now, according to the principle above mentioned, this remainder must be equal to twice the product of 40 and the part of the root still to be found, together with the square of the said part. Now, dividing 425 by 80, (i. e. by twice the part of the root found,) we find for quotient 5; then this part being added to 80, the sum is $80+5$, which being multiplied by 5, the product, $400+25$, or 425, is equal to the remainder of the square, and is evidently equal to twice the product of 40 and 5, together with the square of 5. So, by proceeding according to the Rule, we find the true root, and exhaust the given square. In the same manner the operation may be illustrated in every case in which the root consists of two figures; and it is evident that the process may be carried on farther in the same manner when the root consists of more than two figures, if the given number or square be separated into as many parts as there are figures in the root.

or complete divisors, the quotient figures found by them, (especially the second and third figures in the root,) will sometimes be too great. When any subtrahend exceeds the corresponding dividual, a less figure must be put in the quotient, and the defective divisor must be completed anew, &c.

Note 3.—When it is necessary to bring down more than one period of figures at a time, in order to make the dividual contain the divisor, then a cipher must be put in the quotient, and a cipher annexed to the defective divisor, for every period of figures so brought down more than one.

Note 4.—When all the periods are exhausted, if there be a remainder, the operation may be carried on as much farther as may be thought necessary, by annexing a period of ciphers to each remainder, and continuing the extraction as before;—the ciphers, so annexed, being always considered as places of decimals.—The root will contain just as many places of whole numbers as there are periods of integral figures in the given power, and as many places of decimals as there are periods of decimals used in the operation.

PROOF.—Square the root, when found, and add in the remainder, if any; and the result will be equal to the given number, if the work be right.

EXAMPLES.

1. What is the 2d or square root of 73359.7225?

Operation.

```
 73359.7225(270.85 Ans.
 4
 ----
47)333
   329
   -------
5408)45972
     43264
     --------
54165)270825
      270825
```

Explanation.—1. I first distinguish the given number into periods of two figures each, by putting a point or dot over the unit figure, viz. over 9, and one over every second figure, both ways, from the unit's place; and I have five periods; the first, or left hand period, being 7, the second 33, the third 59, the fourth .72, and the fifth 25.

2. I seek the greatest square number in 7, the left hand period, which I find to be 4; the square root of which, viz. 2, I place to

the right hand of the given number, for the first figure of the required root. I then subtract the said square from the left hand period, and the remainder is 3; to which I annex the next period of figures, viz. 33, and I have 333 for a dividual.

3. I double 2, the figure of the root already found, and it makes 4, which I place at the left hand of the dividual, for a defective divisor.

4. I seek how often 4, the defective divisor, is contained in 33, the dividual exclusive of its right hand figure, and the result or quotient figure is 8, which I set down, on trial, for the second figure of the root, and annex the same figure to the defective divisor, and it makes 48, for a complete divisor. I then multiply 48, the complete divisor, by 8, the quotient figure just found, which gives 384, for a subtrahend. As this subtrahend exceeds the dividual, the quotient figure, taken on trial, must be too great; and therefore I try a less figure, viz. 7, and by making out a subtrahend as before, I find that 7 is the right quotient figure. This subtrahend, viz. 329, I subtract from the dividual, and to the remainder I join the next period of figures, which gives 459, for a new dividual.

5. I double 27, the root already found, which gives 54, for a new defective divisor. I then find that this divisor cannot be had in 45, the new dividual, exclusive of its right hand figure, and therefore I annex a cipher to the root, and another cipher to the defective divisor, and bring down the next period of figures to the dividual. I next seek how often the defective divisor, thus augmented, viz. 540, may be had in 4597, the dividual exclusive of its right hand figure, and the quotient figure is 8. I then complete the divisor, and make out a subtrahend, &c., as before, and continue the operation until all the periods are exhausted.

6. As the given power contains three periods of integers and two of decimals, there must be three integral and two decimal figures in the root; and therefore I point off two figures for decimals, and the answer is 270.85

To prove the operation, I square the root so found; that is, I multiply 270.85 by 270.85, and it gives 73359.7225; which being equal to the given number, I conclude the work is right.

Note 5.—Each defective divisor, except the first one, may be readily found by adding the last quotient figure to the last complete divisor; (or, which is the same thing, bringing down the complete divisor, doubling its right hand figure;) as in the following example.

2. Required the square root of 47.6

```
47.60(6.899+ Ans.
36
——
128)1160
  8 1024
——— ————
1369)13600
   9 12321
———— —————
13789)127900
      124101
      ——————
        3799
```

Here, I annex a cipher to the given number, to make up a full period of decimals. In extracting the root, I find the 2d and 3d defective divisors by adding, in each case, the last quotient figure to the last complete divisor. After having brought down all the figures of the given number, there is a remainder, and I annex a period of ciphers to it, and continue the operation till I have found two more figures in the root; and more figures might be obtained in the same manner. Lastly, I point off three decimal places in the root, because I have used three periods of decimals in the operation.—The work may be proved by multiplying the root by itself, and adding the last remainder to the product.

Note 6.—In extracting the square roots of numbers, it will sometimes happen that the remainder will be equal to, or greater than the corresponding complete divisor, when the quotient figure is correct. Thus, in example 2d, the second remainder, viz. 136, exceeds the corresponding complete divisor, which is 128. But if the second figure of the root, viz. 8, should be increased by 1, the subtrahend would exceed the dividual; and therefore the second figure of the root cannot exceed 8. When the remainder exceeds the complete divisor, you may easily determine whether the quotient figure is right or not, by trying a greater figure; for, always, when the quotient figure is too great, the subtrahend will exceed the dividual.—The same thing will also sometimes happen in extracting the cube roots of numbers by the the following rule, and the true quotient figure may be determined by trial in a similar manner.

More Questions for exercise.

What is the square root of		What is the square root of	
2025?	Ans. 45.	2265.76?	Ans. 47.6
88804?	—— 298.	54.2?	—— 7.36+
10342656?	—— 3216.	.45369?	—— .673+
23049601?	—— 4801.	.005184?	—— .072

CONTRACTION.

When the root is to be extracted to many places of figures, the work may be abbreviated as follows: Proceed in extracting the root according to the foregoing rule, till you find half of the required number of figures, or one figure more; then, take double the root already found, for a divisor, and find the other figures of the root by common division, or by the 2d Contraction in Division of Decimals.*

Ex. 1. To find the square root of 45 to eight places of figures.

In this example, I find the four last figures of the root, (viz. the figures 2039,) by common division.

```
The root.
  45)6.7082039+
  36
  ----
127)9.00
    8 89
  ------
13408)110000
      107264
  ----------
13416)  27360
        26832
        -----
         52800
         40248
         -----
        125520
        120744
        ------
          4776
```

2. Extract the square root of 12 to the 7th decimal figure. Ans. 3.4641016+

3. Extract the square root of 145.037237 to nine places of figures.
Ans. 12.0431406+

* We may always find by this method of contraction, at least as many figures of the root, lacking one, as we previously find by the common method.

PROBLEM II.

To extract the Third, or Cube Root, of any given number.

RULE.

1. Distinguish the given number into periods of three figures each, by setting a point over the place of units, and also over every third figure from thence, to the left in whole numbers, and to the right in decimals, if any; annexing ciphers to the decimals, if necessary, to make out the last period.

2. Find, by trial, the greatest cube number contained in the left hand period, and set its root on the right hand of the given number, for the first figure of the required root. Subtract the cube, thus found, from the left hand period, and annex to the remainder the next period, for a *dividual.*

3. Square the part of the root already found, and multiply this square by 300, for a *defective divisor.*

4. Seek how often the defective divisor may be had in the dividual, and place the result in the quotient for the second figure of the root.

5. Complete the divisor in the following manner; viz. Multiply the root already found, except the last or right hand figure, by 3; to the product annex the last figure of the root; then multiply the result by the said last figure, and call the product the *complement number.* Add this complement number to the defective divisor, and call the sum the *complete divisor.*

6. Multiply the complete divisor by the figure of the root last found, for a *subtrahend;* which subtract from the dividual, and to the remainder join the next period of figures, for a new dividual.

7. A defective divisor of the 2d dividual, (or any subsequent one,) may be very easily found thus: Below the complete divisor last used, write the square of the figure of the root last found; then add together the said square, the complete divisor, and the corresponding complement number; to the sum annex two ciphers, and the result will

be a new defective divisor.* Then find another figure of the root as before, and so proceed.†

Note.—When it is necessary to bring down more than one period of figures at a time, in order to make the dividual contain the divisor; then, for every period of figures so

* Every defective divisor found in this manner, will be equal to 300 times the square of the part of the root found.

† It is somewhat difficult to demonstrate this rule without the aid of Algebra. The following demonstration may be easily understood by those who are acquainted with the first principles of Algebra.

Suppose the root of a cube to consist of two figures; and let a represent the first figure with a cipher annexed, and b, the second figure; then $a+b$ will represent the *root*, and $(a+b)^3$, or $a^3+3a^2b+3ab^2+b^3$, the *cube*. Now, the first term of this cube, viz. a^3, is the cube of the first term of the root; and if we divide the remaining terms of the cube, viz. $3a^2b+3ab^2+b^3$, by $3a^2+3ab+b^2$, the quotient is b, the second term of the root. The first term of this divisor, viz. $3a^2$, represents a *defective divisor*, found according to the 3d article of the Rule; (for, since a represents the first figure of the root with a cipher annexed, $3a^2$ represents 300 times the square of the said first figure;) and the two remaining terms of the said divisor, viz. $3ab+b^2,=(3a+b)\times b$, a *complement number*, made out according to the Rule: and consequently the whole of the said divisor, viz. $3a^2+3ab+b^2$, represents a *complete*, or *true divisor*, found according to the Rule. Hence the reason of the Rule, excepting the last article, is obvious.

That the sum of each complete divisor, complement number, and the square of the last or corresponding quotient figure, is equal to the triple square of the part of the root ascertained, may be shown thus:—

The complete divisor	$=3a^2+3ab+b^2$
The complement number	$= \quad 3ab+b^2$
Square of the last quot. figure=	b^2
Their sum is	$3a^2+6ab+3b^2$; which is

equal to $(a+b)^2\times3$, or 3 times the square of the root; and hence the method of making out the defective divisors, laid down in the 7th article of the Rule, is evidently correct.

brought down, more than one, annex a cipher to the root, and two ciphers to the defective divisor.

The 1st, 2d, and 4th Notes, annexed to the rule for extracting the square root, must also be attended to in extracting the cube root.

PROOF.—Cube the root, when found, and add in the remainder, if any, and the sum will be equal to the given number, if the work be right.

EXAMPLES.

1. Required the cube root of 162784.5

Operation.

		The root.
		162784.500(54.601+
	5×5×5=125	
1st Defective divisor,	7500)	37784=1st dividual.
" Complement no.	616	32464=2d subtrahend.
" Complete divisor,	8116)	5320500=2d dividual.
4×4=	16	5307336=3d subtrahend.
2d Defective divisor,	874800)	13164000000=3d divid.
" Complement no.	9756	8943643801=4th sub.
" Complete divisor,	884556	4220356199 Remainder.
6×6=	36	
3d Defective divisor,	8943480000	
" Complement no.	163801	
" Complete divisor,	8943643801	

5	5	54	5460
5	3	3	3
25	154	1626	163801
300	4	6	1
7500 1st Defective divisor.	616 1st Comp. no.	9756 2d Comp. no.	163801 3d Comp. no.

In this example, I annex two ciphers to the given number, to make up a full period of decimals. In extracting the root, I proceed according to the Rule until I have brought down all the periods; then I annex periods of ciphers to the remainder, and continue the operation till I have found three decimal figures in the root; and the operation might be carried on farther in the same manner. The second defective divisor is found by adding together 16, the square of the second figure of the root, 8116, the first complete divisor, and 616, the corresponding complement number, and then annexing two ciphers to the sum thus obtained. The third defective divisor is found in a similar manner, only two ciphers more than usual are annexed, because one period of figures more than usual is annexed to the remainder, in order to make it contain the divisor. *See the Note annexed to the Rule.*—The work may be proved by raising the root to the 3d power, and adding the last remainder to the result.

In the foregoing example, all the multiplications, performed in making out the several divisors, are exhibited, in order to make the whole operation as intelligible as possible; but, as the multiplier is always a single digit, it is evidently unnecessary to write down all the numbers in the usual form. Each complement number may be readily found as follows; viz. Multiply, *mentally*, the root except the last figure by 3, and place the product to the left hand of the defective divisor, in the next line below; then annex to this product the last figure of the root, and multiply the whole by the said last figure, placing the product below the defective divisor; as in the next following example.

2. Required the cube root of 1953.125

```
           300   1953.125(12.5 The root.
    32×2=   64   1
           ---   ----
           364   )953
             4    728
           ----- ------
           43200)225125
   365×5=   1825 225125
           -----
           45025
```

3.	What is the cube root of 14706125?	Ans.	245.
4.	——— of 122615327232?	——	4968.
5.	——— of 1392915510692$\grave{8}$543?	——	240607.
6.	——— of 78402.752?	——	42.8
7.	——— of 3214?	——	14.757+
8.	——— of 4?	——	1.5874+
9.	——— of 74.128?	——	4.2007+
10.	——— of .15?	——	.5313+
11.	——— of .000684134?	——	.08811+

CONTRACTION.

When the root is to be extracted to many places of figures, the latter part of the process may be very much abbreviated as follows; viz. After having found, by the foregoing Rule, at least one-half of the required number of figures of the root, then take the sum of the last complete divisor and the corresponding complement number for a *constant divisor*, and find the other figures of the root by common division, or by the 2d Contraction in Division of Decimals.

Ex. 1. To find the cube root of 344.02 to ten places.

```
                              The root.
                        344.020(7.006931908+
                        343
                        -------------
             147000000)1020000000
21006×6=       126036  882756216
             ---------- ------------
             147126036 )137243784000
                    36  132543922509
             ----------- ------------
           14725210800 )469986,1491
210189×9=      1891701  4418697
           -----------  ----------
           14727102501   2811644
           -----------   1472899
          1472899,4202   ---------
                         13387459
                         13256091
                         ---------
                          13136810
                          11783192
                          ---------
                           1353618
```

Here, I find the first five figures of the root as in the former examples: Then I take the last remainder, viz. 4699861491, for a dividend, and the sum of the last complete divisor and complement number for a divisor, and find five more figures in the root by the 2d Contraction in Division of Decimals, Rule 2d.

2. Find the cube root of 34567 to nine places of figures.
Ans. 32.5752104+

3. Extract the cube root of 21035.8 to twelve places of figures. Ans. 27.6049105594+

PROBLEM III.

To extract the root of any given power, or to find any proposed root of a given number.

GENERAL RULE.

1. Distinguish the given power, or number, into periods of as many figures each as there are units in the index of the power, that is *two* figures for the second root, *three* for the third root, &c.

2. Begin with the left hand period of the given power, and find, by trial, or by a table of powers, the greatest complete power of the same degree contained in the said period. Place the root of the said power at the right hand of the given number, for the first figure of the required root, and subtract the power from the left hand period; then annex to the remainder the first figure of the next period, for a dividual.

3. Involve the part of the root already found to the next inferior power to that which is given, and multiply this power by the index of the given power, (viz. by 2 for the 2d root, or by 3 for the 3d root, &c.,) for a defective divisor.

4. Find how many times the defective divisor may be had in the dividual, and the result will be another figure of the root, *probably*.

5. Involve the whole ascertained root to the given power, for a subtrahend, which subtract from the two left hand periods of the given number, and to the remainder annex the first figure of the next period, for a new dividual.

6. Find a new defective divisor, and another figure of the root, as before; and so proceed, always subtracting each subtrahend from as many periods of the given number as you have found figures in the root.

Note 1.—As the defective divisors are too small, the quotient figures found by them, (especially the second and third figures,) will sometimes be too great. When the subtrahend exceeds those periods from which it is to be taken, the last quotient figure must be taken less, &c.

Note 2.—When the dividual will not contain the divisor, annex a cipher to the root, and bring down another figure to the dividual; then proceed as before.

Note 3.—When the root is to be extracted to many places of figures; then, after having found, by the foregoing Rule, about half of the required number of figures, or one more than half, you may make out a new dividual and a new defective divisor as usual, and then find the other figures of the root by the 2d Contraction in Division of Decimals.

EXAMPLES.

1. Required the 4th root of 1449832.7281

$1\dot{4}49832\dot{.}728\dot{1}(34.7$ The root.

$3\times3\times3\times3=$ 81 the 1st subtrahend.

$3\times3\times3\times4=108)$ 639 the 1st dividual.

$34\times34\times34\times34=1336336$ the 2d subtrahend.

$34\times34\times34\times4=157216)$ 1134967 the 2d dividual.

14498327281 the 3d subtrahend.

2. Required the 5th root of 371293.

$\dot{3}7129\dot{3}(13$ Ans.

$1\times1\times1\times1\times1=1$

$1\times1\times1\times1\times5=5)27$ dividual.

$13\times13\times13\times13\times13=371293$ subtrahend.

3. Extract the 3d root of 10077.696 by the foregoing general rule. Ans. 21.6

Note 4.—The operations of extracting the roots of high powers by the foregoing general rule, will be very tedious: The following method, when practicable, will be much more convenient.

When the index of the given power, or of the required

root, is a composite number; take any two or more indices whose product is equal to the given index, and extract from the given number a root answering to one of those indices, and then extract from this root, a root answering to another of the indices, and so on to the last; and the root last found will be the answer. Thus, the 4th root is equal to the 2d root of the 2d root, because the index 4 is $=2\times2$; and the 6th root is equal to the 3d root of the 2d root, because $6=3\times2$, &c.

Any power whose index is divisable by 2, may be reduced to a power of half the height, by one extraction of the 2d root; and, any power whose index is divisable by 3, may be reduced to a power of one-third the height, by one extraction of the 3d root; &c. It is, therefore, evident that most powers may either have their roots extracted, or be reduced to lower powers, by extracting the 2d and 3d roots only.

4. Required the 4th root of 20736, by the method laid down in the last Note.

Here, $\sqrt{20736}=144$, and $\sqrt{144}=12$, the Ans.

5. Extract the 18th root of 12 to four places of figures.

Here the index of the given power is 18, which is equal to $2\times3\times3$: therefore, the required root may be found by one extraction of the 2d root and two of the 3d root; as follows: $\sqrt{12}=3.4641016+$; then, $\sqrt[3]{3.4641016}=1.51308+$; then, $\sqrt[3]{1.51308}=1.148+$, the Ans.

6. Extract the 6th root of 21035.8 to four places.

Ans. 5.254+

7. Extract the 9th root of 21035.8 to five places.

Ans. 3.0222+

PROBLEM IV.

To extract the roots of Vulgar Fractions.

RULE.

1. Reduce the given fraction to its lowest terms.

2. If both terms of the fraction be complete powers, extract the root of the numerator, and of the denominator, for the terms of the root required.

3. When the terms are not both complete powers, reduce the vulgar fraction to a decimal, and then extract the

root, as usual: Or, multiply together the numerator and denominator; then the root of this product being made the numerator to the denominator of the given fraction, or the denominator to the numerator of it, will give the root required.

4. Mixed numbers may be reduced to improper fractions and the roots then extracted; or, the vulgar fraction may be reduced to a decimal, then joined to the integer, and the root of the whole extracted.

EXAMPLES.

1. Required the square root of $\frac{4}{25}$.

Here, $\frac{\sqrt{4}}{\sqrt{25}}=\frac{2}{5}$, the Ans.

2. Required the cube root of $\frac{4}{7}$.

Here $\frac{4}{7}=.571428+$, and $\sqrt[3]{.571428}=.829+$ Ans.

3. Required the square root of $2\frac{1}{4}$.

Here $2\frac{1}{4}=\frac{9}{4}$, and $\frac{\sqrt{9}}{\sqrt{4}}=\frac{3}{2}=1\frac{1}{2}$, the Ans.

Or, $2\frac{1}{4}=2.25$, and $\sqrt{2.25}=1.5$, Ans.

4. Required the square root of $\frac{4}{9}$. Ans. $\frac{2}{3}$.

5. Required the square root of $\frac{5}{7}$. Ans. .84515+

6. Required the cube root of $\frac{1}{27}$. Ans. $\frac{1}{3}$.

7. Required the 4th root of $7\frac{58}{81}$. Ans. $1\frac{2}{3}$.

PROPORTIONS AND PROGRESSIONS.

Proportion is *an equality of ratios,* and is either *arithmetical* or *geometrical.* Arithmetical proportion is an equality of arithmetical ratios, and geometrical proportion is an equality of geometrical ratios.* Thus, the numbers

* These two kinds of ratios have been treated of in a preceding part of this work. It may be proper to inform the learner that the *names* of these two kinds of ratios have been adopted arbitrarily, merely for the sake of distinction; ratios of both kinds being applicable both to Arithmetic and Geometry.

There are some other kinds of ratios sometimes mentioned, viz. the following:

A *compound ratio* is the ratio of the product of the corresponding terms of two or more simple geometrical ratios; as in Compound Proportion,

4, 6, 8, 10, are in *arithmetical* proportion, because the *difference* between 6 and 4 is the same as the difference between 10 and 8: And the numbers 2, 6, 4, 12, are in *geometrical* proportion, because the *quotient* of 6 divided by 2, is the same as the quotient of 12 divided by 4.

To denote numbers as being geometrically proportional, a colon is set between the two terms of each couplet, to denote their ratio; and a double colon, or else the sign of equality, between the couplets. We may also denote arithmetical proportionals by separating the couplets in the same manner, and writing a colon, turned horizontally, between the terms of each couplet. So, the geometrical proportionals 4, 2, 6, 3 may be writen thus, 4 : 2 :: 6 : 3, or thus, 4 : 2=6 : 3; and the arithmeticals 5, 2, 7, 4, thus, 5 .. 2 :: 7 .. 4, or thus, 5 .. 2=7 .. 4.

Proportion is distinguished into *continued*, and *discontinued*. When the difference, or the ratio, of the consequent of one couplet and the antecedent of the next couplet, is not the same as the common difference or ratio of the couplets, the proportion is said to be discontinued. So, 2, 5, 6, 9, are in discontinued arithmetical proportion, because 5—2=9—6=3, whereas 6—5=1: and 2, 4, 12, 24, are in discontinued geometrical proportion, because 4÷2 =24÷12=2, but 12÷4=3, which is not the same ratio. But, when the difference, or ratio, of every two succeeding terms is the same quantity, the proportion is said to be continued, and the numbers themselves make a series of continued proportionals, commonly called a *progression*. So, 2, 4, 6, 8, form an arithmetical progression, because the terms

---

or the Double Rule of Three. Thus, the ratio of 12 : 4 is 3, and the ratio of 6 : 3 is 2: The ratio compounded of these is 12×6 : 4×3, or 72 : 12, equal to 6.

The ratio of the *squares* of any two quantities, is called the *duplicate ratio* of the quantities; the ratio of the *cubes* of the quantities is called a *triplicate ratio*; the ratio of their *4th powers*, a *quadruplicate ratio*, &c. Also, the ratio of the *square roots* of two quantities is called a *subduplicate ratio*, and the ratio of their *cube roots* a *subtriplicate ratio*, &c.

Thus, the duplicate ratio of 4 to 2, is the ratio of $4^2 : 2^2$, or of 16 : 4, equal to 4; and the triplicate ratio of 4 to 2 is $4^3 : 2^3$, or 64 : 8, equal to 8, &c. Also, the subduplicate ratio of 4 to 2 is $\sqrt{4} : \sqrt{2}$, and the subtriplicate ratio of 4 to 2 is $\sqrt[3]{4} : \sqrt[3]{2}$, &c.

have all the same common difference; that is, $4-2=6-4=8-6=2$: And 2, 4, 8, 16, form a geometrical. progression, because $4\div2=8\div4=16\div8=2$, all the same ratio.

When the succeeding terms of a progression increase, or exceed each other, it is called an *ascending* progression or series; but, when the terms decrease, the series is said to be a *descending* one. So, 1, 2, 3, 4, &c. is an ascending arithmetical progression or series; but 9, 7, 5, 3, 1, is a descending arithmetical progression. Also, 1, 2, 4, 8, 16, &c. is an ascending geometrical series; and 16, 8, 4, 2, 1, &c. is a descending one.

The first and last terms of any series, are called the *extremes*, and the other terms, the *means*.

ARITHMETICAL PROPORTION AND PROGRESSION.

In Arithmetical Progression the terms have all the same *common difference.*

The most useful part of Arithmetical Proportions is contained in the following theorems.

THEOREM 1.—When four quantities are in arithmetical proportion, the sum of the extremes is equal to the sum of the two mean terms. Thus, of the four, 2, 4, 7, 9; here $2+9=4+7=11$.

THEOREM 2.—In any continued arithmetical progression, the sum of the two extremes is equal to the sum of any two means which are equally distant from the extremes, or equal to double the middle term when there is an uneven number of terms. Thus, in the progression 1, 3, 5, it is $1+5=3+3=6$; and in the series 2, 4, 6, 8, 10, we have $2+10=4+8=6+6=12$.

THEOREM 3.—The difference between the extreme terms of an arithmetical progression, is equal to the common difference of the series multiplied by the number of terms less 1. So, of the six terms, 2, 4, 6, 8, 10, 12, the common difference is 2, and 1 less than the number of terms is 5; then, the difference of the extremes is $12-2=10$, and $2\times5=10$ also.

Consequently, the greatest term is equal to the least term added to the product found by multiplying the common difference by the number of terms less 1.

THEOREM 4.—The sum of all the terms in any arithmetical progression, is equal to half the sum of the extremes multiplied by the number of terms; or the sum of the extremes multiplied by the number of the terms, gives double the sum of the series. This is made evident by setting the terms of any arithmetical progression in an inverted order, under the same series in a direct order, and then adding the terms together in that order.

Thus, in the series 1, 3, 5, 7, 9, 11;
do. inverted 11, 9, 7, 5, 3, 1;

the sums are 12+12+12+12+12+12, which must be double the sum of the single series, and is evidently equal to the sum of the extremes repeated as often as the number of terms.

From the foregoing theorems may be found any *one* of these *five* parts, or things, viz. the *least* term, the *greatest* term, the *number* of terms, the *common difference* of the terms, and the *sum* of the progression, when any *three* of them are given; as in the following

PROBLEMS IN ARITHMETICAL PROGRESSION.

PROBLEM I.—*The extremes and the number of terms being given, to find the sum of all the terms of the progression.*

RULE.—Multiply the sum of the extremes by the number of terms, and half of the product will be the sum of the series: Or, multiply one of those factors by half of the other, and the product will be the answer.

EXAMPLE 1. If the least term of an arithmetical progression be 4, the greatest term 18, and the number of terms 8, what is the sum of the series?

Here 4+18=22, the sum of the extremes: then, 22×8 =176, and 176÷2=88, the Ans.

2. If the extremes be 1 and 312, and the number of terms 15, what is the sum of the series? Ans. 2347½.

3. How many strokes does the hammer of a common clock strike in 12 hours? Ans. 78.

4. What is the sum of the first 100 numbers?

Ans. 5050.

Problem II.—*When one of the extremes, the common difference, and the number of terms are given, to find the the other extreme.*

Rule.—Multiply the number of terms, less 1, by the common difference, and the product will be the difference of the extremes; which being added to the less extreme will give the greater, or subtracted from the greater, will give the less.

Note.—In the same manner may be found any one of the mean terms, when the common difference, one of the extremes, and the number of the required term, (reckoning from the given term or extreme,) are known.

Example 1. The least term of an arithmetical progression is 4, and the common difference is 2; what is the 8th term of the series?

Here the number of terms less 1, is 7: therefore, $7\times2=14$, and $14+4=18$, the 8th term, Ans.

2. If the greatest term be 70, the common difference 3, and the number of terms 21, what is the least term?

Ans. 10.

3. A man puts out $1, at 6 per cent. simple interest, which, in 1 year, amounts to $1.06, in 2 years to $1.12, in 3 years to $1.18, and so on, in arithmetical progression, with a common difference of $.06: what would be the amount in 20 years?

It is obvious that the yearly amounts of any sum, at simple interest, form an arithmetical progression, of which the *principal* is the *first* or *least term*, the *last amount* is the *last* or *greatest* term, the *yearly interest* is the *common difference*, and the *number of years* is 1 less than the *number* of terms. Hence the foregoing question is solved thus: $20\times.06=\$1.20$, and $\$1.20+1=\$2.20$ Ans.

4. Suppose 100 oranges were laid in a row on the ground, 2 yards distant from each other, in a right line, and a basket placed two yards from the first orange; what length of ground must that boy travel over who gathers them up singly, returning with them one by one to the basket?

Ans. 11 miles, 3 fur. 32 rd. 4 yd.

In the last question, the first, or least term, is 4 yards, the common difference is 4 yards, and the number of terms is 100: the *sum* of this series is the answer.

Problem III.—*The extremes and the number of terms being given, to find the common difference.*

Rule.—Divide the difference of the extremes by the number of terms less 1, and the quotient will be the common difference.

Example 1. If the extremes be 4 and 18, and the number of terms 8, what is the common difference of the terms?

Here 7=the number of terms less 1, and 18—4=14=the difference of the extremes: then, 14÷7=2, the common difference of the terms, Ans.

2. A man puts out $100, at simple interest, for 10 years, and receives at the end of said time $160, for the amount of principal and interest: what was the yearly interest, or rate per cent.?

Here the extremes are $100 and $160, and the number of terms is 11; and hence, the *common difference*, or *yearly interest*, is $6, Ans.—*See Ex.* 3d, *Prob.* II.

Problem IV.—*The extremes and the common difference of the terms being given, to find the number of terms.*

Rule.—Divide the difference of the extremes by the common difference of the terms, and the quotient, increased by 1, will be the answer.

Ex. 1. If the extremes be 4 and 18, and the common difference 2, what is the number of terms?

Here 18—4=14, the difference of the extremes; then, 14÷2=7, and 7+1=8, the Ans.

2. In what time will $100 amount to $160, at 6 per cent. per annum, simple interest? Ans. 10 years.

See Ex. 2. *Prob.* III.

GEOMETRICAL PROPORTION AND PROGRESSION.

In Geometrical Progression, the terms have all the same *multiplier*, or *divisor*.

The most useful part of Geometrical Proportion is contained in the following theorems; which are similar to those in Arithmetical Proportion, only *multiplication* and *division* are here used, instead of *addition* and *subtraction*.

THEOREM 1.—When four quantities are in geometrical proportion, (either *continued* or *discontinued*,) the product of the two extremes is equal to the product of the two means. Thus, in the four 2, 4, 3, 6, it is $2\times6=4\times3=12$.

And hence, if the product of the two mean terms be divided by either of the extremes, the quotient will be the other extreme; which is the foundation and reason of the practice in the Rule of Three.

THEOREM 2.—In any continued geometrical progression, the product of the two extremes is equal to the product of any two means that are equally distant from them, or equal to the second power of the middle term when there is an unequal number of terms. Thus, in the series 2, 4, 8, 16, 32, it is $2\times32=4\times16=8\times8=64$.

THEOREM 3.—The quotient of the extreme terms of any geometrical progression, is equal to the common ratio of the series raised to that power whose index is 1 less than the number of terms. Consequently the greatest term is equal to the said power multiplied by the least term. So, of the five terms, 2, 4, 8, 16, 32, the ratio is 2, and 1 less than the number of terms is 4; then, the quotient of the extremes is $32\div2=16$, and the 4th power of the ratio is $2\times2\times2\times2=16$ also.

THEOREM 4.—The sum of all the terms in any geometrical progression, is equal to the greatest term added to the quotient found by dividing the difference of the extremes by the ratio less 1. Thus, the sum of the terms 2, 4, 8, 16, 32, (whose ratio is 2,) is $=32+\frac{32-2}{2-1},=32+30,=62$.

From the foregoing theorems may be found any *one* of these *five* things, the *least* term, the *greatest* term, the *number* of terms, the *common ratio* of the terms, and the *sum of the progression*, when any *three* of them are given; as in the following

PROBLEMS IN GEOMETRICAL PROGRESSION.

PROBLEM I.—*The extreme terms of a geometrical progression, and the common ratio of the terms being given, to find the sum of the series.*

RULE.—Divide the difference of the extremes by the ratio less 1; add the quotient to the greater extreme, and the amount will be the sum of the series.

Example 1. If the least term of a geometrical progression be 2, the greatest term 486, and the ratio 3, what is the sum of all the terms?

Here $(486-2)\div(3-1)=242$, and $242+486=728$, the Ans.

2. If the extremes be 1 and 15625, and the ratio 5, what is the sum of the series? Ans. 19531.

PROBLEM II.—*When one of the extremes, the ratio, and the number of terms are given, to find the other extreme.*

RULE.—Raise the ratio to that power whose index is 1 less than the number of terms: then, multiply this power by the less extreme, to find the greater; or divide the greater extreme by the said power, to find the less.

Note 1.—In the same manner any one of the *means* may be found; only, in this case, the ratio must be raised to that power whose index is less by 1 than the *number* of the *required term*, reckoning from the given term or extreme.

Note 2.—When it is necessary to find a high power of a ratio, it will be convenient to proceed as follows; viz. Write down, in one line, a few of the lower or leading powers of the ratio, and place their indices over them. Add together such indices whose sum shall be equal to the index of the required power; then multiply together the powers belonging to those indices, and their product will be the power required.

Example 1. If the first, or least term, of an ascending geometrical progression be 4, and the ratio 2; what is the 15th term?

I first find the 14th power of the ratio, as follows :—

         +   + +
1, 2, 3, 4, 5, 6, indices.
2, 4, 8, 16, 32, 64, powers of the ratio.
Here $6+5+3=14=$the no. of terms less 1.
And $64\times32\times8=16384$, the 14th power of the ratio.
Then, $16384\times4=65536$, the 15 term, Ans.

N. B. Those indices which are marked with the sign + are added together.

2. If the greatest term of a geometrical progression be 177147, the ratio 3, and the number of terms 11, what is the least term?

1, 2, 3, 4, 5, indices.
3, 9, 27, 81, 243, leading powers of the ratio.

Here $5+5$, or double the index of the 5th power,$=10$, $=$the number of terms less 1; and 243×243, or the square of the 5th power,$=59049$, the 10th power of the ratio. Then, $177147\div59049=3$, the least term, Ans.

3. If the posterity of Noah, which consisted of 6 persons at the time of the flood, increased so as to double their number in 20 years, how many inhabitants were there in the world 2 years before the death of Shem, who died 502 years after the flood? Ans. 100663296.

In the preceding question, the first term is 6, the ratio 2, and the number of terms$=(502-2)\div20,=500\div20=25$: the last term of the series is the answer.

4. Required the 6th term of a descending geometrical progression, whose first or greatest term is 486, and ratio $1\frac{1}{2}$, or 1.5 Ans. 64.

5. What is the amount of $20 for 5 years, at 6 per cent. per annum, compound interest?

Compound interest is that which arises from adding the interest to the principal at the end of each year, and making use of the amount for a new principal; and it is obvious that these several amounts form a geometrical progression, of which the *given principal* is the *least term*, the *number of years* the *number of terms*, the *amount of $1 for one year* the *ratio*, and the *last amount* the *last term*. Thus, in the foregoing question, the least term is $20; the number of terms is 5, and the ratio is $1.06,=the amount of $1 for 1 year; and hence, the answer is $1.06^4\times20=1.26247696\times20=$ $25.2495392

PROBLEM III.—*One of the extremes, the ratio, and the number of terms being given, to find the sum of the series.*

RULE.—Find the other extreme, by Problem II., and then find the sum of the series by Prob. I.

Or, if the less extreme be given, you may find the sum of the series without finding the greater extreme; viz. as follows: Raise the ratio to that power whose index is equal to the number of terms; from which subtract 1; then multiply the remainder by the less extreme; divide the product by the ratio less 1, and the quotient will be the sum of the series.

Ex. 1. The least term of a geometrical progression is 4, the ratio is 2, and the number of terms 18: what is the sum of the series?

Operation by the second method.

1, 2, 3, 4, 5, 6, indices.
2, 4, 8, 16, 32, 64, powers of the ratio.

Here $6+6+6=18$=the number of terms; and $64\times64\times64=262144$, the 18th power of the ratio.

Then, $\frac{(262144-1)\times4}{2-1}=\frac{1048572}{1}=1048572$, Ans.

2. A gentleman, whose daughter was married on a new-year's day, gave her a dollar towards her portion, promising to triple the sum on the first day of every succeeding month in that year: what was the amount of her portion? Ans. $265720.

3. What would 30 yards of broadcloth amount to at 1 pin for the first yard, 2 pins for the second, 4 for the third, &c. supposing the pins to be sold at the rate of 50 for a cent? Ans. $214748.3646

4. A thresher wrought 20 days for a farmer, and by agreement, was to receive 4 kernels of wheat for the first day's work, 12 for the second, 36 for the third, and so on: What did the 20 day's labor amount to, supposing 491520 wheat corns to make a bushel,* and each bushel to be worth $1.25? Ans. $17733.75, rejecting the fractional part of a bushel.

* 7680 wheat or barley corns are supposed to make a pint, or 491520 corns a bushel.

5. If the greatest term of a geometrical progression be 19531250, the ratio 5, and the number of terms 10, what is the sum of the series? Ans. 24414060.

Note.—When the series is a *decreasing one*, and the *number of terms is infinite;* multiply the greatest term by the ratio; divide the product by the ratio less 1, and the quotient will be the sum of the infinite series.

6. Required the value of the interminate decimal .666, &c.

The decimal fraction .666, &c. is evidently equal to the series $\frac{6}{10}+\frac{6}{100}+\frac{6}{1000}$, &c. continued without limit, the ratio of which is 10. Therefore, multiplying $\frac{6}{10}$ (the greatest term in the series) by 10, and then dividing the product by 9, (the ratio less 1,) gives $\frac{60}{90}=\frac{2}{3}$, Ans.

7. Fnd the sum of the infinite series whose greatest term is 100, and ratio 5. Ans. 125.

8. Find the sum of the infinite series 1, $\frac{1}{2}$, $\frac{1}{4}$, $\frac{1}{8}$, &c., the ratio of which is 2. Ans. 2.

9. Supposing a man to walk 20 miles the first day, 19 miles the second, $18\frac{1}{20}$ miles the third, &c., decreasing each day's journey by a ratio of $1\frac{1}{19}$; would he ever arrive at a city 500 miles distant from the place he set out from, if it were possible for him to travel through an infinity of ages and never stop?

Ans. No; the utmost distance he could travel would be but 400 miles.

PERMUTATIONS AND COMBINATIONS.

The *Permutation of Quantities* is the showing how many different ways the order of any given number of things may be varied or changed. This is also called *Variation, Alternation,* or *Changes;* and the only thing to be regarded here, is the order they stand in; for no two parcels are to have all their quantities placed in the same order.

The *Combination of Quantities* is the showing how often a less number of things can be taken out of a greater, and combined together, without considering their places, or the order they stand in. This is sometimes called *Elec-*

tion, or *Choice*; and here every parcel must be different from all the rest, and no two are to have precisely the same quantities or things.

The *Composition of Quantities*, is the taking of a given number of quantities out of as many rows of different quantities, one out of every row, and combining them together. Here no regard is had to their places; and it differs from Combination only as that admits of but *one* row of things.

PROBLEM I.—*To find the number of permutations, or changes, that can be made of any given number of things, all different from each other.*

RULE.*—Find the contined product of the terms in the natural series of numbers, 1, 2, 3, 4, &c. up to the given number of things, and the result will be the answer required.

Example 1. How many changes can be made of the first three letters of the alphabet?

Here the given number of things is 3, and therefore 1×2×3=6, the Ans.

Proof,

1	*a b c*
2	*a c b*
3	*b a c*
4	*b c a*
5	*c b a*
6	*c a b*

2. How many different positions can 7 persons place themselves in at a table? Ans. 5040.

3. How many changes may be rung on 12 bells, and how long would they be in ringing but once over, supposing 15 changes might be rung in one minute?

Ans. The number of changes is 479001600, and the time is 60 Julian years and 261 days.

PROBLEM II.—*Any number of things being given, to find how many changes can be made out of them by taking any given number of quantities at a time.*

RULE.—Take a series of numbers consisting of as many

* The reason of this rule may be shown thus: Any one thing, *a*, is capable of one position only, as *a*. Any two things, *a* and *b*, are capable of two variations only, *ab*, *ba*; the number of which is expressed by 1×2. If there be three things, *a*, *b*, and *c*; then any two of them, leaving out the third, will have 1×2 variations; and consequently when the third is taken in, there will be 1×2×3 variations. If there be four things; then every three of them, without the fourth, will have 1×2×3 variations; and hence the four will admit of 1×2×3×4 variations; and so on, as far as we please.

terms as there are quantities to be taken at a time; beginning at the number of things given, and decreasing continually by 1; and the product of all the terms will be the answer.

Example 1. How many changes may be rung with 3 bells out of 8?

Here the series must begin with the number 8, and consist of three terms: Thus, $8\times7\times6=336$, Ans.

2. How many words can be made with 5 letters out of the 26 in the alphabet, admitting that a number of consonants alone will make a word? Ans. 7893600.

Problem III.—*Any number of things being given, whereof there are several things of one sort, and several of another, &c., to find how many changes can be made out of them all.*

Rule.—1. Take the series $1\times2\times3\times4$, &c. up to the number of things given, and find the product of all the terms.

2. Take the series $1\times2\times3$, &c. up to the number of the given things of the *first* sort, and the series $1\times2\times3$, &c. up to the number of the given things of the *second* sort, &c.

3. Divide the product of all the terms of the first series by the joint product of all the terms of the remaining ones, and the quotient will be the answer.

Example 1. How many different numbers can be made of the following figures, 11222445?

Here the number of things, or figures, is *eight;* there being *two* 1's, *three* 2's, *two* 4's, and *one* 5. Therefore,

$1\times2\times3\times4\times5\times6\times7\times8=40320$, the *dividend.*

1×2(*two* terms, = the no. of 1's) = 2
$1\times2\times3$(*three* terms, = the no. of 2's) = 6
1×2(*two* terms, = the no. of 4's) = 2
1(*one* term, = the no. of 5's) = 1
} The product of these numbers, viz. $2\times6\times2\times1=24$, is the *divisor.*

Then, $40320\div24=1680$, Ans.

2. How many variations can be made of the letters in the word *Bacchanalian?* Ans. 4989600.

Problem IV.—*To find the number of combinations of any given number of things, all different from each other, taken any given number at a time.*

Rule.—1. Take the series 1, 2, 3, 4, &c. up to the numbe to be taken at a time, and find the product of all the terms.

2. Take a series of as many terms, decreasing by 1, from the given number, out of which the election is to be made, and find the product of all the terms.

3. Divide the latter product by the former, and the quotient will be the number sought.

Note.—The operations may often be very much shortened, by the method of contraction made use of in Compound Proportion.

Example 1. How many combinations can be made of 6 letters out of 10?

Operation according to the Rule.

$1\times2\times3\times4\times5\times6=720$ divisor.

$10\times9\times8\times7\times6\times5=151200$ dividend.

Then, $151200\div720=210$ Ans.

Contracted thus: $\dfrac{10\times\overset{3}{9'}\times\overset{2'}{8'}\times7\times6'\times5'}{1\times2'\times3'\times4'\times5'\times6'}=210$ Ans.

2. A butcher bargained with a farmer for a dozen sheep, (at $2 per head,) which were to be picked out of 2 dozen; but being long in choosing them, the farmer told him that if he would give him a cent for every different dozen which might be chosen out of the 24, he should have the whole; to which the butcher readily agreed: What did the sheep cost him? Ans. $27041.56

3. How many locks, whose wards differ, may be unlocked with a key of 6 several wards?

Ans. 63 locks; 6 of which may have one single ward, 15 double wards, 20 triple wards, 15 four wards, 6 five wards, and 1 lock 6 wards.

To solve the last question, find how many combinations can be made of 1, of 2, of 3, &c. to 6 things, out of 6, and the several results will be the number of locks of 1, 2, 3, &c. wards, respectively.

Problem V.—*To find the compositions of any number, in an equal number of sets, the things themselves being all different.*

RULE.—Multiply the number of things in every set continually together, and the product will be the answer.

Ex. 1. Suppose there are 5 companies, each consisting of 9 men; it is required to find how many different ways 5 men may be chosen, one out of each company?

Here we must multiply together as many 9's as there are companies, thus $9\times9\times9\times9\times9=59049$ Ans

2. In how many ways may a man, a woman, and a child, be chosen out of three companies, consisting of 5 men, 7 women and 9 children?

Here $5\times7\times9=315$, the Ans.

POSITION,

Is a method of solving certain kinds of arithmetical questions which cannot be resolved by the common direct rules. It is sometimes called *False Position*, or *False Supposition*, because it makes a supposition of *false* numbers, to work with them as though they were the true ones, and by their means discovers the true numbers sought. It admits of two varieties, viz. *Single Position* and *Double Position*.

In Single Position, the answer to a problem is obtained by *one* assumption or supposition; in Double Position it is obtained by *two*.*

*The rules of Position are applicable to questions, in which the operations to be performed on the numbers sought will not produce any *powers* or *roots* of the said numbers; such, for example, as the questions usually brought to exercise the reduction of simple equations in Algebra.

Those questions in which the results are proportional to their suppositions belong to Single Position: such are those which require the multiplication or division of the number sought by any known number; or in which it is to be increased or diminished, by the addition or subtraction of itself, or of any parts of itself, any given number of times, Questions in which the results are not proportional to their positions, cannot be solved by the Rule of Single Position.

Questions in which different suppositions will give results whose difference is proportional to the difference of the suppositions, may be solved by the Rule of Double Position. This includes all kinds of questions which can be solved by Single Position; and also many others, in which the results are not proportional to their suppositions, viz. those in which the numbers sought are to be increased or diminished, by the addition or subtraction of some given number, which is no *known part* of the number required.

SINGLE POSITION.

RULE.

Assume or take any convenient number, and perform on it the operations mentioned in the question as being performed on the required number: Then, if the result thus obtained be the same with that in the question, the assumed number is the number sought; but if it be not, say, as the result of the operation : is to the true result given in the question :: so is the assumed number : to the number required.*

EXAMPLES.

1. It is required to find a number, to which if one-half, one-third, and one-fourth of itself be added, the sum will be 1500?

Suppose the required number to be	60	*Proof.* The answer=	720
$\frac{1}{2}$ of 60 =	30	$\frac{1}{2}$ of 720 =	360
$\frac{1}{3}$ of 60 =	20	$\frac{1}{3}$ of do. =	240
$\frac{1}{4}$ of 60 =	15	$\frac{1}{4}$ of do. =	180
Result,	125	Sum,	1500

Then, as 125 : 1500 :: 60 : 720, the Ans.

2. A person having about him a certain number of dollars, said that $\frac{1}{2}$, $\frac{1}{3}$, and $\frac{1}{4}$ of them would make 95: How many had he? Ans. $100.

3. A person after spending $\frac{1}{2}$ and $\frac{1}{7}$ of his money, has $50 left: What sum had he at first? Ans. $140.

4. What number is that, which being multiplied by 7, and the product divided by 12, the quotient will be 21? Ans. 36.

5. Seven-eighths of a certain number exceeds four-fifths of it by 6: What is that number? Ans. 80.

6. Divide $440 between A, B, and C, giving B as much

*The results in Single Position, being *proportional to their positions*, the reason of the Rule is obvious. Every question that can be resolved by this rule, may also be resolved by the rule of Double Position; and therefore the latter rule may be employed when it is doubtful whether the question belongs to Single or to Double Position.

as A, and a fifth part more, and C as much as A and B together. Ans. A's part is $100, B's $120, C's $220.

7. Lent a certain sum of money to a friend, to receive simple interest for the same, at the rate of 6 per cent. per annum; and at the end of 10 years I received for principal and interest $400: What was the sum lent?

Ans. $250.

8. Joe Strickland drew a prize in a lottery; the value of which was just 3000 times the price of the ticket. The money he paid for the ticket, if put out at simple interest, at 5 per cent., would not amount to a sum equal to the prize short of 59980 years: Required the price of the ticket, and the value of the prize.

Ans. The prize was $30000, and the ticket cost $10.

DOUBLE POSITION.

RULE.

Assume two different numbers, and perform on them, separately, the operations indicated in the question: Then say, as the difference of the results thus obtained : is to the difference between either of those results and the true result given in the question :: so is the difference of the two assumed numbers : to the difference between the assumed number which gave the result used in the second term of the proportion and the true number required; which difference, (or correction,) being added to that assumed number when it is too little, or subtracted from it when it is too great, will give the number sought.

Note 1.—The results will commonly be too great when the assumed numbers are too great, and too small when the assumed numbers are too small; but sometimes the reverse of this will be true, viz. when the greater of the two assumed numbers gives a less result than the other. When both the results happen to be the same number, (the assumed numbers being one greater and the other less than the true number,) then half the sum of the two assumed numbers will be the number sought.

EXAMPLES.

1. Required to find a number from which if 2 be subtracted, one-third of the remainder will be 16 less than the required number.

First, suppose the required number to be 20, from which take 2, and one-third of the remainder is 6. This being taken from 20, (the assumed number,) the remainder is 14, the *first result*. This result is less than 16, the result in the question, and therefore the assumed number, 20, is less than the number sought.

Secondly, suppose the required number to be 26, from which take 2, and one-third of the remainder is 8, which being taken from 26, the remainder is 18, the *second result*, which is too great. Then,

18—14=4, difference of the first and second results.

18—16=2, difference between the 2d result and the true result given in the question.

26—20=6, difference of the two assumed numbers.

Then, as 4 : 2 :: 6 : 3, the correction, or the difference between the 2d assumed number and the true one; which being subtracted from the said assumed number, the remainder, 23, is the answer required.

2. Suppose the old sea-serpent's head is 10 feet long, and his tail is as long as his head and half the length of his body, and his body is as long as his head and tail; what is the whole length of the monster?

First, suppose the length of his body to be 30 feet; then, adding half of 30 to 10 feet, (the length of his head,) gives 25 feet for the length of his tail. Then, 25+10—30=5, the *first result*, being the number of feet, the supposed length of his body is less than the sum of the lengths of his head and tail. According to the question this result should be 0, and therefore the length of his body must exceed 30 feet.

Secondly, by supposing the length of his body to be 34 feet, and proceeding as before, the *second result* is 3.

Then, as 5—3 : 5—0 :: 34—30 : 10, the correction, to be added to the first assumed number. Therefore the length of his body is 30+10=40 feet, the half of which being added to 10 feet, the length of his head, gives 30 feet for the length of his tail; and consequently his whole length is 10+40+30=80 feet, Ans.

3. Divide $100 among four men, A, B, C and D, so that B may have $4 more than A, C $8 more than B, and D twice as many dollars as C.

Ans. A's share is $12, B's $16, C's $24, D's $48.

4. Two men, A and B, lay out equal sums of money in trade; A gains $126, and B loses $87, and A's money is now double to B's: What sum did each lay out?

Ans. $300.

5. If one person's age be now only 4 times as great as another's, though 7 years ago it was 6 times as great, what is the age of each?

Ans. The age of the younger person is $17\frac{1}{2}$ years, and that of the elder 70 years.

6. A laborer hired to a farmer on the following conditions; viz. for every day he wrought he should receive 50 cents, and for every day he was idle he should pay 20 cents for his boarding. At the end of 50 days he received $14.50, which was then his due. How many days did he work?

Ans. 35.

7. A person being asked the time of day, replied, that the time past noon was exactly equal to $\frac{2}{7}$ of the time to midnight: Required the time of day?

Ans. 40 minutes past 2, P. M.

Note 2.–The foregoing Rule of Double Position depends upon the principle, that the differences between the true and the assumed numbers are proportional to the differences between the result given in the question and the results arising from the assumed numbers. This principle is quite correct in relation to questions in which the numbers sought do not ascend above the first power, but not in relation to any others; and hence, when applied to others, it does not give the *exact* answers. The answers to such questions may, however, be found, or approximated to any assigned degree of exactness, by the following

RULE,

For solving, by approximation, mathematical questions in which the unknown quantities rise above the first power.

Find, by trial, two numbers as near the required number

as can conveniently be found; then, taking these for the two assumed numbers, proceed according to the foregoing Rule of Double Position, and a near approximation to the true answer will be obtained. If, on trial, the answer, thus obtained, should not prove to be sufficiently exact, then, to approximate the required number still more nearly, assume for a second operation, the answer found by the first, and that one of the first two assumed numbers which was nearest the answer, or any other number that may be found nearer, and by proceeding as before an answer will be obtained more correct than the former. In this way, by repeating the process as many times as may be necessary, the true answer may be obtained, or approximated to any degree of exactness required. Each process of approximation will commonly double the number of correct figures in the answer; that is, the answer obtained by each process will contain about twice as many correct figures as there are in each of the assumed numbers.

This rule will serve for extracting the roots of numbers, and for resolving most kinds of compound algebraical equations.

EXAMPLES.

1. Required a number, to which if twice its square be added, the sum will be 100. Ans. 6.82542+

It is easy to see that the required number must be between 6 and 7. These numbers being assumed, therefore, the sum of 6 and twice its square is 78, and the sum of 7 and twice its square is 105. Then, as 105—78 : 105—100 :: 7—6 : .18, the correction; which being taken from 7, the remainder, 6.82, is the required number, nearly. To this number let twice its square be added, and the result is 99.8448, which is less than 100, the true result.

In order to approximate the required number still more nearly, let 6.82 and 7 be the assumed numbers, which give the results 99.8448 and 105; then, as 105—99.8448 : 105—100 :: 7—6.82 : .17458, the correction; which being taken from 7, the remainder is 6.82542, the required number still more nearly; and if the operation were repeated by taking this and the former approximate answer for the assumed numbers, the required number would be found true for eight or nine figures.

2. Required a number, to which if twice its square and three times its cube be added, the sum will be 2000.

Ans. 8.506744+

3. Required to find the cube root of 434 by approximation.

It is easily found, by trials, that the required root lies between 7 and 8. Assuming these two numbers, their results, or cubes, are 343 and 512. Then, as 512—343 : 434—343 :: 8—7 : .53, which being added to 7, gives 7.53 for the root. The cube of 7.53 is 426.957777, which being less than 434, shows that the required root exceeds 7.53. Then, in order to approximate the required root still more nearly, I take 7.53 and 7.54 for the assumed numbers, and proceed as before, and I get 7.5713 for the root; the first four figures of which are correct.

4. A butcher bought a certain number of oxen for $1000. If he had received 5 more oxen for the same money, he would have paid $10 less for each. What was the number of oxen?

Ans. 20.

ALLIGATION,

Teaches how to compound, or mix together, several simples or ingredients, of different qualities or prices, so that the composition may be of some intermediate quality or rate. It consists of two kinds, viz. *Alligation Medial*, and *Alligation Alternate*.

ALLIGATION MEDIAL,

Is the method of finding the mean rate or quality of the composition, from having the quantities and the rates or qualities of the several simples given; which is performed by the following

RULE.

As the sum of the several quantities :
Is to any part of the composition ::
So is their total value :
To the value of the said part.

EXAMPLES.

1. If 6 gallons of wine worth 67 cents a gallon, 7 gallons worth 80 cents a gallon, and 5 gallons worth $1.20 a gallon, be mixed together, what will 2 gallons of the mixture be worth?

Operation.

6 gallons	at 67 cents	=$4.02
7 do.	at 80 do.	= 5.60
5 do.	at 120 do.	= 6.00
18 gallons	are worth	$15.62

Then, as 18gal. : 2gal. :: $15.62 : $1.735+ Ans.

2. A farmer mixed together 15 bushels of rye worth 64 cents a bushel, 18 bushels of Indian corn worth 55 cents a bushel, and 21 bushels of oats worth 28 cents a bushel. I demand the value of a bushel of this mixture?

Ans. 47 cents.

3. A goldsmith melted together 6 oz. of gold of 22 carats fine, 4oz. of 21 carats fine, and 2 oz. of 18 carats fine: I demand the quality or fineness of the composition?

Ans. 21 carats fine.

4. If 5 gallons of spirits at 50 cents a gallon, 2 gallons at $1 a gallon, and 3 gallons of water at 0 a gallon, be mixed together, what will a gallon of the mixture be worth?

Ans. 45 cents.

5. A gallon of proof spirits weighs about 7.73lb. Avoirdupois, and a gallon of clear fresh water about 8.36lb. If 8 gallons of proof spirits, and 4 gallons of water should be mixed together, what would be the weight of one gallon of the mixture? Ans. 7.94lb.

ALLIGATION ALTERNATE,

Is the method of finding what quantities of any number of simples, or ingredients, whose rates are given, will compose a mixture of a given rate: So that it is the reverse of Alligation Medial, and may be proved by it.

CASE I.

When the mean rate of the whole composition, or mixture, and the rates of all the simples or ingredients are given, without any limited quantity, to find the quantity of each ingredient.

RULE.*

1. Reduce the several rates or values of the ingredients to the same denomination, when they are of different denominations; then write them in a column under each other, and place the mean rate, or price, (reduced to the same denomination,) at the left hand.

2. Consider which of the values of the ingredients are greater than the mean rate, and which less; and connect or link together, with a continued line, each *greater value* with one or more *less*, and each *less* with one or more *greater*.

3. See what the difference is between the value of each ingredient and the mean rate, and set down that difference opposite to each value with which such ingredient is connected. Then, if only one difference stands against any value, it will be the quantity belonging to that value; but if more than one, their sum will be the quantity.

Note.—Different modes of linking will produce as many different answers.

---

* *Demonstration.*—When there are only two simples; then, by connecting the less rate with the greater, and placing the differences between them and the mean rate alternately, or one after the other in turn, the quantities resulting are such, that there is precisely as much gained by one quantity as is lost by the other, and therefore the gain and loss, upon the whole, are equal, and are exactly the proposed rate.

In like manner, let the number of simples be what it may, and with how many soever each one is linked, since it is always a less with a greater than the mean price, there will be an equal balance of loss and gain between every two, and consequently an equal balance on the whole. Q. E. D

Questions of this sort admit of a great variety of answers; for, after having found one answer by the above Rule, we may find as many more as we please, by only multiplying or dividing each of the quantities thus found, by 2, 3, 4, &c.: the reason of which is evident; for, if two quantities, of two simples, make a balance of loss and gain, with respect to the mean price, so must also the double or triple, the half or third part, or any other ratio of these quantities, and so on *ad infinitum*.

EXAMPLES.

1. How much oats at 30 cents a bushel, barley at 42cts., Indian corn at 48cts., and rye at 56cts., must be mixed together, so that the compound or mixture may be worth 46 cents a bushel?

Operation.

	cts.	bush.	
cts.	30	10 of oats.	Ans.
46	42	2 of barley.	
	48	4 of corn.	
	56	16 of rye.	

Proof.

bush.	cts.	cts.
10	at 30=	300
2	at 42=	84
4	at 48=	192
16	at 56=	896
32		)1472(46 cts.

Explanation.

1. I first set down, in a column, 30 cents, the value of the oats, 42cts., that of the barley, 48cts., that of the corn, and 56cts., that of the rye; and at the left hand of these I set down 46cts., the mean price, or value of the mixture.

2. I consider which values of the ingredients are greater, and which less than 46c., the value of the mixture, and connect them accordingly, each greater with one less, and each less with one greater; viz. 56, a greater, with 30, a less, and 48, a greater, with 42 a less.

3. I see what the difference is between the value of each ingredient and the mean price; and I find that the difference between 46 and 30 is 16, which I set down opposite to 56, the number with which 30 is connected; the difference between 46 and 42 is 4, which I set down opposite to 48; the difference between 46 and 48 is 2, which I set down opposite to 42; and the difference between 46 and 56 is 10, which I set down opposite to 30.

4. Now, as I have only one difference opposite to each value, these several differences are the quantities required; that is, there must be 10 bushels of oats, 2 of barley, 4 of corn, and 16 of rye.

I prove the answer by Alligation Medial.

Six more answers to the foregoing question may be found, by as many different methods of connecting together the values of the ingredients; viz. as follows:—

c.	c.	bush.	
c. 46	30	2 oats	2d Ans.
	42	10 bar.	
	48	16 corn	
	56	4 rye	

c.	c.	bush.	
c. 46	30	2+10=12 oats	3d Ans.
	42	10 =10 bar.	
	48	16 =16 corn	
	56	16+ 4=20 rye	

c.	c.	bush.	
c. 46	30	10 =10 oats	4th Ans.
	42	2+10=12 barley	
	48	4 = 4 corn	
	56	16+ 4=20 rye	

c.	c.	bush.	
c. 46	30	2+10=12 oats	5th Ans.
	42	2 = 2 barley	
	48	16+ 4=20 corn	
	56	16 =16 rye	

c.	c.	bush.	
c. 46	30	2 = 2 oats	6th Ans.
	42	2+ 10=12 barley	
	48	16+ 4=20 corn	
	56	4 = 4 rye	

c.	c.	bush.	
c. 46	30	2+10=12 oats	7th Ans.
	42	2+10=12 barley	
	48	16+ 4=20 corn	
	56	16+ 4=20 rye	

These seven answers are all that can be found by different modes of linking; but, by multiplying or dividing the quantities in each of these answers, successively, by the numbers 2, 3, 4, &c., we may find as many more answers as we please. Thus, by multiplying each of the quantities in the 7th answer by 2, we have 24 bushels of oats, 24 of barley, 40 of corn, and 40 of rye; which is an answer that will satisfy the conditions of the question: And, by dividing each of the quantities in the 7th answer by 2, we get 6 bushels of oats, 6 of barley, 10 of corn, and 10 of rye; which is another answer; and so on indefinitely.

2. A man would mix two sorts of rum, viz. at 95 cents, and at 1D. 10c. per gallon, with water at 0 per gallon, so that the mixture may be worth 90 cents per gallon: How much of each must the mixture contain?

	c.		gal.	
c.	95		90 of rum at 95cts. per gal.	
90	110		90 of rum at $1.10 do.	Ans.
	0	5+20=	25 of water.	

3. A merchant would mix wines, at 1D. 60c., 1D. 80c., and at 2D. 20c. per gallon, so that the mixture may be worth $2 per gallon: what quantity of each sort must he take?

Ans. 20 gallons at $1.60, 20gal. at $1.80, and 60 gal. at $2.20.

4. A goldsmith has gold of 16, of 18, of 23, and of 24 carats fine: what quantity of each must he melt together, to make a composition of 21 carats fine?

Ans. 3lb. or 3oz. &c. of 16, 2 of 18, 3 of 23, and 5 of 24 carats fine.—Six more answers may be found, by as many different modes of linking; each of which may be proved by Alligation Medial.

5. How much sugar, at 6 cents, at 8c., and at 13c. per lb., must be mixed together, so that the composition may be worth 9c. per lb.? Ans. 4lb. of each sort.

CASE II.

When one of the ingredients is limited to a certain quantity, to find the several quantities of the rest, in proportion to the quantity given.

RULE.

First, find, as in Case I., the quantities of the several ingredients, as though they were all unlimited. Then say, as the quantity, thus found, of that ingredient whose quantity is given in the question : is to the given quantity of said ingredient :: so is the quantity of each of the other ingredients found by Case I. : to the quantity required of each.

Note.—In the same manner questions may be wrought when several of the ingredients are limited to certain quantities, by finding first for one limit, and then for another, &c.—In this Case and the next following one, different modes of linking the values of the ingredients together, will produce different answers; but we cannot find more answers by multiplication and division, as in Case I.

EXAMPLES.

1. A farmer would mix 20 bushels of rye, worth 65 cents a bushel, with Indian corn, worth 50 cents, and oats worth 30 cents a bushel: how much corn and oats must he mix with the 20 bushels of rye, so that the whole mixture may be worth 40 cents a bushel?

Operation.

	c.		bush.
c.	65	10	=10 rye.*
40	50	10	=10 corn.
	30	25+10=35	oats.

Here I find the quantities of the several ingredients by Case 1st, as tho' they were all unlimited.

Then, as 10 : 20 :: { 10 bush. : 20 bush. of corn. / 35 bush. : 70 bush. of oats. } Ans.

Proof, by Alligation Medial.

20	bush. of rye, at 65cts.=	$13.00
20	do. of corn, at 50cts.=	10.00
70	do. of oats, at 30cts.=	21.00
110		$ 110)44.00(.40 mean rate.

2. How much gold of 15, of 17, and of 22 carats fine, must be mixed with 5oz. of 18 carats fine, so that the composition may be 20 carats fine?

Ans. 5oz. of 15, 5oz. of 17, and 25oz. of 22 carats fine.

3. A man being determined to mix 10 bushels of wheat, worth 48 cents a bushel, with rye worth 36c., barley worth 24c., and oats worth 12c. a bushel; I demand how much rye, barley, and oats, must be mixed with the 10 bushels of wheat, that the whole may be worth 28 cents a bushel?

1st Ans. $2\frac{1}{2}$ bush. of rye, 5 of barley, and $12\frac{1}{2}$ of oats.
2d Ans. 40 bush. of rye, 50 of bar., and 20 of oats.
3d Ans. 8 bush. of rye, 10 of bar., and 14 of oats.
4th Ans. $12\frac{1}{2}$ bush. of rye, 5 of bar., and $17\frac{1}{2}$ of oats.
5th Ans. 2 bush. of rye, 14 of bar., and 10 of oats.
6th Ans. 50 bush. of rye, 70 of bar., and 20 of oats.
7th Ans. 10 bush. of rye, 14 of bar., and 14 of oats.

* If the question had limited the quantity of rye to 10 bushels, the quantities of corn and oats found by Case 1st, (viz. 10 bushels of corn and 35 bushels of oats,) would have been the answer; but, since the quantity of rye is to be 20 bushels, it is evident that the quantities of corn and oats found by Case 1st, must, severally, be to the quantities required, as 10 to 20, or as 1 to 2.

N. B. These seven answers arise from as many different ways of linking the rates of the ingredients together.

4. How much linen, at 2s. and at 2s. 5d. a yard, must be taken with 140 yards at 3s. 4d. a yard, that the whole may be worth 2s. 6d. a yard, at an average?

Ans. 200 yards at 2s., and the same quantity at 2s. 5d. a yard.

5. How much land, worth $40 per acre, must be added to a farm containing 50 acres, 2 roods, 20 sq. rods, worth $70 per acre, to make the average value of the whole $50 per acre? Ans. 101 acres, 1 rood.

CASE III.

When the whole composition is limited to a given quantity, to find the quantity of each ingredient.

RULE.

First, find by Case I. the quantities of the several ingredients, as though they were all unlimited. Then say, as the sum of all the quantities thus found : is to the given quantity :: so is the quantity of each ingredient thus found : to the quantity required of each.

EXAMPLES

1. How much gold of 15, 17, and 22 carats fine, must be mixed together, to form a composition of 42oz. of 20 carats fine?

Operation.

20	15	2 oz. of 15 carats fine.	
	17	2 oz. of 17 do.	Here I find the
	22	5+3=8 oz. of 22 do.	proportional quantities by Case I.
		12 oz. the sum.*	

Then, as 12 : 42 :: 2 oz. : 7 oz. of 15 carats fine.
12 : 42 :: 2 oz. : 7 oz. of 17 do. Ans.
12 : 42 :: 8 oz. : 28oz. of 22 do.

* If the question had limited the whole composition to 12oz., the quantities of the several ingredients found by Case 1st, would have been the answer; but since the composition is to consist of 42 oz. it is evident that the quantity of each ingredient found by Case 1st, is to the quantity required of each, as 12 to 42.

2. How much water of no value, and cider worth 20 cents per gallon, must be mixed together, to fill a vessel of 80 gallons, that may be afforded at 18 cents per gallon?

Ans. 8 gallons of water, and 72 of cider.

3. One of the votaries of Bacchus bought a barrel of rum, containing $31\frac{1}{2}$ gallons. After he had purchased the liquor, (which he bought for proof rum,) he suspected that it contained too great a quantity of water; and knowing that a gallon of proof rum weighs 7.73lb. Avoirdupois, and a gallon of water 8.36lb., he weighed the liquor, in order to find how much water it contained more than the usual proportion. The weight of the liquor he found to be 252lb. I demand how much water he bought for rum?

Ans. $13\frac{1}{2}$ gallons.

SIMPLE INTEREST BY DECIMALS.

In Interest *five* quantities are concerned, viz. the *principal*, the *rate*, the *time*, the *interest*, and the *amount*; and any *three* of these, except the principal, the interest, and the amount, being given, the rest may be found. Hence, calculations in interest admit of several problems. The most useful however, and consequently that which claims the greatest degree of attention, is that in which the principal, the time, and the rate are given, to find the interest, or amount.

In Simple Interest by Decimals, the interest of $1 or L1 for one year, at any proposed rate, expressed as the decimal of a dollar or pound, is called the *Ratio*; and it is found by dividing the rate per cent. (or interest of 100 dollars or pounds for one year) by 100. Thus, if the rate per cent. be 6, the ratio is $6 \div 100 = .06$; if the rate be $5\frac{1}{4}$ or 5.25, the ratio is $5.25 \div 100 = .0525$; &c.

PROBLEM I.—*When the principal, time, and rate per cent. are given, to find the interest and amount.*

RULE.*—Multiply the principal, ratio, and time, contin-

* This rule is a contraction of a process in the Double Rule of Three; and the reason of it may be thus illustrated. Let p represent any given

ually together, and the product will be the interest; which add to the principal, and the sum will be the amount.

Note.—When there are parts of a year in the given time, reduce them to the decimal of a year, and annex the decimal to the number of whole years, if any. Also, when the given sum is pounds, shillings and pence, the parts of a pound must be reduced to the decimal of a pound.—This note must be attended to in all the subsequent problems in Simple Interest by Decimals.

Example 1. Required the amount of $211.45, at 6 per cent. per annum, for 5 years.

Here $211.45×.06×5=$63.4350, the interest. Then, $211.45+63.435=$274.885, the amount, Ans.

2. What is the interest of $540, for 4 years, at 7 per cent.?
Ans. $151.20

3. Required the amount of $640, for 12 years and 9 months, at 5½ per cent. Ans. $1088.80

4. Required the amount of 537*l.* 10s. at 6*l.* per cent. per annum, for 5 years. Ans. 698*l.* 15s.

PROBLEM II.—*To find the* PRINCIPAL, *when the amount, time, and rate per cent. are given: Or, to find the* PRESENT WORTH *of a given sum at simple interest for any given time.*

RULE.—Multiply the ratio by the time, and add 1 to the product, for a divisor; by which sum divide the *amount*, or *given sum*, and the quotient will be the *principal*, or *present worth* required.*

Example 1. What principal will amount to $274.885 in 5 years, at 6 per cent. per annum?

.06×5+1=1.30, and $274.885÷1.3=$211.45, Ans.

2. What is the present worth of a debt of $691.20, due

---

principal, *t* the time, *r* the ratio, or interest of $1 for one year, at the given rate, and *i* the interest of the principal for the given time; then, by the Double Rule of Three;—

As $1 principal : *p* }
   1 year : *t* } :: *r* : *i*; that is, $i=\frac{p\times t\times r}{1\times 1}=p\times t\times r$, which is the rule.

The rules for the following problems in Simple Interest are also contractions in the Double Rule of Three, and may be proved by it.

* This rule is a contraction of the rule of Discount, given in page 196.

4 years hence, discounting at the rate of 7 per cent. per annum? Ans. \$540.

3. What principal will amount to \$1088.80 in 12¾ years, at 5½ per cent. per annum? Ans. \$640.

4. What principal will amount to 698*l.* 15s. in 5 years, at 6 per cent.? Ans. 537*l.* 10s.

Problem III.—*When the amount, principal, and time, are given, to find the rate per cent.*

Rule.—Subtract the principal from the amount; divide the remainder by the product of the principal and time, and the quotient will be the ratio; which multiply by 100, and the product will be the rate per cent.

Example 1. At what rate per cent. per annum, will \$211.45 amount to \$274.885, in 5 years?

Here \$274.885—211.45=\$63.435, the interest.

Then, $\frac{63.435}{211.45\times5}$=.06, the ratio, and .06×100=6, the rate per cent., Ans.

2. At what rate per cent. will \$540 amount to \$691.20, in 4 years? Ans. 7 per cent.

3. At what rate per cent. will \$600 amount to \$856.50, in 9½ years? Ans. 4½ per cent.

4. At what rate per cent. will 250*l.* 10s. amount to 400*l.* 16s. in 10 years? Ans. 6 per cent.

Problem IV.—*When the amount, principal, and rate per cent. are given, to find the time.*

Rule.—Subtract the principal from the amount; divide the remainder by the product of the principal and ratio, and the quotient will be the time.

Ex. 1. In what time will \$211.45 amount to \$274.885, at 6 per cent. per annum?

Here $\frac{274.885-211.45}{211.45\times.06}$= 5 years, Ans.

2. In what time will \$540 amount to \$691.20, at 7 per cent.? Ans. 4 years.

3. In what time will \$600 amount to \$856.50, at 4½ or 4.5 per cent.? Ans. 9½ years.

4. In what time will 250*l.* 10s. amount to 400*l.* 16s., at 6 per cent.? Ans. 10 years.

EQUATION OF PAYMENTS BY DECIMALS.

To find the equated time for the payment of several debts due at different times.

RULE.—Find, by Discount, the *present worth* of each debt, at the given or legal rate of interest: then find, by the Rule for Problem IV. in Simple Interest by Decimals, in what time, at the same rate, the sum of all the present worths, thus found, would amount to the sum of all the debts, and the result will be the time required.*

Ex. 1. There are $100 payable in 2 years, and $106 in 6 years: what is the equated time for the payment of the whole, allowing simple interest, at the rate of 6 per cent. per annum?

$	$		$
100÷(.06×2+1)=	89.285+	Present worth of the	100
106÷(.06×6+1)=	77.941+	Do. of the	106
	$167.226+	Do. of	$206

Then, $\frac{206-167.226}{167.226\times.06}$=3.8644+years,=3yr. 10mo. 11da. + Ans.

In the foregoing example, the present values of the debts are found by Prob. II. Simple Interest by Decimals, and then the time is found by Prob. IV.

2. A debt of $1000 is to be paid, one-half in three years, and the other half in six years: what is the equated time for the payment of the whole, allowing 7 per cent. interest?

Ans. 4.3802+ years, or 4 years, 4 months, 16 days+

3. A owes B $200, whereof $40 are to be paid at the end of 3 months, $60 at 6 months, and $100 at 9 months: What is the equated time for the payment of the whole debt, allowing simple interest at the rate of 5 per cent. per annum? Ans. 6 months, 26 da.+

* In this rule, the result depends on the *present* values of the debts; and it is evident that all such transactions and agreements should be regulated in conformity to the *present value* of the money, and the *improvement* of which it is susceptible. This latter principle, which is entirely neglected in the use of the common rule, given in page 199, is acted on in this, in using the rate of interest.

COMPOUND INTEREST BY DECIMALS.

The *ratio*, in Compound Interest, is the amount of $1 or L1 for one year; and it is found by dividing the sum of 100 and the rate per cent. by 100. Thus, if the rate per cent. be 6, the ratio is $(100+6)\div100=1.06$; and if the rate be $5\frac{1}{2}$, or 5.5, the ratio is 1.055, &c.

PROBLEM I.—*When the principal, the rate of interest and the time are given, to find either the amount or interest.*

RULE.—1. Raise the ratio to that power whose index is equal to the number of years: Or, in Table I., in the latter part of this Book, you will find the said power, under the given rate and opposite to the given number of years.*

2. Multiply the power, thus found, by the principal or given sum, and the product will be the *amount* sought; from which if you subtract the given principal, the remainder will be the compound interest.†

Example 1. What is the compound interest of $600 for 3 years, at 6 per cent. per annum?

Here $1.06\times1.06\times1.06=1.191016$, the 3d power of the ratio. Then, $\$1.191016\times600=\$714.6096$, the amount, and $\$714.6096-600=\$114.6096$, the interest, Ans.

In this example, the 3d power of the ratio is the amount of $1 for 3 years at 6 per cent., which may be found in Table I. ready calculated.

---

* The amounts of $1 or L1 in Table I. are so many powers of the amount of $1 or L1 for one year; whose indices are the numbers of years.

† The reason of this rule may be shown thus: The amount of $1 for a year, at 6 per cent., is $1.06, and, by the nature of compound interest, this amount is the principal for the second year. Then, as $1 principal : $1.06 principal :: $1.06 amount : $\$1.06^2$, or $1.1236, the amount at the end of the second year, which is the principal for the third year. Then, as $1 : $1.06 :: $\$1.06^2$: $\$1.06^3$, the amount for the third year, being the principal for the fourth year. By proceeding in this manner, it will be found that the amount of $1, for any given number of years, is equal to that power of the ratio, or amount for one year, which is denoted by the number of years. The amount of $1 being thus determined, it is evident that the amount of any other principal will be obtained by multiplying the amount of $1 by that principal.

The reason of the rules for the other Problems in Compound Interest by Decimals, will easily appear from the foregoing illustration of the rule for Prob. I.

2. Required the compound interest of $2000 for 3 years, at 7 per cent. Ans. $450.086

3. What will $50 amount to in 15 years, at 5 per cent. per annum, compound interest?

The 15th power of $1.05, or the amount of $1 for 15 years, is found by Table I. to be $2.078928; which being multiplied by 50, gives $103.9464, the answer.

Note 1.—When the interest is payable *yearly*, the amount of $1 for half a year will be the square root, for $\frac{1}{3}$ part of a year the cube root, and for $\frac{1}{4}$ part of a year the 4th root, &c. of the amount for one year. Therefore, to find the amount of any given principal for any aliquot part of a year, extract such root of the ratio as is denoted by the said aliquot part; which root multiply by the given principal, and the product will be the amount sought. When the given part of a year is not an *aliquot* part, separate it into two or more parts, one of which shall be an aliquot part of a year, and the rest, either aliquot parts of a year or of one of the other parts; then extract roots out of the amount of $1 for one year, &c. answering to all these parts, and the continued product of these roots will be the amount of $1 for the said parts of a year;* which being multiplied by the given principal, will give the amount required. In all ordinary cases, however, the amount for any part of a year may be found, sufficiently exact, by the rules for computing simple interest; and this method is commonly practised.

4. What will $40 amount to in 6 months, at 6 per cent. per annum, compound interest?

Here $\sqrt{1.06}=1.02956+$ the amount of $1 for 6 months, and $1.02956×40=$41.1824, the Ans.

*We may, in any case, separate the given time into as many parts or portions as we please, and find the amount of $1 for each of these parts, and the continued product of the several amounts, thus found, will be the amount of $1 for the whole time.

The amount of any sum at compound interest may readily be found by the help of a table of the common logarithms of numbers; viz. thus: If there be parts of a year in the given time, reduce them to the decimal of a year; which annex to the number of whole years, if any: then multiply the logarithm of the ratio by the time; add to the product the logarithm of the given principal, and the sum will be the logarithm of the required amount. Then, the natural number answering to this logarithm, being found in the table, will be the answer.

5. What will $500 amount to in 1 year and 4 months, or $1\frac{1}{3}$ years, at 5 per cent. per annum?

Here, the amount of $1 for 1 year is $1.05; its amount for 4 months is $\sqrt[3]{1.05}=1.016396+$, and its amount for 1 year and 4 months is $1.016396×1.05=$1.06721580.

Then, $1.0672158×500=$533.6079, the Ans.

Note 2.—When the interest is payable at intervals of time either longer or shorter than a year; find, by Simple Interest, the amount of $1 or L1 for one of those periods of time, which call the *ratio*. Raise the ratio, thus found, to the power denoted by the number of the said periods, and the result multiplied by the given principal will give the amount sought.

6. What will $100 amount to in 2 years, at the rate of 3 per cent. per *half-year?*

Here the amount of $1 for half a year is $1.03, and its amount for 2 years, or 4 half-years, is $1.03×1.03×1.03×1.03=$1.12550881.

Then, $1.12550881×100=$112.55+, the Ans.

7. What will a principal of $500 amount to in 6 years, at 10 per cent. for every *two years?* Ans. $665.50

Problem II.—*To find the* PRINCIPAL, *when the amount, time, and rate per cent. are given: Or, to find the* PRESENT WORTH *of any given sum at compound interest, for a given time.*

Rule.—Divide the amount, or given sum, by the amount of $1 or L1 for the given time, (found by Prob. I.,) and the quotient will be the *principal*, or *present worth* required.

Ex. 1. What sum must be lent at compound interest, at 5 per cent. per annum, at the birth of a child, so that the amount may be $3000 when the child shall be 21 years old?

I find by Table I. that the amount of $1 for 21 years, at 5 per cent., is $2.785963.

Then, $3000÷2.785963=$1076.82+, the Ans.

2. What principal will amount to $757.486176 in 4 years, at 6 per cent. per annum? Ans. $600.

3. What is the present worth of a debt of $2450.086 due three years hence, discounting at the rate of 7 per cent. per annum, compound interest? Ans. $2000

PROBLEM III.—*To find the rate per cent., when the amount, principal and time are given.*

RULE.—Divide the given amount by the principal, and the quotient will be the amount of $1 for the given time, or the power of the ratio denoted by the time. Then extract such root of this power as is denoted by the number of years, or periods of time at which the interest is payable, and the result will be the ratio; from which the rate per cent. may be easily found.

Or, having found the amount of $1 as directed above, look for it in Table I. in a line with the given time, and over the said amount, at the top of the column, you will find the rate per cent.

Ex. 1. At what rate per cent. per annum will $600 amount to $757.486176, in 4 years?

Here $757.486176÷600=$1.26247696, the 4th power of the ratio, or the amount of $1 for 4 years, at the required rate. Then, $\sqrt[4]{1.26247696}=1.06$, the ratio, or amount of $1 for 1 year. Hence the rate per cent. is 6, Ans.

Or, in Table I. in a line with 4 years is 1.262477, the amount of $1 for the said time, and at the top of the column is 6 per cent., the Ans.

2. At what rate per cent. per annum will $400 amount to $463.05, in three years? Ans. 5 per cent.

PROBLEM IV.—*The amount, principal, and rate per cent. being given, to find the time.**

RULE.—Divide the given amount by the principal, and the quotient will be that power of the ratio denoted by the number of years. Then divide the power, thus found, by the ratio, and the quotient again by the same, and so on till the quotient is a unit, and the number of these divisions by the ratio will be equal to the time sought.

Or, divide the amount by the principal, and then find the quotient in Table I. under the given rate, and in a line with it in the left hand column you will find the time.

---

* Questions of this nature may be easily solved by *logarithms*, as follows: From the logarithm of the given amount subtract the logarithm of the principal; divide the remainder by the logarithm of the ratio, and the quotient will be the time required.

Ex. 1. In what time will $400 amount to $463.05, at 5 per cent. per annum?

Here 463.05÷400=1.157625=the power of the ratio (1.05) denoted by the number of years. Then, 1.157625÷1.05=1.1025, and 1.1025÷1.05=1.05, and 1.05÷1.05=1. Here the number of the divisions is 3, and therefore the time is 3 years, Ans.

Or, by Table I.—Under 5 per cent., in a line with 3 years, is 1.157625, the quotient found by dividing the given amount by the principal; and hence the time is 3 years, as before.

2. In what time will $600 amount to $757.486176, at 6 per cent.? Ans. 4 years.

ANNUITIES, OR PENSIONS, &c.

An *Annuity* is a sum of money payable yearly, either for a certain number of years, or during the life of the pensioner, or forever.

An annuity which is to continue forever, is called a *Perpetuity.*

Annuities are said to be *certain,* when they are to be continued either for some specified time or forever; and *contingent,* when their continuance depends on some contingency, as the life or death of some person, &c.

Sometimes annuities are not to commence till a certain time has elapsed, and the annuities are then said to be in *reversion.*

When the debtor keeps the annuity in his own hands, beyond the time of payment, it is said to be in *arrears.*

The sum of all the annuities remaining unpaid, together with the interest on each, for the time they have been due, is called the *amount.*

The *present worth* of an annuity, is such a sum as being put at interest, would exactly pay the annuity as it becomes due.

ANNUITIES AT SIMPLE INTEREST.

PROBLEM I.—*To find the amount of any annuity or pension in arrears, at simple interest.*

RULE.—1. Multiply the interest of the annuity for one year by the sum of the natural series 1, 2, 3, 4, &c. to the number of years less 1,* and the product will be the whole interest due.

2. Multiply the annuity by the number of years given; add the product to the whole interest due, and the sum will be the amount sought.†

Ex. 1. What is the amount of an annual pension of $100, which has remained unpaid 4 years, allowing 5 per cent. simple interest?

Here the interest of the annuity for one year is $5, and 1+2+3=6=the sum of the natural series of numbers extended to the number of years less 1. Then,

$ 5×6= 30 the whole interest.
$100×4=400 prod. of the annuity and time.

Ans. $430 the amount.

2. If a yearly rent of $250 remain unpaid 7 years, what will then be due, allowing 6 per cent. simple interest?

Ans. $2065.

Note.—If half of the annuity be payable half-yearly, or one-fourth part of it quarterly, &c.; then, for half-yearly payments, take half of the annuity, half of the rate per cent. per annum, and twice the number of years; and for quarterly payments, take a fourth part of the annuity, a fourth part of the rate per cent., and four times the number of years, &c., and then proceed according to the foregoing Rule.

3. If an annuity of $100, payable half of it half-yearly, remain unpaid 4 years; what is the amount, allowing simple interest at the rate of 5 per cent. per annum?

---

* The sum of this series may be found by multiplying the number of years less 1, by half the given number of years.

† It is plain that upon the first year's annuity there will be due so many years' interest as the given number of years less 1, and gradually one year less upon each succeeding year, to that preceding the last, which has but one year's interest, and the last bears none. There is, therefore, due in the whole as many years' interest on the annuity as the sum of the series 1, 2, 3, &c. to the number of years diminished 1. It is evident then, that the whole interest due must be equal to the interest for one year multiplied by the sum of the said series; and that the amount will be found by adding the whole interest to all the annuities, that is, to the product of the annuity and time; which is the Rule.

Here 1+2+3+4+5+6+7=28, the sum of the natural series of numbers extended to the number of half-years less 1; and $50×.025=$1.25, the interest of half of the annuity for half of a year. Then,

$1.25×28= 35 the whole interest due.
$ 50× 8=400 prod. of half of the annuity and the no. of half-years.
——
Ans. $435 amount.

4. If a pension of $100, payable one-fourth part each quarter of a year, be forborne 4 years, what will be the amount, allowing 5 per cent. simple interest?

Ans. $437.50

It may be seen, by comparing the 1st, 3d and 4th examples, that the more frequently the payments are to be made, the greater the amount is.

Problem II.—*To find the present worth of an annuity at simple interest.*

Rule.—Find, as in Discount, the present worth of each payment by itself, allowing discount to the time when it will become due, and the sum of all the present worths, thus found, will be the answer.*

Ex. 1. What is the present worth of a yearly pension of $80, to be continued three years, at 6 per cent. per annum, simple interest?

106 }		75.471, pres. worth of the 1st year's [pension.
112 } : 80 :: 100 :		71.428, do. of the 2d do.
118 }		67.796, do. of the 3d do.

Ans. $214.695+ do. of the whole.

2. What is the present worth of an annuity of $150, to be continued 2 years, at 5 per cent. per annum?

Ans. $279.22+

Note.—The same thing is to be observed as in the first Problem in Annuities, concerning half-yearly and quarterly payments.

3. What is the present worth of a pension of $150, pay-

* The several annuities, or payments, may be considered as so many debts, due 1, 2, 3, &c. years hence, of which the present worths are to be found. Hence the sum of the present worths of all the annuities must be the whole present worth required.

able half each half-year, for 2 years, at 5 per cent. per annum? Ans. $282.546+

4. What is the present worth of a pension of $150, payable one-fourth part each quarter of a year, for 2 years, at 5 per cent. per annum? Ans. $284.229+

By comparing the three last examples, it will be seen that the present worth of half-yearly payments is more advantageous than yearly; and quarterly, than half yearly.

ANNUITIES AT COMPOUND INTEREST.

PROBLEM I.—*To find the amount of any annuity or pension in arrears, at Compound Interest.*

RULE.—1. Make 1 the first term of a geometrical progression, and the amount of $1 or L1 for one year, at the given rate per cent., the ratio.

2. Carry on the series up to as many terms as the given number of years, and find its sum.

3. The sum, thus found, will be the amount of an annuity of $1 or L1 for the given time; which multiply by the given annuity, and the product will be the amount sought.

Or, by Table II.*—Multiply the tabular number, found under the rate and opposite to the time, by the annuity, and the product will be the amount required.

Ex. 1. If a yearly pension of $75, be forborne, or remain unpaid, 4 years, what will it amount to, allowing compound interest at the rate of 6 per cent. per annum?

Here $1+1.06+1.1236+1.191016=$4.374616, the sum of the progression, being the amount of an annuity, of $1 for 4 years, at 6 per cent.

Then, $4.374616×75=$328.0962, Ans.

Or, by Table II. *thus:* Under 6 per cent. and opposite to 4 years is 4.374616, the amount of $1 annuity for the

*Table II. is calculated thus: For 6 per cent, take the first year's amount, which is $1, multiply it by 1.06, (the ratio,) and the product, increased by 1, is 2.06, the second year's amount; which also multiply by 1.06, and the product, plus 1, is the third year's amount, &c. And in this manner proceed in calculating the amounts at other rates.

said time; which being multiplied by 75, gives $328.0962 for the answer, as before.

2. If a person rent a farm at $120 per year, payable yearly, and forbear paying rent for 5 years; how much will he owe to the proprietor at the end of that time, allowing him compound interest at 5 per cent. per annum?

Ans. $663.075+

Note.—When the payments are not to be made *yearly;* then, instead of the amount of $1 for a year, use its amount for the interval between the payments; and instead of the number of years, use the number of payments that would have been made during the time they were remitted, and then proceed as before.—This note must also be attended to in the subsequent problems in Annuities at Compound Interest.

3. If a pension of $100, payable half each half-year, be forborne 2 years, what will it amount to, at 6 per cent. per annum, compound interest?

Here the ratio, or amount of $1 for half a year, is $\sqrt{1.06}$ =1.02956+; and the amount of $1 annuity for 4 half-years is $1+1.02956+1.06+1.0913336=$4.1808936.

Then, $4.1808936×100=$418.08936, the Ans.

4. Let every thing be the same as in the last example, except that compound interest is allowed at the rate of 3 per cent. per half-year, instead of 6 per cent. per year.

In this case the amount of $1 for half a year is $1.03; the amount of $1 annuity for 4 half years is $1+1.03+1.0609+1.092727=$4.183627; and the required amount is $4.183627×100=$418.3627, Ans.

Problem II.—*To find the present worth of annuities, at compound interest.*

Rule.—Divide the annuity by that power of the ratio signified by the number of years, and subtract the quotient from the annuity; then divide the remainder by the ratio less 1, and the quotient will be the present value of the annuity.

Or, by Table III.*—Multiply the tabular number, found

* Table III. is made thus: $1÷1.06=.943396, the present worth of the first year; then, .943396÷1.06=.889996,

under the rate and opposite to the time, by the annuity, and the product will be the present worth required.

Ex. 1. What ready money will purchase an annuity of $150, to continue 4 years, at 6 per cent. per annum, compound interest?

4th power of 1.06=1.26247696)150(118.81404+

Then, from $150
subtract 118.81404

Divisor, 1.06—1=.06) 31.18596

Ans. $ 519.766—

Or, by Table III.: Under 6 per cent. and opposite to 4 years is 3.465105, which being multiplied into $150, gives $519.76575, for the answer.

2. Required the present value of a house, held on a lease, of which three years are unexpired, and bringing a profit rent of $200 per annum, payable yearly, compound interest being allowed at 5 per cent. Ans. $544.649+

3. Let every thing be as in the last question, except that the annuity is payable half each half year, instead of the whole yearly.

Here, the number of half years is 6; and the ratio, or amount of $1 for half a year, is $\sqrt{1.05}$=1.02469507+, the 6th power of which (=the 3d power of 1.05) is 1.157625.

Then, $100÷1.157625=86.3837598+, and

$$\frac{\$100-86.3837598}{1.02469507-1}=\$551.374+, \text{ the Ans.}$$

Problem III.—*To find the present worth of Annuities, Leases, &c., taken in Reversion, at Compound Interest.*

Rule.—Find, by the rule for Prob. II., the present worth of the annuity for the time of its continuance, as if it were to commence immediately; then divide the present

---

which, added to the first year's present worth, is=1.833-392, the second year's present worth; then, .889996÷1.06 =.839619, which being added to the second year's present worth, gives 2.673011, for the third year's present worth, &c.

value, thus found, by that power of the ratio denoted by the time of reversion, (or the time to come before the annuity commences,) and the quotient will be the present worth of the annuity in reversion.

Or, by Table III.—Find the present value of an annuity of $1, at the given rate, for the sum of the time of continuance and time in reversion added together; from which value subtract the present worth of an annuity of $1 for the time in reversion; multiply the remainder by the given annuity, and the product will be the answer required.

Ex. 1. What is the present worth of an annuity of $150, payable yearly, for 4 years; but not to commence till 2 years, at 6 per cent. per annum, compound interest?

4th pow. of 1.06=1.26247696)150.00000(118.81404+
Subtract the quotient,=118.81404

Divide by 1.06—1=.06) 31.18596

$

2d power of 1.06=1.1236) 519.766(462.58+ Ans.

Or, by Table III. *thus:* Under 6 per cent. and opposite to 6 years, (the sum of the time of continuance and time of reversion,) is 4.917324; and opposite to 2 years, (the time of reversion,) is 1.833392: Then, 4.917324—1.833-392=3.083932, the present worth of $1 annuity, in reversion; which being multiplied into $150, gives $462.5898 for the answer.

2. A father leaves to his eldest child for 8 years a profit rent of $400 per annum, payable yearly, and the reversion of it for the 12 years succeeding, to his second child. What is the difference between the present values of these two legacies, and which is worth the most, allowing compound interest on each, at the rate of 6 per cent. per annum?

Ans. The eldest child's portion is worth the most by $379.866+

Problem IV.—*To find the present worth of a Freehold Estate, or an annuity to continue forever, at compound interest.*

Rule.—Subtract a unit from the amount of $1 for a year, or for the interval between the payments; divide the annuity or yearly rent by the remainder, thus found, and the quotient will be the present value required.

Or, as the given rate per cent. : is to the annuity or yearly rent :: so is $100 : to the value required.

Ex. 1. What is the worth of a freehold estate which brings in $180 yearly, allowing 6 per cent. to the purchaser?

Here 1.06—1=.06, and $180÷.06=$3000 Ans.

Or, as $6 : $180 :: $100 : $3000, the Ans.

2. What is the present worth of an annuity of $200, to continue forever, discounting at the rate of 5 per cent.?

Ans. $4000.

3. What is the present value of a perpetuity of $150 per annum, payable each half of it half-yearly, discounting at the rate of 6 per cent. per annum?

Here $\sqrt{1.06}=1.02956301+$, the amount of $1 for half a year; from which subtracting 1, there remains .02956301. Then, $75÷.02956301=$2536.95+ Ans.

Note.—To find what perpetuity can be purchased for a given sum, say, as $100 : is to the given sum :: so is the rate per cent. : to the perpetuity or yearly rent required.

Problem V.—*To find the present worth of a Freehold Estate, or Perpetuity, in Reversion, at compound interest.*

Rule.—Find, (by Prob. IV.,) the present value of the estate, as though it were to be entered on immediately; which value divide by that power of the ratio, (or amount of $1,) denoted by the time of reversion, and the quotient will be the present worth of the estate in reversion.

Or, by Table III.—Find, (as in Prob. II.,) the present worth of the annuity, or rent, for the time of reversion; which subtract from the value of the immediate possession, (found by Prob. IV.,) and the remainder will be the value of the estate in reversion.

Ex. 1. If a freehold estate of $250 per annum, to commence 2 years hence, be for sale, what is it worth, allowing the purchaser 5 per cent. per annum?

First, As 5 : 250 :: $100 : $5000, the value of the immediate possession.

Then, $1.05\times1.05=1.1025$, the 2d power of the ratio, and $5000÷1.1025=$4535.147+, the ans.

Or, by Table III. *thus:* The present worth of $1 annuity for 2 years at 5 per cent. is $1.85941, and $1.85941×

250=$464.85250=the present worth of the annuity for the time of reversion. Then, $5000—464.8525=$4535.1475, the ans.

2. What is the present worth of a perpetuity of $240 per annum, to commence 4 years hence, discounting at the rate of 6 per cent. per annum? Ans. $3168.374+

3. A man has left to his two sons, A and B, an estate which will yield $100 a year forever. A is to enter upon it immediately, and have the use of it 15 years; after which B is to have it forever. Whose portion is the most valuable, and how much the most, discounting at the rate of 5 per cent. per annum, compound interest?

Ans. A's portion is the most valuable by $75.931

PRACTICAL GEOMETRY.

DEFINITIONS.

1. A *point* is that which has position, but not magnitude.

2. A *line* is length, without breadth or thickness.

3. A *surface* or *superficies*, is an extension or a figure, of two dimensions, viz. length and breadth; but it is not considered as having thickness.

4. A *solid* or *body*, has three dimensions, viz. length, breadth and thickness.

5. A plane figure which has three straight sides, or is bounded by three right lines, is called a *triangle*.

6. A *right angled* triangle* has two of its sides perpendicular to each other. The longest side of any right angled triangle is commonly called the *hypothenuse*; the next longest side the *base*, and the shortest side the *perpendicular*.

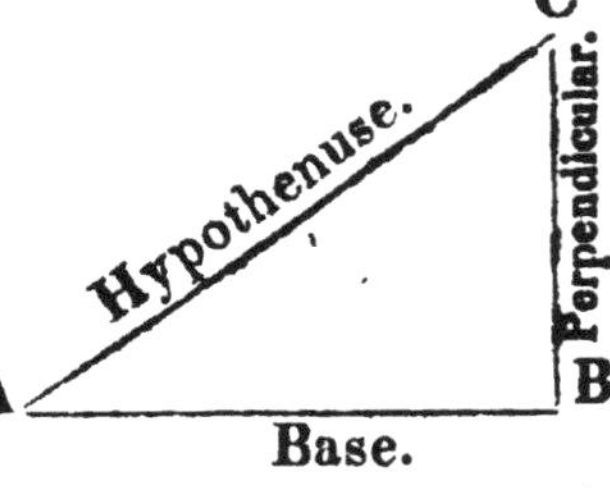

* An *angle* is the inclination or opening of two lines, having different directions, and meeting in a point. Thus, in the above right angled triangle A B C, an angle is formed by the meeting of the two lines B A and C A at A. Every angle is measured by the arch of a circle supposed to be described about the angular point as a centre; the quantity of

7. Triangles which are not right angled, are, in general, called *oblique angled triangles*, or simply *oblique triangles*. The longest side of an oblique triangle is commonly called the *base*, and the other two sides the *legs*.*

8. A figure that is bounded by four straight lines, or sides, is called a *quadrangle*, or a *quadrilateral*.

9. A *parallelogram* is a quadrilateral which has both its pairs of opposite sides parallel; and it takes the following particular names, from the relations of its sides and angles, viz. *rectangle*, *square*, *rhombus*, and *rhomboid*.

10. A *rectangle* is a right angled parallelogram; every two of its adjoining sides being perpendicular to each other. A rectangle is either *oblong* or *square*.

11. An *oblong rectangle*, (improperly called a *long square*,) is a rectangle whose length exceeds the breadth.

12. A *square* is a rectangle whose four sides are all equal.

13. A *rhomboid* is an oblique-angled parallelogram.

14. A *rhombus* is a rhomboid whose four sides are all equal.

15. A *trapezium* is a quadrilateral, whose opposite sides are not equal nor parallel.

---

the angle being reckoned so many degrees, or minutes, &c. as are cut off from the circle by the two lines which form the angle. An angle of 90 degrees, is called a *right angle*: one which is less than 90 degrees, is called an *acute angle*; and one which is greater than 90 degrees, is said to be *obtuse*.—In every triangle, the sum of the three angles is equal to two right angles, or 180 degrees.

Right angled triangles are so called because the angle included between the base and perpendicular is a right angle.—The other two angles of any right angled triangle, whose three sides are known, may be readily found, (when necessary,) to a great degree of accuracy, as follows; viz. As the sum of the hypothenuse and half the longer of the other two sides : is to the shortest side :: so is 86 degrees : to the angle opposite to the shortest side; which being subtracted from 90 degrees, will leave the other angle sought.

* The base of any triangle, or parallelogram, is the side upon which the figure is supposed to stand; and a perpendicular line falling on the base from the opposite angle, is called the *altitude* or *height* of the figure. Any side of a triangle may be considered the base, and the other two sides the legs.

16. A *trapezoid* is a quadrilateral which has two of its opposite sides parallel, but of unequal lengths.

17. A straight line connecting any two opposite angles or corners of a quadrilateral figure, is called a *diagonal.*

18. Plane figures which have more than four sides, are, in general, called *polygons;* and they receive other particular names, according to the number of their sides. Thus, a *pentagon* is a polygon of *five* sides; a *hexigon*, of *six* sides; a *heptagon*, of *seven* sides, &c.

19. A figure which is bounded by straight lines, is called a *regular* figure when all its sides are equal and all the angles equal. When the sides and angles are not all equal, the figure is said to be *irregular.*

20. A *circle* is a plain figure bounded by one curve line, called the *circumference* or *periphery*, every part of which is equally distant from a certain point within the circle, called the *centre.* The *diameter* of a circle is a straight line drawn through the centre, and terminated both ways by the circumference.

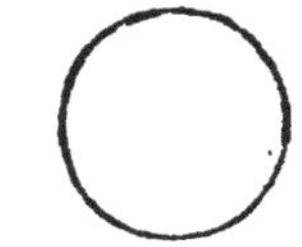

21. An *ellipse*, or *ellipsis*, is an oval figure, resembling a circle, only the length exceeds the breadth. The longest diameter of an ellipse is called the *tranverse* diameter, and the shortest the *conjugate* diameter. These two diameters cross each other, at right angles, in the centre of the ellipse, and each of them divides the ellipse into two equal parts.

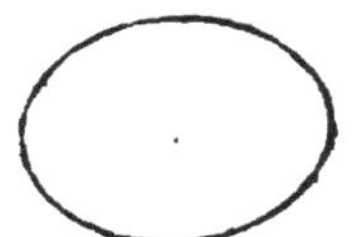

MENSURATION OF PLANE SURFACES.

The *area* or *superficial content* of any plane figure, is the measure of the space contained within its extreme bounds; without any regard to thickness. This area is estimated by the number of little squares it contains, or is equal to;—the side of each of these little measuring squares being an inch, a foot, a yard, or any other fixed quantity: And hence the area is said to be so many square inches, or square feet, or square yards, &c.

Problem I.—*To find the area of any Parallelogram; whether it be an Oblong, a Square, a Rhomboid, or a Rhombus.*

Rule.—Multiply the length by the perpendicular breadth, or height, and the product will be the area.

Note 1.—Since the length and breadth of a square are equal, the area is evidently equal to the square of one of the equal sides.

Note 2.—In finding the contents of superficies, the area will always be of the same name as the dimensions used for finding the same; that is, if the dimensions be *feet*, the area will be *square feet;* or, if the dimensions be *yards*, then the area will be *square yards*, &c.

Example 1. Required the area of a tract of land in form of a parallelogram, whose length is 100 rods, and breadth 80 rods.

$100\times80=8000$ sq. rods, and $8000\div160=50$ acres, Ans.

2. Required the area of a square piece of ground whose side measures 20 chains.

$20\times20=400$ sq. chains, and $400\div10=40$ acres, Ans.

3. How many square yards are contained in a rhombus whose side is 15 feet, and perpendicular breadth 12 feet?

Ans. 20 sq. yd.

Problem II.—*Any two sides of a right angled triangle being given, to find the other side.*

Rule.*—1. When the base and perpendicular are given; square each side separately, and the square root of the sum of these squares will be the length of the hypothenuse.

2. When the hypothenuse and either of the two other sides are given; then the square root of the difference of the squares of the two given sides will be the other side required.

Ex. 1.—Required the hypothenuse of a right angled triangle whose base is 4 chains, and perpendicular 3 chains.

$4\times4+3\times3=16+9=25$, and $\sqrt{25}=5$ chains, Ans.

* The square of the hypothenuse, or longest side of any right angled triangle, is equal to the sum of the squares of the two other sides; whence the reason of the above rule is evident.

2. If the hypothenuse of a right angled triangle be 10 rods, and one of the other sides 6 rods, what is the remaining side? Ans. 8 rods.

PROBLEM III.—*To find the area of a right angled triangle.*

RULE.—Multiply the length of the base by the perpendicular, and half of the product will be the area sought: Or, multiply one of these dimensions by half of the other, and the product will be the area.

Ex. What is the area of a right angled triangle whose base measures 14 rods, and the perpendicular 5 rods?

Here $14\times5=70$, and $70\div2=35$ sq. rods, Ans.

Or, $7\times5=35$ sq. rods, the answer, as before.

PROBLEM IV.—*To find the area of any oblique-angled triangle.*

RULE.—1. When the base and the perpendicular height of the triangle are known, proceed as in Problem III.

2. When the three sides of the triangle are given, the area may be found as follows: From half the sum of the three sides subtract each side separately; then multiply the said half sum and the three remainders continually together, and the square root of the product will be the area of the triangle.

Ex. 1. What is the area of a triangle whose base is 15 feet, and perpendicular height 8 feet?

Here $15\times4=60$ sq. feet, Ans.

2. Required the area of an oblique triangle whose sides are 20, 30, and 40 rods.

Here $20+30+40=90$ rods, the sum of the sides, the half of which is 45 rods. Then, subtracting each side from 45, the remainders are 25, 15 and 5. Then, $45\times25\times15\times5=84375$, and $\sqrt{84375}=290.47+$ sq. rods, Ans.

3. How many acres are contained in a triangle whose sides are 30, 40, and 50 rods? Ans. $3\frac{3}{4}$ A.

PROBLEM V.—*To find the area of a Trapezoid.*

RULE.—Multiply half the sum of the two parallel sides by the perpendicular distance between them, and the product will be the area.

Ex. How many square feet are contained in a board which is 12 feet 6 inches long, 15 inches wide at the greater end, and 11 inches wide at the less end?

Here $(15+11)\div2=13$ inches, the mean width; and 12ft. 6in.=150 inches, the length. Then, $150\times13=1950$ sq. in. $=13\frac{13}{24}$ sq. feet, Ans.

PROBLEM VI.—*To find the area of any Trapezium.*

RULE.—Divide the trapezium into two triangles by a diagonal; then find the areas of these triangles, and add them together for the whole area.

Or thus: Let fall two perpendiculars on the diagonal from the other two opposite angles; then multiply the sum of these two perpendiculars by the diagonal, and half of the product will be the area of the trapezium.

Ex. 1. To find the area of a trapezium whose diagonal is 42 chains, and the two perpendiculars on it, 16 and 18 chains.

Here $16+18=34$, the sum of the two perpendiculars. Then, $42\times34=1428$, and $1428\div2=714$ sq. chains, Ans.

2. A certain tract of land, in form of a trapezium, consists of two triangles—the sides of one triangle are 30, 40, and 50 rods; and the sides of the other, 29, 37, and 50 rods. Required the area of the tract.

Ans. 7 acres, 11.57+ sq. rods.

PROBLEM VII.—*To find the area of a Regular Polygon.*

RULE I.—Multiply together the length of one side, the number of sides, and the length of a perpendicular drawn from the centre of the figure to the middle of one of its sides, and half of the last product will be the area sought.

Ex 1. To find the area of a regular pentagon, each of the 5 sides being 25 feet, and the perpendicular from the centre on each side 17.2048 feet.

Here $17.2048\times5\times25=2150.6000$

And $2150.6\div2=1075.3$ sq. feet, Ans.

RULE II.*—Square one of the sides of the polygon; then multiply that square by the multiplier set against the name

*This rule depends on the principle, that *similar figures are to each other as the squares of their like sides.* The multipliers in the table are the areas of the respective figures to the side 1. Whence the rule is manifest.

of the figure in the following table, and the product will be the area.

No. of sides.	Names.	Multipliers.	No. of sides.	Names.	Multipliers.
3	Trigon*	.43301	8	Octagon	4.82843
4	Tetragon*	1.	9	Nonagon	6.18182
5	Pentagon	1.72048	10	Decagon	7.69421
6	Hexagon	2.59808	11	Undecagon	9.36564
7	Heptagon	3.63391	12	Duodecagon	11.19615

Taking here the same example as before, viz. a pentagon, whose side is 25 feet: Then, $25\times25=625$; and the multiplier for a pentagon being 1.72048, we have $1.72048\times625=1075.3$ sq. feet, the area, as before.

2. What is the area of a trigon, or an equilateral triangle, whose side is 20 rods? Ans. 173.204 sq. rods.

3. What is the area of a hexagon whose side is 5 yards? Ans. 64.952 sq. yards.

PROBLEM VIII.—*To find the area of an Irregular Polygon.*

RULE.--Divide the polygon (or suppose it to be divided) into triangles, by diagonal lines. Then find the areas of all the triangles, separately, and the sum of these areas will be the area of the whole polygon.

Ex. A certain irregular polygon contains three triangles—the base of the first triangle measures 55 rods, and the perpendicular 18 rods; the base of the second triangle is 55 rods, and the perpendicular 13 rods; the base of the third is 44 rods, and the perpendicular is 22 rods: What is the area of the polygon?

Here $55\times18=990$, and $55\times13=715$, and $44\times22=968$. Then, $990+715+968=2673$ sq. rods,=double the area of the polygon. Then, $2673\div2=1336\frac{1}{2}$ sq. rods,=8 acres, 1 rood, $16\frac{1}{2}$ sq. rods, the Ans.

* These two figures are not polygons. A trigon is an equilateral triangle, and a tetragon is a square.

Problem IX.—*The Diameter of a Circle being given, to find the Circumference.*

Rule.—As 7 is to 22, so is the diameter to the circumference, nearly: Or, more exactly, as 113 is to 355, so is the diameter to the circumference.* Or, multiply the diameter by 3.1416,† and the product will be the circumference, very nearly.

Note.—By the converse of this rule the diameter of a circle may be found, when the circumference is known.

Ex. 1. What is the circumference of a circle whose diameter is 226 feet?

As 7 : 22 :: 226 : $710\frac{2}{7}$ ft. Ans.
Or, as 113 : 355 :: 226 : 710ft., the Ans.
Or, 226×3.1416=710.0016ft., the Ans.

2. If the circumference of the earth be 24912 miles, what is its diameter?

As 355 : 113 :: 24912 : $7929\frac{36\frac{1}{5}}{355}$ miles, Ans.

Problem X.—*To find the Area of a Circle.*

Rule I.—Multiply the circumference by the diameter, and $\frac{1}{4}$ of the product will be the area: Or, multiply one of these dimensions by $\frac{1}{4}$ of the other, and the product will be the area.

Rule II.—Multiply the square of the diameter by .7854,‡ for the area.

Rule III.—Multiply the square of the circumference by .07958.

Ex. What is the area of a circle whose diameter is 22.6, and circumference 71?

The answer, by the first rule, is 401.15; by the second rule, 401.150904; and by the third, 401.16278

Note.—A *circular ring*, is the figure contained between the *peripheries* of two concentric circles; and hence the area of a circular ring must be the difference of the areas

* The first of these proportions will give the circumference true to 3 or 4 places of figures, and the second to 7 places.

† This multiplier, correct to fifteen places of decimals, is 3.141592653-589793+; but 3.1416 is sufficiently exact for common use.

‡ This multiplier is 1-4 of 3.1416, being the area of a circle whose diameter is 1.

of the two circles; or equal to the difference of the squares of the diameters of the two circles multiplied by .7854

PROBLEM XI.—*To find the area of an Ellipse.*

RULE.—Multiply the tranverse diameter by the conjugate diameter, and the product by .7854, and the last product will be the area.

Ex. What is the area of an ellipse whose tranverse diameter is 88, and conjugate diameter 72?

Here $88\times72\times.7854=4976.2944$, Ans.

PROBLEM XII.—*To find the area of any long irregular figure.*

RULE.—Take or measure the breadth of the figure in several places, at equal distances: Then divide the sum of these breadths by the number of them, and the quotient will be the mean breadth, nearly: by which multiply the length of the figure, and the product will be the area, near the truth.

Ex. The breadths of an irregular field, at five equidistant places, are 10, 8.1, 9.4, 10.2, and 12.3 rods, and the length is 32 rods: Required the area?

Here $(10+8.1+9.4+10.2+12.3)\div5=10$ rods.

Then $32\times10=320$ sq. rods=2 acres, Ans.

MENSURATION OF SOLIDS.

DEFINITIONS.

1. A *prism*, is a solid of equal size from end to end; the ends being parallel, equal, and like plane figures, and the sides connecting those ends are parallelograms. A prism takes particular names, according to the figure of its base or ends, whether triangular, square, rectangular, or circular, &c.

2. A *parallelopiped*, or *parallelopipedon*, is a prism bounded by six parallelograms, every opposite two of which are equal, alike, and parallel; as a stick of hewn timber of equal bigness from end to end.

3. A *cube*, is a square prism, bounded by six equal square sides, every two adjoining sides being perpendicular to each other; as a square block of wood, whose length, breadth, and thickness, are equal.

4. A *cylinder*, is a round prism, having circles for its ends, or bases; as a round log of equal bigness from end to end.

5. Solids which decrease or taper gradually from the base till they come to a point, are generally called *pyramids;* and they are of different kinds, according to the figure of their bases: Thus; if it has a *square* base, it is called a *square pyramid;* if a *triangular* base, a *triangular pyramid;* if a *circular* base, a *circular pyramid*, or a *cone*. The point where the top of a pyramid ends, is called a *vertix;* a straight line drawn from the vertix to the centre of the base, is called the *perpendicular height* of the pyramid; and a line drawn from the vertix to the middle of one of the sides of the base, is called the *slant height.*

6. A *frustum of a pyramid*, is what remains after the top is cut off by a plane parallel to the base; and is in the form of a stick of timber greater at one end than the other, whether round, or hewn square, &c. If it be *round*, it is usually called the *frustum of a cone.*

7. A *sphere* or *globe*, is a round body, bounded by one curve surface, which is every where equally distant from a certain point within, called the *centre.*

8. A *segment of a globe*, is any part cut off by a plane.

9. A *spheroid*, is a solid resembling an egg in shape, only both its ends are alike.

10. The measure or quantity of a solid, is called its *solidity*, *capacity*, or *content.*

11. Solids are measured by cubes, whose sides are an inch, a foot, or a yard, &c.; and hence the solidity of a body is said to be so many cubic inches, feet, or yards, &c. as will fill its capacity, or another of an equal magnitude.

PROBLEM I.—*To find the solid content of any Prism.*

RULE.—Find the area of one of its ends or bases, by the rule for the figure, whether it be a triangle, a square, a polygon, or a circle, &c.; then multiply this area by the length or height of the prism, and the product will be the solid content.

Note 1.—For a cube, take the cube of its side; and for a parallelopiped, take the continued product of the length, breadth and thickness, for the content.

Note 2.—The solid content will be of the same name as the dimensions which are multiplied together for finding the same; that is, if the dimensions be *inches*, the solidity will be *cubic inches;* or, if the dimensions be *feet*, then the solidity will be *cubic feet*, &c. If you multiply the area of the base in square inches by the length of the prism in feet, and divide the product by 144, the quotient will be the solid content in cubic feet.

Ex. 1. What is the solid content of a cylinder, or round prism, whose diameter is 2 feet, and length 12 feet?

Here $2\times2\times.7854=3.1416$, area of the base.

Then, $3.1416\times12=37.6992$ cub. feet, Ans.

2. If a cubical bin, whose length, breadth and depth, are each 4 feet 2 inches, be filled to the brim with corn, how many bushels will it contain?

4ft. 2in.=50 inches, and $50\times50\times50=125000$ cubic inches, the capacity of the bin. Then, $125000\div2150.4=58.12+$ bushels, Ans.

3. What is the solid content of a stick of hewn timber, which is 12 inches broad, 8 inches thick, and 30 feet long?

This is a parallelopiped; therefore, $12\times8\times30=2880$, and $2880\div144=20$ cub. ft., Ans.

4. If a piece of hewn timber be 22 inches square, and 36 feet long, how many cubic feet does it contain?

Ans. 121.

5. Required the solidity of a triangular prism, whose length is 10 feet, and the three sides of its triangular end, or base, are 3, 4, and 5 feet. Ans. 60 cub. ft.

Note 3.—The superficial content of any prism may be found thus: Multiply the perimeter, or circumference of one end of the prism, by the length, and the product will be the area of all its sides; to which add also the areas of the two ends of the prism, if required.—The area of the whole surface of any cube is equal to 6 times the square of the length of one of its sides.

PROBLEM II.—*To find the solid content of any Pyramid.*

RULE.—Find the area of the base, by the rule for the

figure, whether it be a triangle, a square, a polygon, or a circle; then multiply this area by the perpendicular height of the pyramid, and one-third of the product will be the solid content; or multiply one of those factors by one-third of the other, and the product will be the solidity required.

Ex. 1. If the perpendicular height of a triangular pyramid be 30 feet, and each side of its base 10 feet, what is its solid content?

Here, the base is a trigon, or an equilateral triangle; therefore, $10 \times 10 \times .43301 = 43.301$, the area of the base. Then, $(43.301 \times 30) \div 3 = 433.01$ cubic feet, the solidity, Ans.

2. The largest of the Egyptian pyramids is said to be 693 feet square at its base, and 499 feet high: Required its solid content. Ans. 79881417 cub. ft.

3. Required the solid content of a round stick of timber, which is 30 feet long, 18 inches in diameter at the greater end, and tapers to a point at the other.

Ans. 17.67+ cub. ft.

Note.—The *superficial content* of any pyramid may be found thus: Multiply the perimeter of the base, or sum of all its sides, by the slant height of the pyramid, and half of the product will be the area of all the sides; to which add the area of the base, if required.

PROBLEM III.—*To find the solid content of any Frustum of a Pyramid.*

RULE I.—Add into one sum, the areas of the two ends and the square root of the product of those areas, and one-third of this sum will be the mean area of a section between the two bases; which multiply by the perpendicular height or length of the frustum, and the product will be the solid content.

RULE II.—If it be the frustum of a cone; multiply the diameters of the two bases together; to the product add one-third of the square of the difference of the said diameters; then multiply the sum by .7854, and the product will be the mean area between the two bases; which being multiplied by the height or length of the frustum, will give the solidity.

When the ends or bases of the frustum are any regular right lined figures; then, multiply the side of the greater base by the side of the less; to the product add one-third of the square of the difference of those sides; multiply the sum by the proper multiplier for the figure, found in the table of multipliers for regular polygons, and the product will be the mean area between the bases; which being multiplied by the length of the frustum, will give the solidity.

Ex. 1. If a round log be 20 feet long, its diameter at the greater end 16 inches, and at the less end 10 inches; how many cubic feet does it contain?

This is the frustum of a cone, and its solidity is found by Rule I. as follows: The area of the greater end is $=16\times16\times.7854=201.0624$ sq. inches; that of the less end is$=10\times10\times.7854=78.54$ sq. in.; and the square root of the product of these areas is $\sqrt{(201.0624\times78.54)}=125.664$

Then, $(201.0624+78.54+125.664)\div3=135.0888$ sq. in., the mean area between the bases; and $(135.0888\times20)\div144=18.76+$ cub. feet, Ans.

The same answer may be found more easily by Rule II.

2. Required the solid content of a tapering stick of hewn timber, whose ends are squares; the greater end being 20 inches square, the less end 14 inches, and the length 30 feet.

By Rule II.—$20-14=6$, and $(6\times6)\div3=12$. Then, $(20\times14+12)\times1=292$ sq. inches, the mean area; and $(292\times30)\div144=60\frac{5}{6}$ cubic feet, Ans.

Note.—The *superficial content* of any frustum of a pyramid may be found thus: Add together the perimeters of the two ends; multiply the sum by the slant height, and half the product will be the area of the sides; to which add the areas of the two ends, if required.

PROBLEM IV.—*To find the area of the surface of a Sphere or Globe.*

RULE I.—Multiply the circumference of the sphere by its diameter, and the product will be the area of its surface.

RULE II.—Multiply the square of the diameter of the sphere by 3.1416.

RULE III.—Multiply the square of the circumference by .31831

Ex. Supposing the earth to be a globe, whose diameter is 7930 miles, and circumference 24912 miles, what is the area of its surface?

The answer, found by Rule I. is 197552160 sq. miles.

PRORLEM V.—*To find the solid content of a Sphere or Globe.*

RULE I.—Multiply the area of the surface by the diameter of the sphere, and take one-sixth of the product, for the solidity: Or, which is the same thing, multiply the square of the diameter by the circumference, and take one-sixth of the product.

RULE II.—Multiply the cube of the diameter by .5236, for the solid content.

RULE III.—Multiply the cube of the circumference by .016887

Ex. If the earth be a globe, whose diameter is 7930 miles, and circumference 24912 miles, how many cubie miles are contained in it?

The answer, found by Rule I., is 261,098,104,800 cubic miles.

PROBLEM VI.—*To find the solid content of any Segment of a Globe.*

RULE.—To three times the square of the semidiameter of the segment's base, add the square of its height; then multiply the sum by the height, and the product by .5236, for the content.

Ex. A collier, wishing to burn a quantity of wood into charcoal, piled it up into a stack in the form of a segment of a globe, the height of which was 9 feet, and the diameter of the base 20 feet: how many cubic feet of wood did the pile contain? Ans. 1795.4244 cub. ft.*

PROBLEM VII.—*To find the solid content of a Spheroid.*

---

* The solid content of any *coal pit*, of the usual form, may easily be found as follows: Measure with a rope or chain across the top of the stack of wood, from the bottom on one side to the bottom on the opposite side; then divide the cube of this measure, in feet, by 1895, and the quotient will be the required content, in cords.

Rule.—Multiply the square of the shortest diameter by the longest diameter; then multiply the product by .5236, and the last product will be the solidity.

Ex. If the length, or longest diameter of a spheroid, be 40 inches, and the shortest diameter 30 inches, what is its solidity? Ans. 18849.6 cub. inches.

Problem VIII.—*To find the solid content of any irregular body.*

Rule.—Put the body, or quantity which you would measure, into any vessel of regular shape, and then fill the vessel with water, sand, or any other convenient substance. Then take out the body, and measure the space left empty by its removal, according to the preceding rules, and the content of this space will be the solid content of the irregular body.

MENSURATION OF BOARDS AND TIMBER.

Problem I.—*To find the area or superficial content of any Board.*

Rule.—Multiply the length of the board by the mean breadth, and the product will be the superficial content required.

Note 1.—When the board is broader at one end than at the other; if it tapers regularly, take the breadth in the middle; or else add the breadths of the two ends together, and take half the sum, for the mean breadth: But, if the board does not taper regularly, you may measure the breadth in several places at equal distances, and then divide the sum of these breadths by the number of them, and the quotient will be the mean breadth.

Note 2.—If you multiply the length in feet by the mean breadth in inches, and divide the product by 12, the quotient will be the superficial content in square feet.—In measuring boards, timber, &c. it will be convenient to make use of a measuring rule in which the feet are divided decimally, viz. each foot into 10 equal parts, and these again into 10 parts; for then the dimensions may be taken in feet and decimal parts, and multiplied together like other decimal numbers.

Note 3.—When it is required to measure several boards of the same length; it will be the best way to add all the widths together; (or, if the boards be all of the same width, multiply the width of one board by the number of boards;) then multiply the sum (or product) by the length of one of the boards, and the product will be the contents of all the boards.

Ex. 1. How many square feet are contained in a board which is 13 feet long, and 16 inches wide?

Here $16\times13=208$, and $208\div12=17\frac{1}{3}$ sq. feet, Ans.

2. Required the superficial content of a board which is 12 feet 5 inches long, and 1 foot 6 inches wide, and also its value, at 4 cents per square foot.

The content may be found by several different methods, as follows:—

1st Method.

Length, 149 inches,=12ft. 5in.
Breadth, 18 in.=1ft. 6in.

1192
149

Content, 2682 square inches,=
18 square feet, 90 sq. in.

2d, By Duodecimals.

Ft. ′
12 .. 5
1 .. 6

6 .. 2 .. 6
12 .. 5

Content, 18 .. 7 .. 6

3d, By Practice.

Ft. ′
6in.=½)12 .. 5 ..
6 .. 2 .. 6″

Ans. 18 .. 7 .. 6

4th, By Decimals, thus: 12ft. 5in.= 12.41+ft., and 1ft. 6in.=1.5ft. Then, $12.41\times1.5=18.615$ sq. feet, Ans.

5th, By Vulgar Fractions, thus: 12ft. 5in.=$12\frac{5}{12}$ft., and 1ft. 6in.=$1\frac{1}{2}$ft. Then, $12\frac{5}{12}\times1\frac{1}{2}=\frac{149}{12}\times\frac{3}{2}=\frac{447}{24}=18\frac{5}{8}$sq.ft.

The value is $74\frac{1}{2}$ cents, which may be found by the Rule of Three, or by Practice.

3. How many square feet are contained in a board which is 12 feet long, 14 inches wide at the greater end, and 10 inches wide at the less end? Ans. 12.

4. Having occasion to measure an irregular mahogany plank of 10 feet in length, I found it necessary to measure several breadths, at equal distances from each other, viz.

at every two feet. The breadth of the less end was 8 inches, and that of the greater end 14 inches: the intermediate breadths were 10 inches, 9 inches, 12 inches and 13 inches. How many square feet were contained in the plank?

Ans. $9\frac{1}{8}$.

PoBLEM II.—*Having the breadth of a rectangular board given, to find how much in length will make a square foot, or any other quantity assigned.*

RULE.—Divide the number of square inches in the given area by the width of the board in inches, and the quotient will be the length required in inches.

Ex. From a mahogany plank, 22 inches broad, a square yard is to be cut off: at what distance from the end must the line be drawn?

1sq. yd.=1296sq. in., and $1296 \div 22 = 58\frac{10}{11}$ inches, Ans.

PROBLEM III.—*To find the solid content of any piece of timber.*

RULE I.—If the stick, or piece of timber, be a *prism*, or a *pyramid*, or the *frustum of a pyramid*, its solid content may be found by the rule for the figure, as in the Mensuration of Solids.

RULE II.—The customary method of measuring timber, is to gird the stick round the middle with a string; then the square of one-fourth of this girt being multiplied by the length of the stick, gives the solid content, nearly.—This Rule is commonly called the *quarter-girt rule.*

Note 1.—If the piece of timber be very irregular, gird it in several places equally distant from each other, and divide the sum of these circumferences by the number of them, for the mean circumference.

Note 2.—If the length of the piece of timber be taken in feet, and the other dimensions in inches; then you may multiply the area of the base in square inches, or the square of the quarter girt, in inches, by the length in feet, and divide the product by 144, and the quotient will be the solid content in cubic feet.

EXAMPLES, *performed by Rule* II.

1. If the circumference of a stick of hewn timber be 60 inches, and the length 20 feet, how many cubic feet does it contain?

60÷4=15 inches, the quarter girt. Then, 15×15×20=4500, and 4500÷144=31¼ cub. ft. Ans.

2. If the circumference of the greater end of a round stick of timber be 136 inches; that of the less end 32 inches; (or which is the same thing, if the quarter girt in the middle be 21 inches;) and the length 48 feet; how many cubic feet does it contain? Ans. 147.

3. How many cubic feet are contained in the trunk of a tree, whose length is 17¼ feet, and its circumference in five places, at equal distances, as follows; viz. 9.43 feet, 7.92 feet, 6.15 feet, 4.74 feet, and 3.16 feet?

Ans. 42.51+cub. ft.

Problem IV.—*Having given the size of a piece of timber of equal bigness from end to end, to find how much in length will make a given number of cubic inches, or feet, &c.*

Rule.—Divide the given number of cubic inches by the area of one end of the stick, in square inches, and the quotient will be the length required, in inches.

Ex. 1. From a stick of hewn timber, 8 inches wide and 6 inches thick, I would cut off a solid foot: what length of the stick must I cut off?

Here 8×6=48 sq. inches, the area of one end.
Then, 1728÷48=36 inches=3 feet, Ans.

2. If a round stick of timber be 10 inches in diameter, how much in length will make 4 solid feet?

Ans. 7 ft. 4 in.+

Problem V.—*To find how many cubic feet a round stick of timber will contain, when hewn square.*

Rule.—Multiply the square of the diameter of the less end of the stick, in inches, by half the length in feet; divide the product by 144, and the quotient will be the answer.

Ex. If the diameter of a round stick of timber be 22 inches, and its length 12 feet, how many solid feet will it contain when hewn square?

Here (22×22×6)÷144=20⅙ cub. feet, Ans.

Problem VI.—*To find how many feet of square edged*

boards, of a given thickness, can be sawn from a log of a given diameter.

RULE.—Find (by Prob. 5th) the solid content of the log, when made square: Then say, as the thickness of the board including the saw calf : is to 12 inches :: so is the number of cubic feet contained in the log when made square : to the number of square feet of boards.

Ex. How many feet of square edged boards, $1\frac{1}{4}$ inch thick, including the saw calf, can be sawn from a log 16 feet long, and 24 inches in diameter?

$(24\times24\times8)\div144=32$ cub. ft. Then, as $1\frac{1}{4}$in. : 12 in. :: 32 : $307\frac{1}{5}$ sq. ft., the Ans.

MENSURATION OF ARTIFICERS' WORKS.

Artificers compute the contents of their works by several different measures; as glazing and masons' flat work, by the square foot; painting, plastering, paving, &c. by the square yard; flooring, partitioning, roofing, tiling, &c. by the square of 100 square feet; and brick-work, either by the square yard or square rod.

All works, whether superficial or solid, are computed by the rules proper to the figure of them, whether it be a triangle, a rectangle, a parallelopiped, or any other figure.

ARTICLE 1.—*Of Glaziers' and Masons' flat work, estimated by the square foot.*

Ex. 1. There is a house with three tiers of windows, three in a tier; the height of the first tier is 7 feet 10 inches; that of the second 6 feet 8 inches; that of the third 5 feet 4 inches; and the breadth of each window is 3 feet 10 inches: What will the glazing come to, at 15 cents a square foot?

Ft. in.	
7 .. 10	The
6 .. 8	heights,
5 .. 4	added.
19 .. 10	
3	No. of windows in a tier.
59 .. 6	Height of all the windows.

Then, by Multiplication of Duodecimals:

Ft. in.
59 .. 6
3 .. 10
49 .. 7 .. 0
178 .. 6
228 .. 1 .. 0

Then, $228\frac{1}{12}\times.15=\$34.2125$, the Ans.

2. What is the price of a marble slab whose length is 5 feet 7 inches, and breadth 1 foot 10 inches, at 20 cents a square foot? Ans. $2.047+

ART. 2.—*Of Pavers', Painters', Plasterers', and Joiners' works, estimated by the square yard.*

Ex. 1. What will the paving of a court yard come to at 50 cents per square yard, if the length be 27 feet 6 inches, and the breadth 14 feet 9 inches?

27ft. 6in.=27.5ft., and 14ft. 9in.=14.75ft. Then, 27.5 ×14.75=405.625 sq. ft.=45.06+sq. yards, and 45.06×.50 =$22.53, the Ans.

2. If a room be 7 yards long, 6 yards wide, and 3 yards high, what is the expense of plastering its sides and upper part, at the rate of 10 cents per square yard?

(7+6)×2=26 yards, the circumference of the room. Then, 26×3=78 sq. yards, the area of the sides, and 7×6= 42 sq. yards, the area of the upper part. Then 78+42= 120 sq. yards, the superficial content of the whole wall, which, at 10 cents per square yard, amounts to $12, Ans.

3. How many square yards are contained in a piece of wainscotting, which is 8 feet 3 inches long, and 6 feet 6 inches broad?

Ans. 5.95+sq. yards; or 5 sq. yd. 8 sq. ft. 7′..6″.

ART. 3.—*Measuring by the square of* 100 *square feet; as Flooring, Partitioning, Roofing, Tiling, &c.*

Ex. 1. How many squares of 100 square feet, are contained in the floor of a room which is 32 feet 6 inches long, and 24 feet 3 inches wide?

Ans. 7.88125 squares; or 7 squares, and 88sq. ft. 1′..6″.

2. Suppose a house is 44 feet 6 inches in length, and 18 feet 3 inches in breadth, and the roof is of a true pitch; what will the roofing amount to, at $2.40 per square of 100 square feet? Ans. $29.2365

Note.—The roof of a house is said to be of a true pitch, when the width of the roof, on each side, or the length of each rafter, is equal to three-fourths of the breadth of the building: then the superficial content of the whole roof is found by multiplying the length of the house by $1\frac{1}{2}$ times the breadth; or by multiplying the length by the breadth, and then increasing the product by $\frac{1}{2}$ of itself.

Art. 4.—*Of Bricklayers' Work.*

Bricklayers estimate their work either by the square yard, (of 9 square feet,) or the square rod, (of $272\frac{1}{4}$ square feet,) and always at the rate of a brick and a half thick. So that if a wall be more or less than this standard thickness, it must be reduced to it, by multiplying the superficial content of one side of the wall by the number of half-bricks in the thickness, and then dividing the product by 3.

Note.—The whole length of a wall which forms the sides of any building may be found by measuring the length on the outside and on the inside of the building, and taking half the sum of these two measures: Or, if the base of the building be a rectangle, the whole length of the wall may be found by subtracting 4 times its thickness from twice the sum of the length and breadth of the house.

Ex. 1. How many square rods of standard brick-work are contained in a wall, which is 68 feet, 1 inch, in length, 24 feet high, and $2\frac{1}{2}$ bricks, or 5 half-bricks, thick?

Here $68\frac{1}{12}\times24=1634$ sq. feet,$=6+$sq. rods.

Then, $(6\times5)\div3=10$ standard square rods, Ans.

2. How many square yards of standard brick-work are contained in a wall, 62 feet 6 inches long, 14 feet 8 inches high, and 2 bricks thick?

Ans. 135.8+sq. yards, or 135sq. yd. 7sq. ft. 2′ .. 8″.

Art. 5.—*To find how many bricks, of a given size, will be necessary to build a wall, of any given dimensions.*

Rule.—As the solid content of one brick : is to that of the wall :: so is 1 : to the number of bricks required.

Ex. A man is determined to build a brick house of the following dimensions; viz. length 40 feet; breadth 30 feet; height of the eaves, or lower part of the roof, 20 feet; perpendicular height of each triangular gable end 16 feet: the walls are to be a foot thick, and there are to be left through them 30 places for windows, each 6 feet high and 4 feet wide, and 5 places for doors, each 7 feet high and 5 feet wide. How many bricks 8 inches long, 4 wide, and $2\frac{1}{2}$

thick, must be taken to build the walls of the house?

(40+30)×2=140 feet, from which subtracting 4 times the thickness of the wall, the remainder, 136 feet, is the whole length of wall. Then, 136×20×1=2720 cub. feet, the solid content of the wall, exclusive of the two gable ends, and 30×16×1=480 cub. feet, the contents of both the gable ends; and hence the whole content of the wall, including the places for windows and doors, is 2720+480=3200 cubic feet.

Then, (6×4×30)+(7×5×5)=720+175=895 cubic feet, to be deducted for the windows and doors; which being taken from 3200 cubic feet, the remainder, 2305 cubic feet,= 3983040 cubic inches, is the solid content of all the bricks.

Now, one brick contains 8×4×2½=80 cub. inches: Therefore, as 80 cub. in. : 3983040 cub. in. :: 1 brick : 49788 bricks, Ans.

GAUGING.

Gauging is the art of measuring and finding the contents or capacities of all sorts of vessels; such as casks, brewers' vessels, &c. &c.

The content of any vessel in the shape of any of the solids mentioned in the "Mensuration of Solids," may be found in gallons, or bushels, &c. by the following

RULE.

Measure the inside of the vessel, according to the rule for the figure, and calculate its content or capacity, in cubic inches, as though it were a solid. Then the content, thus found, may be reduced to wine gallons, by dividing by 231, the number of cubic inches in a wine gallon; or to beer gallons, by dividing by 282; or to bushels, by dividing by 2150.4, &c.

Ex. 1. There is a bin, in form of a parallelopiped, 60 inches long, 40 inches wide, and 50 inches deep: How many bushels of corn will it hold?

Here (60×40×50)÷2150.4=55.8+bushels, Ans.

2. If the length of a cylindrical wooden bottle be 6 inch-

es, and the diameter 7 inches, how much wine will it hold?

Ans. 1 gallon, nearly.

3. The diameter of a round mash tub is 36 inches at the bottom, within, and 42 inches at the top; and its perpendicular height or depth is 48 inches: Required its content in wine and beer gallons.

Ans. 248.71+wine gal., or 203.73+beer gal.

N. B. The tub mentioned in the last question being the frustum of a cone, its content, in cubic inches, is found by the rule for solids of that form.

To find the capacity of a Cask of the usual form.

RULE I.

1. Take the dimensions of the cask, within, in inches, viz. the length of the cask, and the diameters at the bung and head.

2. To the head diameter add two-thirds of the difference between the head and bung diameters, and the sum will be the mean diameter; but, if the staves be but little curving, add only six-tenths of the said difference.

3. Square the mean diameter, thus found, and multiply the square by the length; then divide the product by 294 for wine gallons, or by 359 for beer gallons.

Note.—There is a difference in the thickness of the heads of different casks, for which proper allowances must be made in taking the length. The head diameter may be taken on the outside, close to the chimes; to which add, for small casks, three-tenths of an inch; for casks of 40 or 50 gallons, four-tenths; and for larger casks, 5 or 6 tenths, and the sum will be very nearly the head diameter within.

RULE II.

With any straight rod, take the diagonal of the cask, from the centre of the bung hole to the end or head of the cask on the opposite side; then multiply the cube of this diagonal in inches by .00272 for wine gallons, or by .002228 for beer gallons.

Note.—The diagonal should be measured both ways from the bung hole; and, if these two measures differ, take half their sum, for a mean diagonal.

Ex. 1. Required the content, in wine and beer gallons, of a cask, whose bung diameter is 28 inches, head diameter 22 inches, and length 40 inches.

By Rule I. thus: $28-22=6$, the difference between the bung and head diameters, two-thirds of which is 4; and hence the mean diameter is $22+4=26$ inches. Then,

$$\left.\begin{array}{l}(26\times26\times40)\div294=91\frac{143}{147} \text{ wine gallons.}\\(26\times26\times40)\div359=75\frac{115}{359} \text{ beer gallons.}\end{array}\right\} \text{Ans.}$$

2. How many wine gallons in a cask whose diagonal is 34 inches? Ans. 106.90688 gal.

To find a Ship's burthen, or to Gauge a Ship.

There is such a diversity in the forms of ships, that no general rule can be applied to answer all varieties; however, the following rules are practised.

Rule I.—Multiply continually together, the length of the keel, the breadth of the ship at the main beam, and the depth of the hold, in feet; divide the product by 95, and the quotient will be the ship's burthen, in tons.

Rule II.—Multiply together the length of the keel, the breadth of the beam, and half the said breadth; divide the last product by 95, and the quotient will be the tonnage.

Rule III.—The weight of a ship's burthen is half the weight of the water she can hold.

Ex. What is the tonnage of a ship, whose length is 72 feet, breadth 24 feet, and depth 12 feet?

By Rule I. thus: $(72\times24\times12)\div95=218\frac{26}{95}$ tons, Ans.

The same answer may be found by Rule II.

MISCELLANEOUS PROBLEMS.

Problem I.—*Having the sum and difference of any two numbers given, to find those numbers.*

Rule.—From the sum of the required numbers take their difference, and half of the remainder will be the less number; to which add the difference, and the amount will be the greater number.

Or, half the difference of the required numbers added to half their sum will give the greater number, and the said half difference subtracted from the half sum will give the less number.

Example 1. Required to find two numbers whose sum is 25, and difference 11?

Here, $(25-11)\div2=14\div2=7$, the less number; and $7+11=18$, the greater number.

Or, $25\div2=12\frac{1}{2}$=half the sum, and $11\div2=5\frac{1}{2}$=half the difference. Then, $12\frac{1}{2}+5\frac{1}{2}=18$, the greater number; and $12\frac{1}{2}-5\frac{1}{2}=7$, the less number.

2. Divide 100 dollars between A and B, so that A may have 14 dollars more than B.

Ans. A $57, and B $43.

PROB. II.—*Having the sum of two numbers and their product given, to find those numbers.*

RULE.—From the square of their sum subtract 4 times their product; then extract the square root of the remainder, and it will be the difference of the two numbers. You will then have the sum and difference of the two required numbers, from which you may find the numbers by Prob. I.

Ex. 1. The sum of two numbers is 28, and their product is 147: what are those numbers?

$(28\times28)-(147\times4)=784-588=196$, and $\sqrt{196}=14$, the difference of the required numbers. Then, by Prob. 1st, $(28-14)\div2=7$, the less number; and $7+14=21$, the greater number.

2. A certain tract of land, in form of a right angled parallelogram, is 160 rods in circuit, and contains 9 acres, 1 rood, and 20 square rods. I demand the length and breadth of the tract? Ans. Length 50 rods, breadth 30 rods.

PROB. III.—*Having the difference of two numbers, and their product given, to find those numbers.*

RULE.—To the square of their difference add 4 times their product, and the square root of the amount will be the sum of the two required numbers. Then the numbers may be found by Prob. I.

Ex. 1. What are those two numbers whose difference is 7, and product 44? Ans. 4 and 11.

2. It is required to lay out 47 acres 2 roods and 16 square rods of land, in form of a parallelogram, the length of which shall exceed the breadth by 80 rods.

Ans. 136 rods and 56 rods.

PROB. IV.—*Having the sum of two numbers, and the sum of their squares given, to find those numbers.*

RULE.—From twice the sum of their squares subtract the square of their sum, and the square root of the remainder will be the difference of the two numbers. Then find the numbers by Prob. I.

Ex. 1. The sum of two numbers is 50, and the sum of their squares is 1700: what are those numbers?

Ans. 10 and 40.

2. The length of the hypothenuse of a right angled triangle is 5 chains, and the sum of the base and perpendicular is 7 chains: required the length of each of the two last mentioned sides.

Ans. The base is 4 chains, and the perpendicular is 3 chains.

Note.—The square of the hypothenuse, or longest side, of any right angled triangle, is equal to the sum of the squares of the base and perpendicular; and consequently the square of either of the two last mentioned sides is equal to the difference of the squares of the other two sides.

PROB. V.—*Having the difference of two numbers, and the sum of their squares given, to find those numbers.*

RULE.—From twice the sum of their squares subtract the square of their difference, and the square root of the remainder will be the sum of the two numbers. Then find the numbers by Prob. I.

Ex. The hypothenuse of a right angled triangle measures 20 rods, and the difference of the other two sides is 4 rods—these two sides are required.

Ans. 12 rods and 16 rods.

PROB. VI.—*Having the sum of two numbers, and the difference of their squares given, to find those numbers.*

RULE.*—Divide the difference of the squares by the sum of the numbers, and the quotient will be the difference of the two numbers. Then find the numbers by Prob. I.

Ex. 1. One-hundred dollars so divide,
Between two worthy men,
That when each part is fairly squar'd,
The difference is but ten.

Ans. The greater part is $50.05, the less $49.95.

2. A liberty pole 100 feet high, standing on a plain, breaks and hangs on the stump, so that the top rests on the ground at the distance of 10 feet from the upright part: At what height from the ground did it break?

Ans. $49\frac{1}{2}$ feet.

PROB. VII.—*Having the difference of two numbers, and the difference of their squares given, to find the numbers.*

RULE.—Divide the difference of the squares by the difference of the numbers, and the quotient will be the sum of the numbers. Then find the numbers by Prob. I.

Ex. 1. Between my brother's age and mine,
Are just a dozen years;
And forty dozen are between
Their second pow'rs, or squares.
What are our ages? Ans. 26 yr. and 14 yr.

2. The base of a right angled triangle is 15 chains, and the difference between the hypothenuse and perpendicular is 2.5 chains—these two sides are required.

Ans. 46.25 ch. and 43.75 ch.

PROB. VIII.—*To find an arithmetical mean between any two given numbers.*

RULE.—Divide the sum of the given numbers by 2, and the quotient will be the arithmetical mean required.

Ex. Find an arithmetical mean between the two numbers 4 and 10. Ans. 7.

* The product of the sum and difference of any two numbers is equal to the difference of the squares of the numbers; whence the reason of the rules for Problems 6th and 7th is obvious.

PROB. IX.—*To find any assigned number of arithmetical means between two given terms or extremes.*

RULE.—Subtract the less term or extreme from the greater; divide the remainder by 1 more than the number of means required to be found; that is, divide by 2 for 1 mean, by 3 for 2 means, &c., and the quotient will be the common difference of the terms; which being added continually to the least term, or subtracted from the greatest, will give the mean terms required.

Ex. Required to find 5 arithmetical means between the numbers 2 and 14.

$(14-2)\div6=2$, the common difference of the terms, which being added continually to the least term, gives 4, 6, 8, 10, and 12, for the means required.

PROB. X.—*To find a geometrical mean proportional between any two given numbers.*

RULE.—Multiply the two numbers together, and extract the square root of their product, which will give the mean proportional sought.

Ex. Find a geometrical mean between 4 and 9.

Here $\sqrt{(9\times4)}=6$, Ans.

PROB. XI.—*To find any assigned number of geometrical means between two given numbers or extremes.*

RULE.—Divide the greater number by the less, and extract such root of the quotient whose index is 1 more than the number of means required; that is, extract the 2d root for one mean, the 3d root for two means, and so on; and that root will be the common ratio of all the terms: Then, with the ratio, multiply continually from the least term, or divide continually from the greatest, and you will have the mean terms required.

Ex. Required to find three geometrical means between the numbers 2 and 162.

$162\div2=81$, and $\sqrt[4]{81}=3$, the ratio. Then, $2\times3=6$, and $6\times3=18$, and $18\times3=54$.

Or, $162\div3=54$, and $54\div3=18$, and $18\div3=6$.

So, 6, 18, and 54, are the means required.

PROB. XII.—*Having given the base and perpendicular of*

*any triangle, to find the side of a square inscribed in the same.**

RULE.—Divide the product of the base and perpendicular by their sum, and the quotient will be the side of the inscribed square.

Ex. If the base of a triangle be 30 rods, and the perpendicular 20 rods, what is the side of a square inscribed in the triangle?

Here $(30\times20)\div(30+20)=12$ rods, Ans.

PROB. XIII.—*The diameter of a circle being given, to find the side of a square inscribed in the circle.*

RULE.—Multiply the diameter of the circle by .707106, and the product will be the side of the inscribed square.

Ex. Required the side of a square inscribed in a circle whose diameter is 100 feet. Ans. 70.7106 feet.

PROB. XIV.—*To find the side of a square equal in area to any given superficies whatever.*

RULE.—Extract the square root of the number of square rods, or feet, &c. contained in the given superficies, and this root will be the side of the square sought, in the same name with the area.

Ex. 1. If the area of a triangle be one acre, what is the side of a square equal in area thereto?

$\sqrt{160}=12.64+$rods, Ans.

2. Find the side of a square which is equal in area to a parallelogram 100 rods long and 25 rods wide.

Ans. 50 rods.

PROB. XV.—*To lay-out a given quantity of land in form of a parallelogram, having the length to the breadth in a given ratio.*

RULE.—As the greater number of the given ratio, is to the less; so is the given area, to a fourth term, whose square root is the breadth. Then, as the less number of the given ratio, is to the greater; so is the breadth, to the length: Or, the length may be found by dividing the area by the breadth.

* One right-lined figure is *inscribed* in another, or the latter circumscribes the former, when all the angular points of the former are placed in the sides of the latter.

Ex. 1. It is required to lay out 25 acres of land in a rectangular form, having the length to the breadth in the ratio of 8 to 5.

25 acres=4000 square rods.
As 8 : 5 :: 4000 sq. rd. : 2500 sq. rd.
Then, $\sqrt{2500}$=50 rods, the breadth. } Ans.
As 5 : 8 :: 50 : 80 rods, the length. }

2. Suppose I would set out an orchard of 600 trees, so that the number of trees in length shall be to the number in breadth as 3 to 2, and the distance of each tree from the next 7 yards; how many trees must the orchard be in length, and how many in breadth, and how many square yards of ground will it contain?

Ans. 30 trees in length and 20 in breadth; and, if no space be left on the outside of the orchard, it will contain 26999 sq. yards of ground.

PROB. XVI.—*To find the diameter of a circle equal in area to an ellipsis, (or oval,) whose tranverse and conjugate (or longest and shortest) diameters are given.*

RULE.—Find, by Prob. 10th, a geometrical mean between the two given diameters of the ellipsis, and that mean will be the diameter of a circle equal to the ellipsis.

Ex. Find the diameter of a circle equal in area to an ellipsis whose tranverse diameter is 81 feet, and conjugate diameter 64 feet. Ans. 72 feet.

PROB. XVII.—*To find the side of a cube equal to any given solid.*

RULE.—The cube root of the number of cubic inches, or feet, &c. contained in the given solid, will be the side of the cube sought.

Ex. The statute bushel contains 2150.4 cubic inches: I demand the side of a cubic box that will hold that quantity?

$\sqrt[3]{2150.4}$=12.9+ inches, Ans.

PROB. XVIII.—*The solid content of any frustum of a cone, its height or length, and the diameter of one end, being given, to find the diameter of the other end.*

RULE.—Multiply the height or length of the frustum by the decimal .2618, and divide the solid content by that product: From the quotient subtract three-fourths of the square of the given diameter, and from the square root of the remainder subtract half of the given diameter; and the last remainder will be the diameter required.

Note.—If the given dimensions and the solid content of the frustum are not all of the same name, they must be reduced to the same: Thus, if the dimensions be in inches, then the solid content must be in cubic inches, &c.—This note must also be attended to in solving other problems of the like nature.

Ex. A cooper would make a tub in the form of a conic frustum, to hold just 10 barrels, wine measure; and would have the inside diameter at the bottom 40 inches, and the perpendicular height or depth 50 inches. What must be the diameter at the top?

10 barrels contain 72765 cubic inches. Half of 40 is 20, and $\frac{3}{4}$ of the square of 40 is 1200. Then,

$$\sqrt{\left(\frac{72765}{50\times.2618}-1200\right)}-20=46+\text{ inches, Ans.}$$

Prob. XIX.—*When the dimensions of any surface and its area are given, to find the dimensions of any larger or smaller surface of similar shape.**

Rule.—As the area of the surface whose dimensions are known, is to that of the other; so is the square of any one of the dimensions of the former, to the square of the corresponding or like dimension of the latter; the square root of which is the dimension required.

Or, if the surface whose dimensions are required be $\frac{1}{2}$, $\frac{1}{3}$, or twice, or three times, &c. as large as the other; then the square root of the like part, or multiple, of the square of any dimension of the latter, will be the corresponding dimension of the former.

Note.—When there are several dimensions to be found; then, after having found one of them by the above rule, the rest may be found thus: As the dimension of the given surface which corresponds with the known dimension of the other, is to the said known dimension; so is any other dimension of the former, to the corresponding dimension of the latter.

* The areas of similar surfaces are to each other as the squares of their like linear dimensions; and the contents of similar solids are proportional to the cubes of their like dimensions; whence the reason of the rules for Problems 19th and 20th is obvious.

Ex. 1. The three sides of a triangle are 6, 8, and 10 rods, and the area is 24 square rods: I demand the sides of a similar triangle, whose area is 96 square rods?

As 24 : 96 :: 6×6 : 144. Then √144=12 rods, the length of the shortest side. Then I find the other sides thus: As 6 : 12 :: 8 : 16 rods; and, as 6 : 12 :: 10 : 20 rods.—Ans. The three sides are 12, 16, and 20 rods.

2. The side of a certain square is 8 rods: I demand the side of another square, which is $\frac{1}{4}$ as great?

Here (8×8)÷4=16, and √16=4 rods, Ans.

3. The diameter of a certain circle is 10 yards: I demand the diameter of a circle 9 times as large?

(10×10)×9=900, and √900=30 yards, Ans.

4. The base of a triangle measures 4 rods, and the area of the triangle is 6 square rods: I demand the area of a similar triangle whose base measures 8 rods?

As 4×4 : 8×8 :: 6 sq. rd. : 24 sq. rd. Ans.

Prob. XX.—*When the dimensions of any solid and its content are given, to find the corresponding dimensions of another solid of similar shape, but either greater or less.*

Rule.—As the content of the solid whose dimensions are known, is to that of the other; so is the cube of any dimension of the former, to the cube of the corresponding dimension of the latter; the cube root of which will be the dimension required.

Or, if the content of the solid whose dimensions are required, is any part or multiple of the content of the other solid, then the cube root of the like part or multiple of the cube of any dimension of the latter, will be the corresponding dimension of the former.

Note 1.—Observe the same Note here that is annexed to the rule for Prob. 19th.

Ex. 1. If the length of a cylindrical gallon bottle be 6 inches, and the diameter 7 inches; what are the dimensions of a quart bottle of similar shape?

(6×6×6)÷4=54, and ∛54=3,7+ inches, the length of the quart bottle. Then, as 6 : 3.7 :: 7 inches : 4.3+ inches, the diameter.

2. If a bullet of 3 inches diameter weigh 4lb. what must

be the diameter of a bullet, made of the like metal, to weigh 32lb.?

4lb. : 32lb. :: 3×3×3 : 216. Then, $\sqrt[3]{216}=6$ inches, Ans.

3. The diameter of the earth is about 7930 miles, and that of the sun 883246 miles: Required the ratio of the magnitudes of these two bodies.

$$\text{As } 7930^3 : 883246^3 :: 1 : 1381737+$$

Ans. The sun is 1381737 times as large as the earth.

4. If a ship of 300 tons burthen be 75 feet long in the keel, how long is the keel of a similar built ship whose burthen is 600 tons? Ans. 94.4+ft.

5. If a ship of 300 tons burthen be 75 feet long in the keel, what is the burthen of another ship of the same form, whose keel is 100 feet long? Ans. $711\frac{1}{9}$ tons.

Note 2.—The strength of cables, and consequently the weights of their anchors, are as the cubes of the diameters or peripheries of the cables.

6. If an anchor of $2\frac{1}{4}$cwt. require a cable 6 inches in circumference, what must be the circumference of a cable for an anchor of 18cwt.? Ans. 12 inches.

PROB. XXI.—*If the dimensions of any surface be measured by a chain, or a measuring rule, &c. that is either too long or too short, and the area of the surface be computed from the erroneous measure; then, to find the true area.*

RULE.—As the square of the length of a true chain, or measuring rule, &c., is to the square of the length of the chain, &c. used; so is the area computed from the erroneous measure or survey, to the true area required.

Ex. Suppose a field, measured by a two rod chain 3 inches too long, is found (from the erroneous measure) to contain 41 acres 1 rood and 33 square rods; what is the true area of the field?

33ft. 3in.=399in. Length of the chain used.

33ft.=396in. True length of a 2 rod chain.

		A. R. sq. rd.		A. R. sq. rd.	
Then, 396^2 : 399^2	::	41 .. 1 .. 33	:	42 .. 0 .. 13+	Ans.

PROB. XXII.—*When the dimensions of any solid have*

been measured by a measuring rule, &c. that is either too long or too short, and the content of the solid computed from the erroneous measure; then, to find the true content.

RULE.—As the cube of the length of a true measuring rule, &c., is to the cube of the length of that used; so is the solid content computed from the erroneous measure, to the true content.

Ex. If a stick of timber be measured by a measuring rule supposed to be 2 feet long, but which is found to be only 23.8 inches long, and the solid content of the stick, as calculated from the erroneous measure, be 40 cubic feet; what is the true content? Ans. 39+ cub. ft.

PROB. XXIII.—*To find the true weight of any thing by a pair of scales, the beam of which is unequally divided.*

RULE.—Weigh the given body, or quantity, in each scale; then, the mean proportional between these two weights, (found by Prob. 10th,) will be the true weight required.

Ex. Suppose a bar of silver weighs 10 ounces in one scale of a false balance, and 12 ounces in the other; what is the true weight?

Here $\sqrt{(12\times10)}=10.954+$ oz. Ans.

PROB. XXIV.—*The quantities of water that may run through two or more pipes of different diameters, in any given time, are directly, and the times in which any given quantity of water may pass through them, are inversely, as the squares of the diameters of the pipes.*

Ex. 1. If 120 gallons of water will run through a pipe of 2 inches diameter in 10 minutes, how many gallons will run through another pipe, of 3 inches diameter, in the same time? As $2\times2 : 3\times3 :: 120$ gal. : 270 gallons, Ans.

2. If a pipe, 1 inch in diameter, will fill a certain cistern in an hour, in what time will a pipe of 2 inches diameter fill the same cistern?

As $2\times2 : 1\times1 :: 60$ min. : 15 minutes, Ans.

3. If 100 gallons of water run through a pipe of 2 inches diameter in 8 minutes, what is the diameter of a pipe that will discharge the same quantity in 2 minutes?

As 2min. : 8min. :: 2×2in. : 16in., the square of the diameter. Then, $\sqrt{16}=4$ inches, Ans.

PROB. XXV.—*To estimate the distance of objects on level ground, or at sea, having only the height given.*

RULE.—1. To the earth's diameter, (viz. 7930 miles, or 41870400 feet,) add the height of the eye, and multiply the sum by that height; then, the square root of the product will be the distance, at which an object on the surface of the earth or water, can be seen by an eye so elevated.

2. To find the distance between two elevated objects, when a right line joining them touches the earth's surface; work for each object separately, as above directed, and the sum of the distances, thus found, will be the distance sought.

Ex. Chimborazo, the highest peak of the Andes, and the highest mountain in America, is, in height, about 4 miles above the level of the sea, and may be seen at a great distance, by a spectator on the Pacific ocean. How far may this lofty summit be seen by a spectator on the Pacific, supposing the eye of the observer to be elevated 20 feet above the surface of the water, and no other mountain to intervene?

$$\sqrt{(\overline{41870400+20}\times20)}=28938\text{ft.}=5.48+ \text{ miles.}$$
$$\sqrt{(\overline{7930+4}\times4)}=178.14+ \text{ miles.}$$

Ans. 183.62+ miles.

PROB. XXVI.—*To estimate the height of objects on level ground, or at sea, having only the distance given.*

RULE.—From the given distance, take the distance which the elevation of your eye above the surface will give, found by Prob. 25th. Divide the square of the remainder by the diameter of the earth, and the quotient will be the height required.

Ex. Just rising from the sea I've seen,
When station'd in the shrouds,
The lofty peak of Teneriffe,
That penetrates the clouds—
Just fifty leagues from it was I,
My reck'ning being true,
And from the water to my eye,
The feet were eighty-two.

Suppose these observations just,
To make the question brief,
Above the level of the sea,
How high is Teneriffe?

$\sqrt{(41870400+82\times82)}=58595+$ft.$=11.097+$ miles.

Now, 50 leagues=150 miles, and $150-11.097=138.903$. Then, $(138.903\times138.903)\div7930=2.433+$ miles, the height of the peak of Teneriffe, Ans.

Prob. XXVII.—*To measure the height of a tree, or other object.*

Rule.—Set up a pole perpendicularly, the length of which above the ground is known. Go to the foot of the tree, and make a mark in it at the height of your eye above the ground, and make a mark in the pole at the same height. Then go backward till you find such a station that your eye shall be exactly in a range with the top of the pole and the top of the tree, and also in a range with the marks in the pole and tree. Measure the distance from that station to the foot of the pole, and also to the foot of the tree. Then, as the distance from your station to the foot of the pole, is to the distance from the said station to the foot of the tree; so is the height of the pole above the mark, to the height of the tree above the mark. Then, add to the height so found, the distance from the mark in the tree to the ground, and the sum will be the true height of the tree.

Note.—If the tree is not perpendicular, but leans, let the pole be placed parallel to it, and the same process will give its length.

Ex. What is the height of a tree, when, if you set up a perpendicular pole 20 feet above the ground, and take such a station that your eye is in a range with the top of the tree and the top of the pole, your eye is 5 feet from the ground, and your station 10 feet from the pole, and 64 feet from the tree?

As 10 : 64 :: 15ft. : 96ft. Then, $96+5=101$ft. Ans.

Prob. XXVIII.—*When the magnitudes and densities of two or more bodies are given, to find their relative weights, or quantities of matter.*

Rule.—Multiply together the numbers which denote

the magnitude and density of each body, and the products will show the relative quantities of matter which they contain.

Ex. If the magnitude of the earth be to that of the moon as 50 to 1, and the density of the former to that of the latter as 4 to 5, what is the ratio of their quantities of matter?

Here 50×4=200, and 1×5=5; therefore the quantity of matter in the earth is to that in the moon, as 200 to 5, or as 40 to 1, Ans.

PROB. XXIX.—*Motion of bodies, with their Velocities.*

If the quantities of matter in any two or more bodies, put in motion, be equal, the forces by which they are moved will be in proportion to their velocities, or swiftness of motion.

If the velocities of the bodies be equal, the forces will be directly as the quantities of matter in the bodies, that is, as their weights.

If both the quantities of matter and the velocities be unequal, the forces with which the bodies are moved, will be in a ratio compounded of their quantities of matter and velocities.

Ex. 1. Suppose the battering ram of Vespasian weighed 60000lb.; that it was moved at the rate of 24 feet per second of time, and that this was sufficient to demolish the walls of Jerusalem: with what velocity must a cannon ball of 42lb. weight have been moved, to have done the same execution?

Here the forces are equal, and therefore the velocities are inversely as the weights of the bodies; viz. as 42lb. : 60000lb. :: 24ft. : 34285$\frac{5}{7}$ feet per second, Ans.

2. There are two bodies put in motion, one of which weighs 100lb., the other 60lb.; but the less body is impelled by a force 6 times as great as the other: The proportion of the velocities wherewith these bodies move is required.

As 60lb. : 100lb.
1 force : 6 force } :: 1 velocity : 10 velocity.

Ans. The velocity of the greater body is to that of the less as 1 to 10.

PROB. XXX.—*Of Gravity.*

All bodies possess the attraction of gravitation in proportion to the quantities of matter which they contain. This principle of attraction causes the weight of bodies on the earth's surface.

As the form of the earth is very nearly that of a globe, it is evident that a body placed at the earth's centre would be attracted equally in every direction, because it would be surrounded on all sides by equal portions of the earth; and consequently the body would there have no weight.

From the earth's centre to its surface, the weight of bodies *increases*, in proportion to the distance from the centre; and above the surface their weights *decrease*, in a duplicate ratio, or as the squares of their distances from the earth's centre increase.

Ex. 1. If a body weigh 100lb. at the earth's surface, what would it weigh, if placed 793 miles below the surface, supposing the semidiameter of the earth to be 3965 miles? As 3965 : 3965—793 :: 100lb. : 80lb. the Ans.

2. If a body weigh 4 lb. at the surface of the earth, what would it weigh at the distance of 3965 miles above the earth's surface, or 2 of the earth's semidiameters from the centre?

3965+3965=7930 miles, the distance from the centre.

Then, as 7930×7930 : 3965×3965 :: 4lb. : 1lb. Ans.

Or, as 2×2 the square of 2 semidiameters : 1×1 the square of 1 semidiameter :: 4lb. : 1lb. the ans., as before.

3. If a body on the surface of the earth weigh 180lb., at what distance from the earth's centre must it be placed, to make its weight only 20lb.?

Ans. Either at 3 semidiameters, or $\frac{1}{9}$ of the earth's semidiameter from the centre.

Note 1.—The weight of a body at the surfaces of two different planets, is as the products of the diameters and densities of the planets.

4. If a stone weigh 1lb. at the surface of the earth; required its weight at the surfaces of the sun and the several planets, whose diameters and densities are given below.

	Sun.	Jupiter.	Saturn.	Earth.	Moon.
Proportional densities,	100	78.5	36	392.5	464
Diameters in miles,	883246	89170	79042	7930	2180

$$7930\times392.5 : \left\{\begin{array}{r}883246\times100\\ 89170\times78.5\\ 79042\times36\\ 2180\times464\end{array}\right\} :: \begin{array}{c}\text{lb.}\\ 1\end{array} : \left\{\begin{array}{r}\text{28.37lb. at the Sun.}\\ \text{2.24lb. at Jupiter.}\\ \text{.91lb. at Saturn.}\\ \text{.32lb. at the Moon}\end{array}\right.$$

Note 2.—The forces with which two planets attract each other, are directly as the quantities of matter in the planets, and inversely as the squares of their distances.

5. If the quantity of matter in the earth be to that in the moon as 40 to 1, and the distance of the centres of these planets be 240000 miles; whereabouts between them are their attractions equal to each other?

This question may be solved as follows: As the sum of the square roots of their quantities of matter : is to the square root of the quantity of matter in the earth :: so is the distance of their centres : to the required distance from the earth's centre.

Now, $\sqrt{40}=6.324555$, and $\sqrt{1}=1$. Then, as 6.324555 +1 : 6.324555 :: 240000 miles : 207233 miles, the distance from the earth's centre, Ans.

6. If the attraction of the moon raise a tide on the earth 5 feet, what will be the height of a tide raised by the earth on the surface of the moon, under similar circumstances, supposing their diameters and quantities of matter as before stated?

The effects which the attractions of these bodies have on each other's surfaces, are directly as the quantities of matter contained in the bodies, and inversely as their diameters. Therefore,

$$\text{As } \left.\begin{array}{r}1 : 40\\ 7930 : 2180\end{array}\right\} :: \text{5ft. : 55 feet nearly, Ans.}$$

PROB. XXXI.—*Of the effects of Light and Heat.*

The effects or degrees of light and heat are reciprocally proportional to the squares of their distances from the centre whence they are propagated.

Ex. If the distance of the planet Mercury from the Sun be 36 millions of miles, and the Earth's distance from the Sun 95 millions, the degree of light and heat received by Mercury, compared with that received by the Earth, is required.

As 36×36 : 95×95 :: 1 : 7 nearly.—Ans. As 7 to 1.

Prob. XXXII.—*To measure any height, by the time in which a heavy body will fall from it to the ground.*

Rule.*—Multiply the number of seconds in the given time by 4, and the square of the product will be the space fallen through, in feet.

Note.—The reverse of this rule will give the time in which a heavy body will fall through a given space.

Ex. 1. I dropped a stone from the top of a perpendicular precipice, and found that it descended to the bottom in three seconds: Required the height of the precipice?

Ans. 144 feet.

2. Ascending bodies are retarded in the same ratio that descending bodies are accelerated; therefore, if a ball, discharged from a gun, return to the earth in 12 seconds, how high did it ascend?

The ball being half the time, or 6 seconds, in its ascent; therefore, $6\times4=24$, and $24\times24=576$ feet, Ans.

3. There is a steeple in Salisbury, in England, which is about 400 feet high: In what time would a bullet fall from the top of this lofty spire to the bottom?

$\sqrt{400}=20$, and $20\div4=5$ seconds, Ans.

Prob. XXXIII.—*To find the velocity acquired by a falling body, per second, (or by a stream of water, having the perpendicular descent given,) at the end of any given period of time.*

Rule.—Multiply the perpendicular space fallen through by 64, and the square root of the product is the velocity required.

Or, the velocity acquired at the end of any period of time, is equal to twice the mean velocity with which it has passed during that period.

---

* It has been ascertained, by experiments, that heavy bodies, near the surface of the earth, fall 1 foot in the first quarter of a second, 3 feet the second quarter, 5 feet the third, and so on; and hence the velocities acquired by bodies in falling, are found to be proportional to the squares of the times in which they fall.

Ex. 1. If a ball fall through a space of 484 feet in $5\frac{1}{2}$ seconds, with what velocity will it strike?

Here $\sqrt{(484\times64)}=176$ feet per second, Ans.

Or, $(484\div5.5)\times2=176$ feet, the answer, as before.

2. There is a sluice, or flume, one end of which is a foot lower than the other: What is the velocity of the stream per second? Ans. 8 feet.

PROB. XXXIV.—*The weight of a body, and the space fallen through, being given, to find the force with which it will strike.*

RULE.—Multiply the weight of the body by the velocity with which it will strike, (found by Prob. 33d,) and the product will be the momentum, or force with which it will strike.

Ex. If a rammer, used for driving the piles of a bridge, weighs 4500lb. and falls through a space of 9 feet, with what force does it strike the pile?

$\sqrt{(9\times64)}\times4500=108000$lb. momentum, Ans.

PROB. XXXV.—*To find the quantity of pressure against a sluice, or bank, which pens water.*

RULE.—Multiply the area of the sluice, under water, by the depth of the centre of gravity, (which is equal to half the depth of the water,) in feet, and that product again by $62\frac{1}{2}$, (the number of Avoirdupois pounds in a cubic foot of fresh water,) or by 64.4lb. (the Avoirdupois weight of a cubic foot of salt water,) and the product will be the number of pounds required.

Ex. 1. Suppose the length of a sluice or flume to be 30 feet, the width at the bottom 3 feet, and the depth of the water 4 feet; what is the pressure against the side of the sluice?

$30\times3=90$ square feet, the area of the bottom. Then, 90×2 (the depth of the centre of gravity) gives 180 cubic feet, and $180\times62.5=11250$lb. Ans.

Note.—The perpendicular pressure of fluids on the bottoms of vessels, is estimated by the area of the bottom multiplied by the altitude of the fluid.

2. Suppose a vessel is 3 feet wide, 5 feet long, and 4 feet

high; what is the pressure on the bottom, it being filled with fresh water to the brim?

3×5=15 square feet, the area of the bottom. Then, 15×4=60 cubic feet, and 60×62.5=3750lb. Ans.

Prob. XXXVI.—*To find the length of a pendulum that will swing in any given time.*

Rule.—Multiply the square of the number of seconds in the given time by 39.2, and the product will be the length required, in inches.

Note.—The converse of this rule will give the time in which a pendulum of any given length will swing.

Ex. 1. Required the length of a pendulum that will vibrate in half a second?

.5×.5=.25, and 39.2×.25=9.8 inches, Ans.

2. There are two pendulums; one of which will swing once in a second, and the other once in two seconds: Required the length of each?

Ans. 39.2 inches, and 156.8 inches.

3. How often will a pendulum that is 9.8 inches in length vibrate?

$\sqrt{}(9.8\div39.2)$=.5 sec. Ans. Once in half a second.

Prob. XXXVII.—*To find what weight may be raised or balanced by any given power with a lever.*

Rule.—As the distance between the body to be raised or balanced and the fulcrum or prop, is to the distance between the prop and the point where the power is applied; so is the power, to the weight which it will balance.

Note.—No allowance is here made for the weight of the lever, which ought to be done in order to obtain the *exact* answer.

Ex. 1. If a man, whose weight is 160lb., rest on the end of a lever, 100 inches long, what weight will he balance on the other end, if the prop be 10 inches from the said weight? As 10in. :: 90 : 160lb. : 1440lb. Ans.

2. At what distance from a weight of 1440lb. must a prop be placed, so that a weight of 160lb., applied 90 inches from the prop, will balance it?

As 1440lb. : 160lb :: 90in. : 10 inches, Ans.

3. If a lever be 100 inches long, what weight lying on the lever, 10 inches from one end resting on a pavement, may be moved by means of a force of 160lb. lifting at the other end of the lever?

As 10in. : 100in. :: 160lb. : 1600lb. Ans.

4. What weight, hung at 30 inches distance from the centre of motion of a steel-yard, will balance a barrel of pork weighing 240lb., freely suspended at 2 inches distance from the said centre, on the contrary side? Ans. 16lb.

PROB. XXXVIII.—*To find what weight may be raised by any given power, with a wheel and axle.*

RULE.—The proportion for the wheel and axle, (in which the power is applied to the circumference of the wheel, and the weight is raised by a rope, which coils about the axle as the wheel turns round,) is, as the diameter of the axle : is to the diameter of the wheel :: so is the power applied to the wheel : to the weight suspended by the axle.

Ex. 1. Suppose the diameter of the wheel to be 60 inches, and that of the axle 6 inches; what weight at the axle will balance 1lb. at the wheel?

As 6in. : 60in. :: 1lb. : 10lb. Ans.

2. A mechanic would make a windlass in such a manner, as that 2lb. applied to the wheel, shall just balance 20lb. suspended from the axle; now, supposing the diameter of the axle to be 4 inches, what must be the diameter of the wheel? As 2lb. : 20lb. :: 4in. : 40 inches, Ans.

PROB. XXXIX.—*To find what weight may be raised by any given power with a screw.*

RULE.—As the distance between the threads of the screw, is to the circumference of a circle described by the power applied at the end of the lever; so is the power, to the weight required.

Note 1.—At a medium about one-third part of the effect of the machine is destroyed by friction.

Note 2.—The reverse of the above rule, will give the power that must be applied to the end of the lever, to raise a given weight.

Ex. 1. There is a screw whose threads are an inch asunder; the lever by which it is turned is 30 inches long, and the weight to be raised is 2240lb.: What power or force must be applied to the end of the lever, sufficient to turn the screw, that is, to raise the weight?

The length of the lever being the semidiameter of the circle, the diameter is 60 inches, and the circumference $3.1416 \times 60 = 188.496$ inches: Therefore, as 188.496in. : 1in. :: 2240lb. : 11.88lb.+ Ans.

2. Let the lever be 30 inches long, (which will describe a circle 188.496 inches in circumference,) the threads of the screw an inch asunder, and the power 11.88lb.: Required the weight to be raised?

As 1in. : 188.496in. :: 11.88lb. : 2240lb. nearly, Ans.

3. Let the power be 11.88lb., the weight 2240lb., and the threads of the screw an inch asunder, to find the length of the lever.

As 11.88lb. : 2240lb. :: 1in. : 188.5+in. Then, as 355 : 113 :: 188.5in. : 60 inches, nearly, the diameter of the circle; and hence the length of the lever is 30 inches, Ans.

PROB. XL.—*To find whether any given year is a leap year, or not, according to the present style.*

RULE.—Divide the year of our Lord by 4, and if the division terminates without a remainder, the year is a leap year; otherwise it is a common year, and the remainder shows what year it is after bissextile or leap year.

But when the given number of years has two or more ciphers at the right hand, you must reject two ciphers at the right hand before you divide by 4; and then the remainder, if any, will show what century it is after a centesimal leap year.

Ex. Required to find whether the years 1800 and 1820 were leap years, or not.

4)18|00(4	4)1820(455
Rem. 2	0

So, it is found that the year 1800 was a common year, of 365 days, and the year 1820 a leap year, of 366 days.

In like manner it is found that the year 1817 was the first, and the year 1830 the second after bissextile,

PROB. XLI.—*To find the Dominical Letter* for any given year, according to the present style.*

RULE.—1. *To find the Dominical Letter for any year of the present century:* Add to the year of our Lord, a fourth part of itself, omitting fractions, if any; then divide the sum by 7, and the remainder will be the index of the Dominical Letter for that year; the indices, or numbers of the letters, being as given below.

2. *To find the Dominical Letter for any year of the Christian Era:* Divide the number of whole centuries by 4, and set down what remains as well as the quotient. To 3 times the said quotient add the remainder, if any; then subtract the sum from the given year of the Christian Era; to the remainder add a fourth-part of the said year, omitting fractions; divide the sum by 7, and the remainder will be the index of the required letter.

Note.—In every leap year there are two Dominical Letters; that is, the one found by the above rule, will be the Dominical Letter in March, April, and the subsequent months, and the next following letter, according to the order of the alphabet, in January and February.

Letters,	A,	B,	C,	D,	E,	F,	G.
Indices,	0, (7)	6,	5,	4,	3,	2,	1.

Ex. 1. To find the Dominical Letter for A. D. 1830.

Ans. C.

$$\begin{array}{r} 4)1830 \\ +457 \\ \hline 7)2287 \\ \hline 326..5 \text{ Rem.} \end{array}$$

2. Required the Dominical Letters for the year 1796, this being a *leap year.*

* The *Dominical Letters,* are the first seven letters of the alphabet. The letter A is supposed to be annexed to the first day of the year, in the Calendar; B to the second day, and so on; and the letter which happens to fall on the first Sabbath in the year, is called the *Dominical Letter* for that year. But, as each month invariably begins with a certain letter, there are two Dominical Letters in every leap year; the Dominical Letter in January and February being different from that for the subsequent months.

4)17 No. of whole centuries.

4 ... 1 Remainder.
3

13 To be subtracted from 1796.

Ans. B and C.

4)1796
−13

1783
+449

7)2232

318 ... 6

Prob. XLII.—*To find what year of the Julian Period corresponds with any given year.*

Rule.—Add 710 to the year of the World, or 4713 to the year of the Christian Era, and the sum will be the answer.

Ex. 1. What year of the Julian Period answers to A. D. 1831? Here 1831+4713=6544, Ans.

2. The Israelites departed from Egypt in the year of the world 2513: what year of the Julian Period was that?

Ans. The 3223d.

Prob. XLIII.—*To find the Cycle of the Sun, Cycle of the Moon, or Golden Number, and Indiction, for any given year.*

Rule.—Find (by Prob. 42d) the year of the Julian Period which corresponds with the given year; which divide by 28 for the Solar Cycle; by 19 for the Golden Number, and by 15 for the Indiction, and the several remainders will be the numbers required; but, if nothing remain, the divisor will be the number sought.

Ex. Required the several cycles in A. D. 1831.

1831+4713=6544, the year of the Julian Period.

28)6544(233	19)6544(344	15)6544(436
Rem. 20	8	4
Solar Cycle.	Golden Number.	Indiction.

Prob. XLIV.—*To find the Epact for any given year, according to the present style.*

Rule.—Divide the given year of the Christian Era by

19; multiply what remains by 11, and if the product does not exceed 30 it is the epact required; but if it exceeds 30, divide it by 30, and the remainder will be the epact.

Ex. 1. Find the epact for the year 1831.

```
19)1831(96          7
   171             11
   ---             --
   121          30)77
   114             --
   ---              2 . . . 17 The epact, Ans.
Rem. 7
```

2. What was the epact in A. D. 1824? Ans. 0.

PROB. XLV.—*To find the times of the New and Full Moon, and of the First and Last Quarters, in any given month.*

RULE.—To the epact for the given year add the number annexed to the name of the given month in the table below; subtract the sum from 30, for the time of *new-moon;* from 15, for the *full;* from 7, for the *first quarter*, and from 22, for the *last quarter*; and the several remainders will be the days of the month in which the respective changes or phases happen.*

For Jan. add	0	For May, add	2	For Sept. add	7
Feb.	2	June,	4	Oct.	8
March,	0	July,	4	Nov.	9
April,	2	August,	6	Dec.	10

Note.—If the subtrahend (or number to be subtracted) be greater than the minuend; add as many 30's to the minuend as may be necessary to make it exceed the subtrahend, and then subtract as before.

Ex. For May 1830.

The epact for the year 1830 is 6, and the number to be added to it for May is 2, which makes 8. Then,

30—8=22d day, the time of new-moon.
15—8= 7th day, do. of full-moon.
7+30—8=29th day, do. of the 1st quarter.
22—8=14th day, do. of the last quarter.

* This method of calculating the moon's phases is by no means exact It will, however, generally give the *days* in which the changes happen.

MISCELLANEOUS QUESTIONS FOR EXERCISE.

1. The difference of two numbers is 88, and the less number is 114: what is the greater number? Ans. 202.

2. What number subtracted from 1000 will leave 210 for the remainder? Ans. 790.

3. The product of two numbers is 22440, and one of them is 120: what is the other number? Ans. 187.

4. If it be supposed that as many persons die in the world in $33\frac{1}{3}$ years as are equal to the entire population, how many die each hour, at an average, supposing the whole population to be 650 millions, and the year to consist of 365 days, 6 hours? Ans. 2224+

5. How much time, in the course of 12 years, does a person, who rises at 5 o'clock in the morning, gain over another who continues in bed till 7, supposing both to retire to rest at the same hour? Ans. 1 year.

6. If the human heart beat 70 times in a minute, and each pulsation transmit 4 Avoirdupois ounces of blood, and the whole blood be one-twentieth part of the weight of the body, in what time will the whole blood of a man, whose weight is 140lb., circulate through the heart?

Ans. 24 seconds.

7. The latitude of St. Petersburgh, the capitol of Russia, is 59 degrees, 56 minutes, north; that of London is 51° 31' N., and that of Montreal, in Lower-Canada, is 45° 31' N. How many statute miles is St. Petersburgh north of the latitude of London, and how many miles is London north of the latitude of Montreal?

Ans. St. Petersburgh is $582\frac{13}{30}$ miles north of the latitude of London, and London is $415\frac{1}{5}$ miles north of the latitude of Montreal.

8. An American silver dollar weighs $15\frac{187}{875}$ drams, Avoirdupois, and a cent weighs just half as much as a dollar. What is the value of an Avoirdupois pound of copper at that rate? Ans. 33 cents, $6\frac{7}{13}$ mills.

9. In the United-States, a Troy ounce of standard gold is worth \$$17\frac{7}{9}$, and a Troy ounce of standard silver \$$1\frac{2}{13}$. Required the value of an Avoirdupois ounce of each of these metals?

Ans. { Value of an Avoirdupois ounce of gold \$$16\frac{11}{54}$.
Do. of silver \$$1\frac{43}{832}$.

10. In the year 1827, the British government owed about L900,000,000, sterling, and the pound sterling is equal to 3oz. 17pwt. 10gr. of silver. How many Avoirdupois tons of silver would this sum make, and how long a string of wagons would it take to carry the whole, a ton at a load, allowing 3 rods to each wagon and horses? I would also know how long it would take to count the money, in dollars, reckoning without intermission 12 hours a day, at the rate of 60 dollars a minute, and $365\frac{1}{4}$ days to the year?

Ans. { The weight would be 106645 tons, 8cwt. $18\frac{2}{7}$lb.
The distance 999 miles, 6 furlongs, 16 rods+
The time 253 years, 184 days, 4h. 6 min. 40 sec. }

11. From 8 leagues, subtract 20 miles, 20 furlongs, 20 rods, 20 yards, 20 feet, 20 inches and 20 barley-corns, without reducing any of the numbers to other denominations.* Ans. 1 mile, 3fur. 14rd. 5yd. 1ft. 9in. 1b. c.

12. Suppose I would put 520 bushels of apples into casks, containing 3 bushels and 1 peck each; how many casks must I procure? Ans. 160.

13. If a cow yield 15 quarts of milk a day, for 240 days, and 20 quarts of milk make 1 lb. of butter; how much butter may be thus obtained in the season, and what does it amount to at 15 cents a lb.?

Ans. Weight 180lb.; value $27.

14. A gallon, Dry Measure, contains $268\frac{4}{5}$ cubic inches, and a gallon, Wine Measure, 231 cubic inches. How many bushels of corn can be put into a cask that will hold just a barrel, or $31\frac{1}{2}$ gallons, of wine?

Ans. 3bush. 1pk. $4\frac{9}{32}$qt.

15. A butcher laid out $900 for cattle: he bought oxen at $40 each, cows at $15, steers at $14, and calves at $6; and an equal number of each. How many of each sort did he purchase? Ans. 12.

16. A farmer carried a load of produce to market: he

---

* In performing compound subtraction, when we have occasion to borrow, we usually borrow as many of the denomination which we would subtract as make 1 of the next higher; but, when it is necessary, we may borrow as many of the lower denomination as make 2, 3, or 4, &c. of the next higher, and then carry accordingly to the said higher denomination. Thus, in example 11th, in order to subtract the 20 barley corns, borrow 21 barley-corns, which are equal to 7 inches; then subtract, and carry 7 to the 20 inches, and so proceed.

sold 527lb. of pork, at 7 cents a pound; 175lb. of cheese, at 8 cents a lb., and 215lb. 12oz. of butter, at 15 cents a lb.: in pay he received $14\frac{1}{2}$ yards of cloth, at $87\frac{1}{2}$ cents a yard; 40lb. of sugar, at 9 cents a lb.; 10 bushels of salt, at $62\frac{1}{2}$ cents a bushel; and the balance in money: how much money did he receive? Ans. $60.715

17. A merchant bought a quantity of goods in New-York for $500, and paid $43 for their transportation; he sold them so as to gain 24 per cent. on the whole cost: for how much did he sell them? Ans. $673.32

18. Bought a book, the price of which was marked $2.-25, but for cash the bookseller sold it at $33\frac{1}{3}$ per cent. discount: what was the cash price? Ans. $1.50

19. What is the sum of the third and half third of $1? Ans. 50 cents.

20. What is the difference between six dozen dozen and half a dozen dozen? Ans. 792.

21.* From ninety take forty,
From forty take ten;
Subtract six from sixty,
And what remains then? Ans. 134.

22. What number is that, which being divided by $\frac{2}{5}$, the quotient will be $12\frac{13}{16}$? Ans. $5\frac{1}{8}$.

23. A person who owned $\frac{2}{5}$ of a vessel, sold $\frac{3}{4}$ of his share for $1200: what was the value of the vessel, at that rate? Ans. $4000.

24. What part of 10 cents is $\frac{2}{3}$ of 4 cents? Ans. $\frac{4}{15}$.

25. If the third of six be three,
What will the fourth of twenty be? Ans. $7\frac{1}{2}$.

26. If 2 be 3, and 3 be 5, and $5\frac{1}{2}$ be 11;
What is the $\frac{1}{2}$ of 26, and the $\frac{1}{3}$ of 27?

$$\left.\begin{array}{r} \text{As } 2 : 3 \\ 3 : 5 \\ 5\frac{1}{2} : 11 \end{array}\right\} \left.\begin{array}{l} :: \frac{26}{2} : 65 = \frac{1}{2} \text{ of } 26, \text{ according to the ques.} \\ \text{Then, as } 13 : 65 :: \frac{27}{3} : 45 = \frac{1}{3} \text{ of } 27. \end{array}\right\} \text{Ans.}$$

---

* The poetical questions in this collection, are taken from other books. It will be seen that some of these questions are very ordinary compositions of poetry; but, as they are in other respects ingenious questions, calculated to instruct, as well as amuse the student, I have thought proper to insert them.

27. Whereas an eagle and a cent
Just three score yards did buy,
How many yards of that same cloth
For ninety dimes had I? Ans. 53.94+

28. If from a staff just four long,
A shadow five is made,
What is the steeple's height, in yards,
That's ninety feet in shade? Ans. 24yds.

29. If the forward wheels of a waggon be each 14 feet, 6 inches, in circumference, and the hind wheels 15 feet, 9 inches, how many more times will the former turn round than the latter, in running from Boston to New-York, the distance being 248 miles? Ans. $7167\frac{97}{609}$ times.

30. A ship has a leak which will fill it so as to make it sink in 10 hours; it has also a pump which will clear it in 15 hours: now if the crew begin to pump when the ship begins to leak, in what time will it sink?

In 1 hour the ship would be $\frac{1}{10}$ filled by the leak, but in the same time it will be $\frac{1}{15}$ emptied by the pump; therefore, it will be $\frac{1}{10}-\frac{1}{15}=\frac{1}{30}$ filled in 1 hour; and consequently it will sink in 30 hours, Ans.

31. A person, looking on his watch, was asked the time of day, who answered, it is between 5 and 6 o'clock; but a more particular answer being required, he said that the hour and minute hands were then exactly together. What was the time?

The velocities of the two hands of a watch, or a clock, are to each other as 12 to 1; the minute hand performing 12 revolutions in 12 hours, and the hour hand but one; and consequently the former performs, in 12 hours, 11 revolutions more than the latter. Now, it is obvious that whenever the hour and minute hands are exactly together, except at 12 o'clock, the number of *entire* hours past 12, must be equal to the difference between the numbers of the revolutions performed by the two hands since 12 o'clock: Therefore, as the difference 11 : the dif. 5 :: 12 hours : 5 hours, $27\frac{3}{11}$ minutes, the time required.

32. As I was hunting, on the forest grounds,
Up sprang a hare before my two grey-hounds;
The dogs, being swift on foot, did fairly run,
Unto her fifteen rods just twenty-one;

The distance which she started up before,
Was four-score thirteen rods, just, and no more:
Now, this I'd have you unto me declare,
How far they ran before they caught the hare?

Ans. $325\frac{1}{2}$ rods.

33. A and B are on opposite sides of a circular field 268 rods in circumference; they begin to go round it, both the same way, at the same instant of time; A goes 22 rods in 2 minutes, and B 34 rods in 3 minutes: how many times will they go round the field before the swifter will overtake the slower? Ans. A $16\frac{1}{2}$ times, and B 17 times.

34. Some sportsmen, having placed a fox 100 yards distant from some hounds, let them all start together; and the hounds ran $2\frac{1}{2}$ times as fast as the fox: I demand how far the fox ran before the hounds overtook him?

As $2\frac{1}{2}$—1yd. gained by the hounds : 100yd. the whole distance to be gained :: 1yd. run over by the fox : $66\frac{2}{3}$ yards, the answer.

35. A frigate pursues a ship at 8 leagues distance, and sails twice as fast as the ship: how far must the frigate sail before she will come up with the ship?

Ans. 16 leagues.

36. A can perform a certain piece of work in 3 days; B can do the same in 4 days, and C in 5 days: In what time will all three finish it, working together?

It is evident that A can perform $\frac{1}{3}$, B $\frac{1}{4}$, and C $\frac{1}{5}$, of the required work in 1 day; and consequently all of them together can perform $\frac{1}{3}+\frac{1}{4}+\frac{1}{5}=\frac{47}{60}$ of the work in 1 day. Therefore, as $\frac{47}{60}$ of the work : 1, the whole :: 1 day : $1\frac{13}{47}$ day, Ans.

37. If a saddler can make a saddle in 1 day, his journeyman 2 saddles in 3 days, and his apprentice 1 saddle in 3 days, how long will it take all of them together to make a dozen saddles?

$1+\frac{2}{3}+\frac{1}{3}=2$, the number of saddles that all of them can make in one day. Then, as 2 saddles : 12 saddles :: 1 day : 6 days, Ans.

38. A and B, together, can do a certain piece of work in 20 days; with the assistance of C they can do it in 12 days: In what time could C do it by himself?

It is evident that C can do as much in 12 days as A and

B, together, can in 8 days: Therefore, as 8 days : 20 days :: 12 days : 30 days, Ans.

39. If 6 men, having each 6 degrees of strength and 4 degrees of activity, build a wall 20 feet long, 6 high, and 4 thick, in 2 days, when the day is 10 hours long; how many men, having each 4 degrees of strength and 8 degrees of activity, must be employed to build another wall, 200 feet long, 8 high, and 6 thick, in 10 days, when the day is 12 hours long? Ans. 15 men.

40. If 14 men in 15 days build 16 rods of wall,
How many men must added be to do it in 2, that's all?
Ans. 91.

41. If 50 cents buy 15 pounds of bread that's made of rye,
How many loaves of 6 pounds each will four score eagles buy? Ans. 4000.

42. If 30 cents and 90 dimes buy 50 pints of wine,
What is the cost of 60 quarts in current Federal coin?
Ans. \$22.32

43. If 18 dollars and a dime, for 13 weeks supply
For meat and drink, that is my board, how much a day give I? Ans. 19$\frac{81}{91}$ cents.

44. If 5 pounds of ginger cost 58 cents,
How many pounds, in New-England, for 92 pence?
Ans. 11$\frac{4}{261}$ lb.

45. My sister's years are just to mine,
As forty-six to sixty-nine—
And if you multiply them o'er,
The product's six times sixty-four;
Show thou our ages unto me,
And I thy wedded wife will be—
All other proffers are in vain,
No other shall that favor gain.
Ans. 16 years the younger, 24 the elder.

46. When first the marriage knot was tied
Between my wife and me,
My age, in years, was to my bride's,
As three times three to three;
But now, when ten and half ten years,
We man and wife have been,
My age to her's exactly bears
As three times six to nine:

Now tell, I pray, from what I've said,
What were our ages when we wed?

Ans. { Thy age, when married, must have been
Just forty-five; thy wife's fifteen.

47. If a half and a third to my age added be,
With a complement number of twice twenty-three,
One hundred and one the sum would then be—
The age of the Author pray show unto me?

48. A market woman bought a certain number of eggs at 2 for a cent, and as many more at 3 for a cent—the eggs being mingled, she sold them all out at 5 for 2 cents, and by so doing lost 5 cents. How many eggs had she?

Ans. 300.

49. The yearly interest of Harriet's money, at 6 per cent., exceeds one-twentieth of the principal by $100, and she does not intend to marry any man who is not scholar enough to tell her fortune; pray what is it?

Ans. $10000.

50. A man left to his three sons, A, B, and C, whose ages were 13, 15, and 17 years, $4500, to be divided in such manner that the three parts being put out at simple interest, at 6 per cent., should amount to equal sums when the boys were respectively 21 years of age: I demand what each part must be?

Solved by Single Position, as follows: Suppose that when the sons were respectively 21 years of age, the share of each, together with the interest on the same, amounted to $1480; or, which is the same thing, suppose the youngest son's share was $1000. Then, by Prob. II. Simple Interest by Decimals, the two other shares are found to be $1088.235+ and $1193.548+. The sum of these three shares is $3281.783. Then,

$		$		$		$		
As 3281.783	:	4500	:: {	1000	:	1371.20+	A's share.	} Ans.
				1088.235	:	1492.19+	B's do.	
				1193.548	:	1636.59+	C's do.	

51. Old John, who had in credit lived,
Tho' now reduced, a sum received;
This lucky hit's no sooner found,
Than clam'rous duns came swarming round;

To the landlord—baker—many more,
John paid, in all, pounds ninety-four.
Half what remained a friend he lent,
On Joan and self one-fifth had spent;
And when of all these sums bereft,
One-tenth of th' sum received had left:
Now show your skills, ye learned youths,
And by your work the sum produce. Ans. L141.

52. If my horse and saddle be worth $135, and the horse be worth 8 times as much as the saddle, what is the value of the horse? Ans. $120.

53. Two shepherds met upon their way,
And thus one said, ('twas true,)
"Give half thy flock of sheep to me,
Then I'll have eighty-two."
"Nay friend," the other soon replied,
"Add but a third to mine,
Of thy own sheep, then I shall have
One hundred twenty-nine."
His answer being strictly true,
No scholar will impeach;
Then by your knowledge show to me
How many sheep had each?

Solved by Double Position, as follows: First, suppose the less flock consisted of 18 sheep; then, by the question, $(82-18)\times2=128$=the number of sheep in the greater flock; to which add $\frac{1}{3}$ of 18, and the sum, 134, is the first result; which ought to be 129. Secondly, suppose the less flock consisted of 30 sheep, and proceed as before, and the second result is 114. Then, as 134—114 : 134—129 :: 30—18 : 3, the correction, which being added to 18, gives 21, the number of sheep in the smaller flock; and hence, $(82-21)\times2=122$=the number of sheep in the larger flock.

54. Divide $40 between A and B, so that B's part shall be to A's as A's is to $40.

Ans. A's part is $24.721+, and B's $15.279—

Note.—The last question may be solved by the rule given at pages 288 and 289 of this Book. The answers to questions 55th and 56th may also be found by the same rule, or by trials; and as these answers, when found, may be easily proved, I shall not insert them.

55. A number find, the cube of which,
Increas'd by half the square,
Five hundred taken from this sum
Will leave just forty-four.

56. A farm containing 100 acres, and valued at $3600, is to be divided between two men, A and B. Now, the values of the two shares are to be equal, but A is to have his part set off where the land is worth $15 more per acre than the rest of the farm. How many acres will each part contain?

57. A country clown addressed a charming belle,
Who in both wit and learning did excel;
The youth, unskilled in numbers, as will show,
Desirous was the lady's age to know—
When she replied with a majestic air,
With piercing words peculiar to the fair,
"My age is such if multiplied by three,
Two-sevenths of that product tripled be—
The square root of two-ninths of that is four,
Now tell my age or never see me more."

Ans. 28 years.

Note.—Question 57th, and others of a similar nature, may be solved by taking the last, or final result, and reversing all the operations mentioned in the question.

58. An army fought three battles; in the first battle, half the men were killed, and half a man more; in the second, half the remainder were killed, and half a man more; and in the third they lost half of what remained, and half a man more, when their number was reduced to 275. How many were there at first? Ans. 2207.

59. The third part of an army was killed, the fourth part taken captive, and 1000 fled: How many were killed, and how many taken captive?

Ans. 800 were killed, and 600 taken prisoners.

60. In a river, supposing two boats start at the same time from places 300 miles apart; the one proceeding up stream is retarded by the current 2 miles per hour, while that moving down stream is accelerated the same; if both be propelled by a steam engine, which would move them 10 miles per hour in still water, how far from each starting place will the boats meet?

It is evident that the boat which goes down stream must

move 12 miles per hour, and the other boat 8; and that each hour will bring the boats 20 miles nearer each other. Therefore,

As 20 : 300 :: { 12m. : 180 miles from the upper place. } Ans.
{ 8m. : 120 miles from the lower place. }

61. Sold a quantity of goods for $475, and by so doing lost 5 per cent., whereas in trading I ought to have gained 25 per cent. How much were the goods sold under their real value? Ans. $150.

62. A man sold cloth at 75 cents per yard, by which he cleared $\frac{1}{5}$ of the money; but finding the cloth better than any in market at that price, he raised his price to 90 cents per yard. What did he gain per cent. by the latter price?

Ans. 50 per cent.

63. Suppose 220 apples were laid on the ground, $3\frac{1}{2}$ yards distant from each other, in a straight row, and a basket placed $16\frac{3}{4}$ yards from one end of the row, in a right line with it; how far must a person travel to gather the apples one by one into the basket? Ans. 100 miles.

64. One Sessa, a native of India, having invented the game of chess, showed it to his Prince, who was so delighted with it that he promised him any reward he should ask; on which Sessa requested that he might be allowed one kernel of wheat for the first square on the chess board, 2 for the second, 4 for the third, and so on, doubling continually to 64, the whole number of squares. Now, supposing 491520 of these wheat corns to make a bushel, what is the amount of all the wheat?

Ans. 37,529,996,894,754$\frac{131071}{983040}$ bushels.

65. A crafty young fellow, in numbers well skill'd,
Once agreed with a farmer to work in his fields,
And for thirty years' service to take no reward
But the wheat that should from one kernel be rear'd.
Ten kernels from one, the first year, did grow—
Those kernels, that season, the farmer did sow—
Next season, these kernels one hundred did yield—
One thousand next summer he got from the field.
From season to season, as each harvest came,
For thirty fine summers, the increase was the same;
The last of these harvests the laborer had,
And thus, for his labor, most amply was paid.

If a bin should be made, of a cubical form,
Large enough to receive all the kernels of corn,
The depth of this bin, and its content pray tell,
If one thousand nine hundred wheat corns make a gill?

Ans. Depth of the bin 25903.596+ miles; its content 2,055,921,052,631,578,947,368,421$\frac{1}{19}$ bushels.

66. If one cent had been put out at compound interest at the commencement of the Christian era, at such a rate per cent. that the sum should have doubled once in 15 years, what would the amount have been in A. D. 1815? And if the amount should be in a solid globe of gold, what would be the diameter of this globe, supposing a cubic foot of gold to be worth $306056?

Ans. Amount, $13292279957849158729038070602803-445.76; diameter of the globe of gold 825972.7611+ miles; and hence, this globe must be more than 1129997 times as large as the earth.

Note.—In the last question, the ratio is 2, and the number of terms 121. The last term of this series is the amount of the money in cents; and this term is the 120th power of 2, or the 40th power of 8, which may be found in Table IV. To find the diameter of the globe of gold, divide the cubic feet of gold by .5236, (or divide the value of the gold, in dollars, by 160250.9216, the product of 306056 and .5236,) and the cube root of the quotient will be the diameter in feet.

67. Hiero, king of Sicily, ordered his jeweller to make him a crown, containing 63 ounces of gold. The workman thought that substituting part silver was only a proper perquisite; which taking air, the king requested the famous Archimedes to examine the crown; who, on putting it into a vessel of water, found it raised the fluid 8.2245 cubic inches: and having discovered that the cubic inch of gold weighed 10.36 ounces, and that of silver but 5.85 ounces, he found by calculation what part of the king's gold had been changed. Required the quantities of gold and silver in the crown.

63÷8.2245=7.66+oz., the weight of a cubic inch of the crown. Then, by Alligation Alternate, Case III.—

oz.	oz.	cub. in.
7.66	10.36	1.81
	5.85	2.70
	Sum,	4.51

As 4.51 : 8.2245 :: { 1.81 : 3.3007+cub. in. of gold. / 2.70 : 4.9237+ do. of silver. }

Then, 3.3007×10.36=34.195252 oz. of gold } Ans.
And 4.9237× 5.85=28.803645 oz. of silver }

The answer may also be obtained by Double Position.

68. How many different whole numbers can be expressed by the nine digits, without having the same figure more than once in the same number? Ans. 986409 numbers.

Note.—To solve the last question, find by Prob. II. in Permutations and Combinations, how many different numbers of one figure—of two figures—of three figures—and so on, to numbers of nine figures, can be made out of the nine digits, and the sum of the several results, thus found, will be the answer.

69. The largest of the Egyptian pyramids is 499 feet high, and its base is 693 feet square; the next less pyramid is 655 feet square at the base, and 398 feet high. How much ground is covered by each of these wonderful monuments?

Ans. The largest covers 11 acres, 4 square rods, and the other 9 acres, 3 roods, 15$\frac{225}{1089}$ square rods.

70. What is the area of a circular ring 3 rods broad, the inner circumference of which measures 355 rods?

Ans. 6.83298 acres.

71. A man has a square garden, each side of which measures 20 rods, and a circular fish pond in the centre, the diameter of which is 10 rods; how much ground has he?

Ans. 2 acres, 1.46 sq. rods.

72. What length of cart tire will it take to band a wheel 5 feet in diameter? Ans. 15 feet, 8$\frac{1}{2}$ inches, nearly.

73. The earth is a globe or ball, of about 7930 miles diameter, and the atmosphere, which surrounds it on every side, is supposed to extend about 45 miles from the earth's surface. How many cubic miles does the atmosphere contain?

The diameter of the earth is 7930 miles, and that of the earth and atmosphere is 8020; the difference between the cubes of these numbers being multiplied by .5236, gives 8991442983.6, the number of cubic miles in the atmosphere.

74. Required the ratio of the surfaces, and of the solidities, of the earth and moon; the earth's diameter being 7930 miles, and the moon's 2180.

Similar superficies are proportional to the *squares*, and similar solids to the *cubes* of their like dimensions: Therefore, the magnitude of the earth is to that of the moon as the cube of 7930 to the cube of 2180, or as 48.1 to 1, nearly; and the area of the earth's surface is to the moon's as the square of 7930 to the square of 2180, or as 13.2 to 1, nearly.

75. Required the length of a prop that will support a weak place in the perpendicular side of a building, 15 feet from the ground, and have the foot of the prop stand 6 feet from the bottom of the building.

Ans. 16 feet, 2 in. nearly.

76. A certain house, standing on a plain, is 28 feet wide and 35 feet high, the top of the roof being 15 feet higher than the feet of the rafters. It is required to find the length of a ladder that will just reach from the ground to the top of the house, and lie on the roof, exactly parallel to it. Ans. 47.8+feet.

77. If a cubic foot of iron were drawn into a bar a quarter of an inch square, what would be its length, provided there were no waste of metal? Ans. 768 yards.

78. A horse in the midst of a meadow suppose,
Made fast to a stake by a line from his nose;
How long must this line be, that feeding all round,
Permits him to graze just an acre of ground?

Ans. 117 ft. 9 in.+

79. On the 4th of July a pole was erected,
Compos'd of six pieces, and nicely connected;
Two feet and six inches it measured around,
On the place where it stood at the top of the ground;
The form was a cone in surface complete,
The height of the same was twice sixty feet:
What length of inch riband, procured at the shop,
Will wind round the pole from bottom to top,

And have it lie smooth and plain to be seen,
By leaving a space of five inches between?

Ans. 100 yards.

It is evident that the riband will cover a sixth part of the surface of the pole; and since the width of the riband is 1 inch, a sixth part of the number of square inches in the said superficial content is the length of the riband in inches.

80. Suppose a half bushel is exactly round—
That the depth of the same eight inches is found;
If the breadth be thirteen and two-fifths inches, quite,
The measure is *legal*, or very near right;
But, suppose I would make one of another frame,
And have ten inches the depth of the same;
Now, pray of what length must the diameter be,
That it may with the former in measure agree?

Ans. 11.7+ inches.

81. I agreed with a tinker, whose name was Doolittle,
To make for my wife a flat-bottomed kettle—
Twelve inches exactly the depth of the same,
And twenty-five gallons of beer to contain—
The number of inches across on the top
Was twice on the bottom, when new at the shop;
How many inches across must the bottom then be,
Likewise on the top, pray show unto me?

Solution.—282×25=7050 cubic inches, the capacity of the kettle; and, since the kettle is the frustum of a cone, 7050÷12=587.5 sq. inches=the mean area between the bases. Now, suppose the top diameter to be 30 inches; then, by the question, the bottom diameter will be 15 inches, and (by the rule for measuring a frustum of a cone) the mean area of a section between the bases will be 412.335 sq. inches. But similar superficies are proportional to the squares of their like dimensions; therefore, as 412.335 sq. in. : 587.5 sq. in. :: 30×30 : 1282.33, the square of the real top diameter. Then, √1282.33=35.8+ inches, the top diameter, and 35.8÷2=17.9 inches, the bottom diameter, *Ans.*

82. A globe—a cube—a hemisphere,
All three, *in surface*, equal are,
Each surface being six square feet—
Pray tell the cubic feet in each?

Ans. The cube contains 1 cubic foot, the globe 1.38+ cub. ft., and the hemisphere 1.06+cub. ft.

Note.—A *hemisphere* is half of a sphere. The area of the whole surface of a hemisphere is equal to three-fourths of the superficial content of a globe of the same diameter.

83. A tract of land is to be laid out in form of a square, and to be enclosed with a post and rail fence, 5 rails high; so that each rod of fence shall contain 10 rails. How large must this square be, to contain just as many acres as there are rails in the fence that encloses it?

This question, or any other of a similar nature, may be solved thus: Suppose any number of rails you please; then, as the area that a fence consisting of the said number of rails will enclose : is to the area which it ought, by the question, to enclose :: so is the supposed number of rails in the fence : to the true number required. Suppose that 40 rails will fence the square; then, its perimeter is 4 rods, its side is 1 rod, and its area 1 square rod. But, by the question, the area ought to be as many acres as there are rails in the fence, viz. 40 acres, or 6400 sq. rods; therefore, as 1 sq. rd. : 6400 sq. rd. :: 40 rails : 256000, the number of rails in the fence, or the number of acres in the square. Hence, the perimeter of the square is 25600 rods or 80 miles, and each side is 20 miles, *Ans.*

84. A landed man two daughters had,
And both were very fair;
To each he gave a piece of land,
One round, the other square.
At forty dollars the acre, just,
Each piece its value had;
The dollars which encompass'd each,
For each exactly paid.
If 'cross a dollar be an inch,
And just a half inch more,
Which did the better portion have,
That had the round, or square?

Ans. The one who had the square had the better portion, by \$239310$\frac{18}{113}$.

Note.—In solving the last question, I have supposed the diameter of a circle to be to the circumference as 113 to 355.

85. A well was dug, once, very deep,—
Its form a cylinder—
Its depth was just one hundred feet,

By ten diameter.
A wall was built within the well,
To line its sides around,
Its height, in feet, I need not tell,
It measur'd with the ground.
The thickness of this wall increas'd
From bottom to its top,
Three feet at first, and four at last,
And slanted truly up.
I now propose to have you tell,
How many cubic feet
Of stone were put within the well,
To make the wall complete.

Ans. 7121 cub. feet, nearly.

86. I placed a bowl into the storm,
To catch the drops of rain—
A half a globe was just its form,
Two feet across the same:
The storm was o'er, the tempest past,
I to the bowl repaired—
Six inches deep the water stood,
It being measur'd fair.
Suppose a cylinder, whose base
Two feet across within,
Had stood exactly in that place,
What would the depth have been?

Ans. $2\frac{1}{2}$ inches.

87. A cooper made for me a cask,
Just forty inches long;
Across each head two feet its width,
But three feet at the bung.
The cooper then did nicely make
A second cask for me,
Just like the other cask in shape,
But larger every way.
This second cask, when fill'd, did hold
A third more than the first—
The dimensions of the first I've told;
Required those of the last.

Ans. Length 44+ inches; bung diameter 39.6+ inches, and head diameter 26.4+ inches.

88. I drop'd a ball from Jackson's bridge,
Its height above the stream to tell—
A pendulum one foot in length,
Made six vibrations while it fell.
Can you in numbers fairly show
How high this lofty bridge must be
Above the stream that glides below,
The feet and inches show to me?
Ans. 176ft. 3in.+

89. How far from the end of a stick of timber, 30 feet long, must a lever be placed, so that two men may each carry just as much at the lever as a third man at the end of the stick? Ans. 5 feet.

Divide the length of the stick by the number of men, and half of the quotient will be the length required.

90. Suppose two men carry a burden of 200lb. weight between them, hung on a pole, the ends of which rest on their shoulders; how much of this load will be borne by each man, if the whole length of the pole be 4 feet, and the weight hang six inches from the middle? Ans. The man nearest the weight will carry 125 lb., and the other 75lb.

91. Five men own a grindstone of 50 inches diameter; I demand how much of its diameter each man must grind off to have equal shares, if one first grind his share, and then the next, and so on, until the stone is ground away, making no allowance for the eye? Ans. The first must grind off, or diminish the diameter, 5.2786+ inches, the second 5.9915+ inches, the third 7.1071—inches, the fourth 9.2621—inches, and the fifth 22.3607—inches.

Note.—The last question may be solved as follows: Square the diameter of the stone; then, the square root of $\frac{4}{5}$ of this square is the diameter of the stone after the first man has ground off his share, and the square root of $\frac{3}{5}$ of the said square is the diameter after the second man has ground off his part, &c. Then by subtracting the several diameters from each other, you will find the several parts which the men grind off.

TABLE I.—*Showing the amount of $1, at Compound Interest.*

Yrs.	3 per cent.	4 per cent	5 per cent.	6 per cent.	7 per cent.
1	1.030000	1.040000	1.050000	1.060000	1.070000
2	1.060900	1.081600	1.102500	1.123600	1.144900
3	1.092727	1.124864	1.157625	1.191016	1.225043
4	1.125509	1.169859	1.215506	1.262477	1.310796
5	1.159274	1.216653	1.276282	1.338226	1.402552
6	1.194052	1.265319	1.340096	1.418519	1.500730
7	1.229874	1.315932	1.407100	1.503630	1.605781
8	1.266770	1.368569	1.477455	1.593848	1.718186
9	1.304773	1.423312	1.551328	1.689479	1.838459
10	1.343916	1.480244	1.628895	1.790848	1.967151
11	1.384234	1.539454	1.710339	1.898299	2.104852
12	1.425761	1.601032	1.795856	2.012196	2.252192
13	1.468534	1.665074	1.885649	2.132928	2.409845
14	1.512590	1.731676	1.979932	2.260904	2.578534
15	1.557967	1.800944	2.078928	2.396558	2.759032
16	1.604706	1.872981	2.182875	2.540352	2.952164
17	1.652848	1.947900	2.292018	2.692773	3.158815
18	1.702433	2.025817	2.406619	2.854339	3.379932
19	1.753506	2.106849	2.526950	3.025600	3.616528
20	1.806111	2.191123	2.653298	3.207135	3.869684
21	1.860295	2.278768	2.785963	3.399564	4.140562
22	1.916103	2.369919	2.925261	3.603537	4.430402
23	1.973587	2.464716	3.071524	3.819750	4.740530
24	2.032794	2.563304	3.225100	4.048935	5.072367
25	2.093778	2.665836	3.386355	4.291871	5.427433
26	2.156592	2.772470	3.555673	4.549383	5.807353
27	2.221289	2.883369	3.733456	4.822346	6.213868
28	2.287928	2.998703	3.920129	5.111687	6.648838
29	2.356566	3.118651	4.116136	5.418388	7.114257
30	2.427262	3.243398	4.321942	5.743491	7.612255
31	2.500080	3.373133	4.538039	6.088101	8.145113
32	2.575083	3.508059	4.764941	6.453386	8.715271
33	2.652335	3.648381	5.003189	6.840590	9.325340
34	2.731905	3.794316	5.253348	7.251025	9.978114
35	2.813862	3.946089	5.516015	7.686087	10.676581
36	2.898278	4.103933	5.791816	8.147252	11.423942
37	2.985227	4.268090	6.081407	8.636087	12.223618
38	3.074783	4.438813	6.385477	9.154252	13.079271
39	3.167027	4.616366	6.704751	9.703507	13.994820
40	3.262038	4.801021	7.039989	10.285718	14.974458

TABLE II.—*Showing the amount of an Annuity of $1.*

Yrs.	3 per cent.	4 per cent.	5 per cent.	6 per cent.
1	1.000000	1.000000	1.000000	1.000000
2	2.030000	2.040000	2.050000	2.060000
3	3.090900	3.121600	3.152500	3.183600
4	4.183627	4.246464	4.310125	4.374616
5	5.309135	5.416322	5.525631	5.637092
6	6.468409	6.632975	6.801912	6.975318
7	7.662462	7.898294	8.142008	8.393837
8	8.892336	9.214226	9.549108	9.897467
9	10.159106	10.582795	11.026564	11.491315
10	11.463879	12.006107	12.577892	13.180794
11	12.807795	13.486351	14.206787	14.971642
12	14.192029	15.025805	15.917126	16.869941
13	15.617790	16.626837	17.712982	18.882137
14	17.086324	18.291911	19.598631	21.015065
15	18.598913	20.023587	21.578563	23.275969
16	20.156881	21.824531	23.657491	25.672528
17	21.761587	23.697512	25.840366	28.212879
18	23.414435	25.645412	28.132384	30.905652
19	25.116868	27.671229	30.539003	33.759991
20	26.870374	29.778078	33.065954	36.785591
21	28.676485	31.969201	35.719251	39.992726
22	30.536780	34.247969	38.505214	43.392290
23	32.452883	36.617888	41.430475	46.995827
24	34.426470	39.082604	44.501998	50.815577
25	36.459264	41.645908	47.727098	54.864512
26	38.553042	44.311744	51.113453	59.156382
27	40.709633	47.084214	54.669126	63.705765
28	42.930922	49.967582	58.402582	68.528111
29	45.218850	52.966286	62.322711	73.639798
30	47.575415	56.084937	66.438847	79.058186
31	50.002678	59.328335	70.760789	84.801677
32	52.502758	62.701468	75.298829	90.889778
33	55.077841	66.209527	80.063770	97.343164
34	57.730176	69.857908	85.066959	104.183754
35	60.462081	73.652224	90.320307	111.434779
36	63.275944	77.598313	95.836322	119.120866
37	66.174222	81.702246	101.628138	127.268118
38	69.159449	85.970336	107.709545	135.904205
39	72.234232	90.409149	114.095023	145.058458
40	75.401259	95.025515	120.799774	154.761965

TABLE III.--*Showing the present value of an Annuity of $1.*

Yrs.	3 per cent.	4 per cent.	5 per cent.	6 per cent.
1	.970874	.961538	.952381	.943396
2	1.913470	1.886094	1.859410	1.833392
3	2.828612	2.775090	2.723248	2.673011
4	3.717099	3.629894	3.545950	3.465105
5	4.579708	4.451821	4.329476	4.212363
6	5.417192	5.242136	5.075691	4.917324
7	6.230284	6.002054	5.786372	5.582381
8	7.019693	6.732744	6.463211	6.209793
9	7.786110	7.435331	7.107820	6.801691
10	8.530204	8.110895	7.721733	7.360086
11	9.252625	8.760476	8.306412	7.886874
12	9.954005	9.385073	8.863249	8.383843
13	10.634956	9.985647	9.393570	8.852682
14	11.296074	10.563122	9.898638	9.294983
15	11.937936	11.118387	10.379655	9.712248
16	12.561103	11.652295	10.837767	10.105894
17	13.166119	12.165668	11.274064	10.477258
18	13.753514	12.659296	11.689585	10.827602
19	14.323800	13.133938	12.085319	11.158115
20	14.877476	13.590325	12.462208	11.469920
21	15.415025	14.029159	12.821150	11.764075
22	15.936918	14.451114	13.163000	12.041580
23	16.443610	14.856840	13.488571	12.303377
24	16.935544	15.246961	13.798639	12.550356
25	17.413150	15.622078	14.093942	12.783355
26	17.876845	15.982767	14.375183	13.003165
27	18.327034	16.329584	14.643031	13.210533
28	18.764111	16.663061	14.898125	13.406163
29	19.188457	16.983712	15.141071	13.590720
30	19.600444	17.292031	15.372448	13.764830
31	20.000431	17.588491	15.592807	13.929085
32	20.388768	17.873549	15.802673	14.084042
33	20.765794	18.147643	16.002546	14.230228
34	21.131839	18.411195	16.192901	14.368140
35	21.487222	18.664610	16.374191	14.498245
36	21.832254	18.908279	16.546848	14.620986
37	22.167237	19.142576	16.711284	14.736779
38	22.492463	19.367861	16.867889	14.846018
39	22.808217	19.584482	17.017037	14.949074
40	23.114774	19.792771	17.159083	15.046296

Table IV.—*Powers of the numbers* 2, 3, 4, 5, 6, 7, 8, 9.

Indices.	Powers of 2.	Powers of 3.	Indices.
1	2	3	1
4	16	81	4
7	128	2187	7
10	1024	59049	10
13	8192	1594323	13
16	65536	43046721	16
19	524288	1162261467	19
22	4194304	31381059609	22
25	33554432	847288609443	25
28	268435456	22876792454961	28
31	2147483648	617673396283947	31
34	17179869184	16677181699666569	34
37	137438953472	450283905890997363	37
40	1099511627776	12157665459056928801	40

Indices.	Powers of 4.	Powers of 5.	Indices.
1	4	5	1
4	256	625	4
7	16384	78125	7
10	1048576	9765625	10
13	67108864	1220703125	13
16	4294967296	152587890625	16
19	274877906944	19073486328125	19
22	17592186044416	2384185791015625	22
		298023223876953125	25
		37252902984619140625	28
		4656612873077392578125	31
		582076609134674072265625	34
		72759576141834259033203125	37
		9094947017729282379150390625	40

Indices.	Powers of 4.
25	1125899906842624
28	72057594037927936
31	4611686018427387904
34	295147905179352825856
37	18889465931478580854784
40	1208925819614629174706176

Table IV. *continued.*

Indices.	Powers of 6.	Powers of 7.	Indices.
1	6	7	1
4	1296	2401	4
7	279936	823543	7
10	60466176	282475249	10
13	13060694016	96889010407	13
16	2821109907456	33232930569601	16
19	609359740010496	11398895185373143	19
		3909821048582988049	22
		1341068619663964900807	25
		459986536544739960976801	28
		157775382034845806615042743	31
		54116956037952111668959660849	34
		18562115921017574302453163671207	37
		6366805760909027985741435139224001	40

Indices.	Powers of 6.
22	131621703842267136
25	28430288029929701376
28	6140942214464815497216
31	1326443518324400147398656
34	286511799958070431838109696
37	61886548790943213277031694336
40	13367494538843734067838845976576

Ind.	Powers of 8.	Powers of 9.	Ind.
1	8	9	1
4	4096	6561	4
7	2097152	4782969	7
10	1073741824	3486784401	10
13	549755813888	2541865828329	13
16	281474976710656	1853020188851841	16
		1350851717672992089	19
		984770902183611232881	22
		717897987691852588770249	25
		523347633027360537213511521	28
		381520424476945831628649898809	31
		278128389443693511257285776231761	34
		202755595904452569706561330872953769	37
		147808829414345923316083210206383297601	40

TABLE IV. *continued.*

Indices.	Powers of 8.
19	144115188075855872
22	73786976294838206464
25	37778931862957161709568
28	19342813113834066795298816
31	9903520314283042199192993792
34	5070602400912917605986812821504
37	2596148429267413814265248164610048
40	1329227995784915872903807060280344576

Note.—The foregoing table contains only one-third of the powers below the 40th; the rest being omitted to save room. When any power below the 40th, which is not in the table, is wanted, it may readily be found, either by multiplying the next lower power, or dividing the next higher power by the root or first power. Thus, the 17th power of 2 may be found by multiplying the 16th power by 2; and the 18th power may be obtained by dividing the 19th power by 2, &c.—In this table, each power of 4 is the square of the corresponding power of 2; and all the powers of 9 are the squares of the corresponding powers of 3. Also, each power of 8 is the cube of the like power of 2, and the product of the corresponding powers of 2 and 4. Therefore, the 1st, 4th, 7th, 10th, &c. powers of 4, are the 2d, 8th, 14th, 20th, &c. powers of 2; the 1st, 4th, 7th, &c. powers of 8, are the 3d, 12th, 21st, &c. powers of 2, and the 1st, 4th, 7th, &c. powers of 9, are the 2d, 8th, 14th, &c. powers of 3; and consequently the table contains powers of 2 as high as the 120th, and powers of 3 up to the 80th. Lastly, since the multiplication of powers answers to the addition of their indices, and the division of powers, to the subtraction of their indices, it is evident that the foregoing table contains a great number of products, quotients, powers and roots of numbers; and hence, teachers may take from this table, for their pupils, as many questions in Multiplication, Division, Involution and Evolution, as they please. The table will also be very useful in solving questions in Geometrical Progression.

TABLE V.—*For measuring Wood.*

1 ft.	2 ft.	3 ft.	4 ft.	Width.	5 ft.	6 ft.	7 ft.	8 ft.
Cd.	Cd.	Cd.	Cd.	Ft. in.	Cd.	Cd.	Cd.	Cd.
.0010	.0020	.0029	.0039	1 .. 6	.0049	.0059	.0068	.0078
10	21	31	41	7	52	62	72	82
11	22	33	43	8	54	65	76	87
11	23	34	46	9	57	68	80	91
12	24	36	48	10	60	72	84	95
12	25	37	50	11	62	75	87	100
13	26	39	52	2 .. 0	65	78	91	104
14	27	41	54	1	68	81	95	109
14	28	42	56	2	71	85	99	113
15	29	44	59	3	73	88	103	117
15	30	46	61	4	76	91	106	122
16	31	47	63	5	79	94	110	126
16	33	49	65	6	81	98	114	130
17	34	50	67	7	84	101	118	135
17	35	52	69	8	87	104	122	139
18	36	54	72	9	90	107	125	143
18	37	55	74	10	92	111	129	148
19	38	57	76	11	95	114	133	152
20	39	59	78	3 .. 0	98	117	137	156
20	40	60	80	1	100	120	141	161
21	41	62	82	2	103	124	144	165
21	42	63	85	3	106	127	148	169
22	43	65	87	4	109	130	152	174
22	44	67	89	5	111	133	156	178
23	46	68	91	6	114	137	160	182
23	47	70	93	7	117	140	163	187
24	48	72	95	8	119	143	167	191
24	49	73	98	9	122	146	171	195
25	50	75	100	10	125	150	175	200
25	51	76	102	11	127	153	178	204
26	52	78	104	4 .. 0	130	156	182	208
27	53	80	106	1	133	160	186	213
27	54	81	109	2	136	163	190	217
28	55	83	111	3	138	166	194	221
28	56	85	113	4	141	169	197	226
.0029	.0058	.0086	.0115	5	.0144	.0173	.0201	.0230

Table V. *continued.*

9 ft.	10 ft.	11 ft.	12 ft.	Width.	13 ft.	14 ft.	15 ft.	16 ft.
Cd.	Cd.	Cd.	Cd.	Ft. in.	Cd.	Cd.	Cd.	Cd.
.0088	.0098	.0107	.0117	1 .. 6	.0127	.0137	.0146	.0156
93	103	113	124	7	134	144	155	165
98	109	119	130	8	141	152	163	174
103	114	125	137	9	148	160	171	182
107	119	131	143	10	155	167	179	191
112	125	137	150	11	162	175	187	200
117	130	143	156	2 .. 0	169	182	195	208
122	136	149	163	1	176	190	203	217
127	141	155	169	2	183	197	212	226
132	146	161	176	3	190	205	220	234
137	152	167	182	4	197	213	228	243
142	157	173	189	5	205	220	236	252
146	163	179	195	6	212	228	244	260
151	168	185	202	7	219	235	252	269
156	174	191	208	8	226	243	260	278
161	179	197	215	9	233	251	269	286
166	184	203	221	10	240	258	277	295
171	190	209	228	11	247	266	285	304
176	195	215	234	3 .. 0	254	273	293	313
181	201	221	241	1	261	281	301	321
186	206	227	247	2	268	289	309	330
190	212	233	254	3	275	296	317	339
195	217	239	260	4	282	304	326	347
200	222	245	267	5	289	311	334	356
205	228	251	273	6	296	319	342	365
210	233	257	280	7	303	327	350	373
215	239	263	286	8	310	334	358	382
220	244	269	293	9	317	342	366	391
225	250	275	299	10	324	349	374	399
229	255	280	306	11	331	357	382	408
234	260	286	313	4 .. 0	339	365	391	417
239	266	292	319	1	346	372	399	425
244	271	298	326	2	353	380	407	434
249	277	304	332	3	360	387	415	443
254	282	310	339	4	367	395	423	451
.0259	.0288	.0316	.0345	5	.0374	.0403	.0431	.0460

To find, by the help of the foregoing table, the solid content of any pile or load of wood, the dimensions of which are known.

RULE.

1. *When the length of the pile or load is an exact number of feet;* find the length at the top of the table, and the width in the middle column; and in a line with the width and under the length you will find the content of a section of the pile one inch high or thick. Multiply the tabular content, thus found, by the height of the pile, in inches; point off four decimal places at the right hand of the product, and you will have the content required, in cords and decimal parts.

2. *When the length of the pile consists of feet and inches;* then, to the tabular content answering to the feet in length, add a twelfth part of the tabular content answering to as many feet in length as there are inches besides the number of entire feet; and the sum will be the content of a section of the pile an inch thick; which multiply by the height of the pile, in inches, and the product will be the content required.

Note 1.—The ciphers, which, according to decimal rules, ought to be prefixed to the decimal numbers in the table, in order to make each fraction contain four places of figures, are omitted, except in the top and bottom numbers in each column. It will not be necessary, in using these decimal numbers, to prefix the ciphers and decimal points, but care must always be taken to point off the proper number of decimal places in the answer.

Note 2.—When the length of the pile exceeds the limits of the table, then take $\frac{1}{2}$, or $\frac{1}{3}$, &c. of the length, and find the content of that part of the pile; which content multiply by 2 when $\frac{1}{2}$ the length is used, or by 3 when $\frac{1}{3}$ of the length is used, &c., and the product will be the content of the whole pile. In a similar manner the content of a pile may be found, when the width exceeds the limits of the table; viz. by first finding the content of $\frac{1}{2}$, or $\frac{1}{3}$, &c. of the width, and then multiplying the content thus found by 2, or 3, &c.

EXAMPLES.

1. Required the content of a pile of wood 12 feet long, 3 feet 6 inches wide, and 5 feet high; and also the value of the wood, at $3 a cord.

.0273 Cd. Tabular content.
60 inches=5ft. The height of the pile.

1.6380 Cd. Content of the pile. } Ans.
Then, 1.638×3=$4.914, Value of the wood. }

2. Bought a load of wood, 8 feet 9 inches long, 2 feet 10 inches wide, and 2 feet 6 inches high, at the rate of 16 shillings, New-England currency, per cord: How much is the amount in federal money?

Here, the tabular content answering to the given width and 8 feet in length is .0148Cd.; and that answering to the said width and 9 feet (9 being the number of odd inches) in length, is .0166; one-twelfth of which is .0166÷12=.0013+. Then, .0148+.0013=.0161Cd. the content of a section of the load an inch thick; which being multiplied by 30, the height in inches, gives .4830Cd., the content of the whole load. Then, to find the value of the wood; .483×16= 7.728 shillings, and 7.728÷6=$1.288, Ans.

3. Bought a pile of wood 22 feet long, 4 feet wide, and 4 feet 6 inches high, at 16s. 8d. per cord: what did it cost?

Here, because the length of the pile exceeds the limits of the table, I take $\frac{1}{2}$ of the length, which is 11 feet; and I find, by calculation, that the content of a pile 11 feet long, 4 feet wide, and 4 feet 6 inches high, is 1.5444Cd.; which being multiplied by 2, gives 3.0888Cd., the content of the whole pile. Then, since 16s. 8d.=200 pence, we have 3.0888×200=617.7600 pence=2*l.* 11s. 5$\frac{3}{4}$d.+, the Ans.

TABLE VI. *Of the Specific Gravities of several solid and fluid bodies;* the second column of which shows the weight, in Aveirdupois ounces, of a cubic foot of each substance.

	oz.		oz.
Platina, (pure,) - -	23000	Chalk, - - - - -	1793
Fine Gold, - - -	19400	Alum, - - - - -	1714
Standard Gold, - -	18888	Lignum vitæ, - -	1327
Mercury, - - - - -	14000	Coal, - - - - -	1250
Lead, - - - - - -	11325	Mahogany, - - -	1063
Fine silver, - - -	11091	Cows' milk, - -	1034
Standard silver, -	10535	Box wood, - - -	1030
Copper, - - - -	9000	Sea water, - - -	1030
Fine brass, - - -	8350	Common water, -	1000
Steel, - - - - -	7850	Red wine, - - -	993
Iron, - - - - - -	7645	Proof spirits, - -	925
Pewter, - - - - -	7471	Dry oak, - - -	925
Cast iron, - - -	7425	Ice, - - - - - -	908
White lead, - - -	3160	Living Men, - - -	891
Marble and hard stone,	2700	Spirits of turpentine,	864
Green glass, - -	2620	Alcohol, - - - - -	850
Clear glass, - - -	2600	Dry ash and maple,	800
Flint stone, - - -	2570	Ether, - - - - -	732
Common stone, - -	2520	Dry elm and fir, -	600
Brick, - - - - -	2000	White pine, - -	569
Common earth, - -	1984	Cork, - - - - -	240
Dry ivory, - - -	1825	Common air, - - -	$1\frac{1}{4}$
Sulphur, - - - -	1810	Inflammable air, -	$\frac{1}{8}$

The use of this table may be illustrated by a few examples.

Ex. 1. Required the solid content of an irregular piece of marble, which weighs 100lb.
As 2700oz. : 100lb. :: 1 cubic foot : 1024 cub. inches, Ans.

2. What weight of sheet lead, $\frac{1}{120}$ of a foot thick, will roof a house, the area of the roof being 6000 square feet?

$6000\times\frac{1}{120}$=50 cubic feet, the solid content of the lead. Then, as 1 cub. ft. : 50 cub. ft. :: 11325 oz. (the weight of a cubic foot of lead,) : 566250 oz.=35390lb. 10 oz. Ans.

3. If the weight of a counterfeit coin, consisting of gold and silver, be 265 grains, and its specific gravity be 16000, what quantities of standard gold and silver does it contain?
Ans. 173.3+ grains of gold, and 91.6+ grains of silver.

1st Month January.	2d Month February.	3d Month March.	Dominical Letters.							4th Month April.	5th Month May.	6th Month June.
			A	B	C	D	E	F	G			
1 8 15 22 29	5 12 19 26	5 12 19 26	1	7	6	5	4	3	2	2 9 16 23 30	7 14 21 28	4 11 18 25
2 9 16 23 30	6 13 20 27	6 13 20 27	2	1	7	6	5	4	3	3 10 17 24	1 8 15 22 29	5 12 19 26
3 10 17 24 31	7 14 21 28	7 14 21 28	3	2	1	7	6	5	4	4 11 18 25	2 9 16 23 30	6 13 20 27
4 11 18 25	1 8 15 22 29	1 8 15 22 29	4	3	2	1	7	6	5	5 12 19 26	3 10 17 24 31	7 14 21 28
5 12 19 26	2 9 16 23	2 9 16 23 30	5	4	3	2	1	7	6	6 13 20 27	4 11 18 25	1 8 15 22 29
6 13 20 27	3 10 17 24	3 10 17 24 31	6	5	4	3	2	1	7	7 14 21 28	5 12 19 26	2 9 16 23 30
7 14 21 28	4 11 18 25	4 11 18 25	7	6	5	4	3	2	1	1 8 15 22 29	6 13 20 27	3 10 17 24

7th Month July.	8th Month August.	9th Month September.	Dominical Letters.							10th Month October.	11th Month November.	12th Month December.
			A	B	C	D	E	F	G			
2 9 16 23 30	6 13 20 27	3 10 17 24	1	7	6	5	4	3	2	1 8 15 22 29	5 12 19 26	3 10 17 24 31
3 10 17 24 31	7 14 21 28	4 11 18 25	2	1	7	6	5	4	3	2 9 16 23 30	6 13 20 27	4 11 18 25
4 11 18 25	1 8 15 22 29	5 12 19 26	3	2	1	7	6	5	4	3 10 17 24 31	7 14 21 28	5 12 19 26
5 12 19 26	2 9 16 23 30	6 13 20 27	4	3	2	1	7	6	5	4 11 18 25	1 8 15 22 29	6 13 20 27
6 13 20 27	3 10 17 24 31	7 14 21 28	5	4	3	2	1	7	6	5 12 19 26	2 9 16 23 30	7 14 21 28
7 14 21 28	4 11 18 25	1 8 15 22 29	6	5	4	3	2	1	7	6 13 20 27	3 10 17 24	1 8 15 22 29
1 8 15 22 29	5 12 19 26	2 9 16 23 30	7	6	5	4	3	2	1	7 14 21 28	4 11 18 25	2 9 16 23 30

To find the day of the month by this Almanac: Under the Dominical Letter for the current year find the day of the week, and in a line with it in the calendar you will find the day of the month.

A Table *for finding the Dominical Letter, or Letters, for any year, according to the present style.*

Centuries.				*Odd Years.*																	
20	21	22	23																		
16	17	18	19																		
A	C	E	G	0	6		17	23	28	34		45	51	56	62		73	79	84	90	
G	B	D	F	1	7	12	18		29	35	40	46		57	63	68	74		85	91	96
F	A	C	E	2		13	19	24	30		41	47	52	58		69	75	80	86		97
E	G	B	D	3	8	14		25	31	36	42		53	59	64	70		81	87	92	98
D	F	A	C		9	15	20	26		37	43	48	54		65	71	76	82		93	99
C	E	G	B	4	10		21	27	32	38		49	55	60	66		77	83	88	94	
B	D	F	A	5	11	16	22		33	39	44	50		61	67	72	78		89	95	

To find, by the foregoing table, the Dominical Letter for any given year; seek first at the top for the number of whole centuries, and then at the right hand for the odd years, and in a line with the odd years, under the centuries, you will find the Dominical Letter for that year. Thus, the Dominical Letter for the year 1830 is C, which stands under 18, the number of centuries, and in a line with 30, the odd years.—In every leap year there are two Dominical Letters; the letter standing against the odd years being the Dominical Letter in the 3d, 4th, and following months; and the letter standing against the blank immediately above the odd years, or below those of the preceding year, is the Dominical Letter in the 1st and 2d months. Thus, the Dominical Letters in the year 1820 were B and A; and in the year 1828, F and E.

Method of assessing or apportioning Taxes.

In assessing town taxes, an inventory of the property of the whole town, both *real* and *personal*, and also of the whole number of polls, is first to be obtained: And, if any part of the tax is to be averaged on the polls, or otherwise, we must deduct what the polls, &c. amount to; and then with the remainder, we may find the tax on $1, and by it make a table containing the taxes on 1, 2, 3, &c., to 9 dollars; then on 10, 20, 30, &c., to 90 dollars; and then on 100, 200, &c., to 900 dollars; and so on as far as may be necessary. Then, knowing the inventory of an individual, we may readily find the tax on his property by addition; as in the following

EXAMPLE.

A certain town, valued at $125700, raises a tax of $2000; there are 650 polls, which are taxed 75 cents each, and the remainder is to be assessed on the property: What is A's tax, whose *real* estate is valued at $2150.70, *personal* property at $784.25, and who pays for 2 polls?

650×.75=$487.50, amount of the poll taxes, and $2000—487.50=$1512.50, to be assessed on property. Then, as $125700 : $1512.50 :: $1 : $.0120326, tax on $1.

TABLE.

	$	$		$	$		$	$
Tax on	1 is	.012	Tax on	10 is	.120	Tax on	100 is	1.203
	2	.024		20	.241		200	2.407
	3	.036		30	.361		300	3.610
	4	.048		40	.481		400	4.813
	5	.060		50	.602		500	6.016
	6	.072		60	.722		600	7.220
	7	.084		70	.842		700	8.423
	8	.096		80	.963		800	9.626
	9	.108		90	1.083		900	10.829

Tax on $1000=$12.033; on $2000=$24.065

Now, to find A's tax, his real estate being $2150.70, I find, by the table, that

The tax on	$2000 is	$24.065
—— on	100	1.203
—— on	50	.602
—— on	0.70	.008
Tax on	$2150.70 is	$25.878
In like manner I find the tax on his personal property to be		$9.437
2 polls at $.75 each,		$1.500
Ans. A's tax is		$36.815

Note.—To find, by the foregoing table, the tax on any given number of cents, remove the decimal point in the tax on that number of dollars two places to the left. Thus, the tax on 70 cents is found to be $.00842, or $.008 nearly.

What will B's tax amount to, whose inventory or list is

$1500 *real*, and $875 *personal* property, and who pays for 3 polls? Ans. $30.827

Method of calculating School Bills.

Every teacher of a common school, ought to keep a daily account of the number of scholars sent to the school by each proprietor, and also an account of what each pays, or contributes towards defraying the expenses of the school, by boarding the teacher, or finding wood for fuel, &c. When the school is terminated, then to find how the accounts on the school bill stand, with respect to each proprietor, observe the following

RULE.

1. Find the number of days' schooling charged to each proprietor, and add all these numbers into one sum total.

2. Find the total amount of all the expenses of the school. Then, the shortest method of finding each proprietor's share of the expenses is as follows:

3. Divide the amount of the expenses by the total amount of the days, and the quotient will be the share for 1 day. But, if the district draws any money from a public fund, which is to be applied towards defraying the expenses of the school, subtract it from the amount of the expenses; then divide the remainder by the whole number of days, and the quotient will be the share for 1 day, as before. Then make a little table, by multiplying the share for 1 day by 1, 2, 3, &c., to 9. By the help of this table the share for any greater number of days may be readily found by addition.

4. Calculate, by the table, each proprietor's share of the expenses, in proportion to the number of days' schooling charged to him; from which subtract the amount of his credit, if any, and the remainder will be what is due *from* him; or, if his amount of credit exceeds his share of the expenses, then subtract the latter from the former, and the remainder will be what is due *to* him.

Note.—If any children go to a school in a district to which they do not belong, and their parents, or guardians, (on account of their being nonresidents,) are not allowed any shares of the public school money; or, if any indigent persons in a district, are, by consent of the district, released from paying their shares of *some part* of the expenses of the school; then, a separate bill must first be made out for such persons, and the amount of what is due from them, together with what they have paid or contributed towards supporting the school, must be deducted from the whole amount of the expenses; and then the remainder must be apportioned among the other proprietors, as directed in the foregoing rule.

EXAMPLE.

Suppose the summary of the accounts on a school bill to be as follows:

The School District No. 2, in the Town of ——, State of Conn., Dr. to S. M. for teaching a school in said District 5 months, at \$14 per month, - - - - - - \$70.00

Expense of boarding Teacher $21\frac{2}{3}$ weeks, at \$1.50 per week, - - - - - - - - - - - 32.50

Wood used for fuel, 5 cords, at \$3 per cord, 15.00

Other expenses of the school, - - - - - - 2.25

Amount of the expenses, \$119.75

Money received from the school fund, deducted, 39.25

Remainder, \$80.50

Proprietors.	Dr.	Days.	Cr. By Wood, &c.	
A,	To schooling children,	145	A,	\$5.58
B,	Do.	99		
C,	Do.	346	C,	10.38
D,	Do.	560	D,	11.84
E,	Do.	608	E,	14.00
F,	Do.	153	F,	7.95
Whole number of days,		1911	Amount,	\$49.75

This school bill is calculated as follows:—

\$80.50÷1911=\$.042124+, the share for 1 day. I next make out the following table by multiplying \$.042124 by 1, 2, 3, &c., to 9.

Days.		$
1	- - -	.04212
2	- - -	.08425
3	- - -	.12637
4	- - -	.16850
5	- - -	.21062
6	- - -	.25274
7	- - -	.29487
8	- - -	.33699
9	- - -	.37912

Note.—Removing the decimal point in each of these tabular numbers one place to the right, will give the shares for 10, 20, 30, &c. days; and removing the point two places to the right, gives the shares for 100, 200, 300, &c. days.

I find each proprietor's share of the expenses as follows:—

The number of days' schooling charged to A is 145, and I find, by the table, that

The share for 100 days is			$4.212
——	for 40	do.	1.685
——	for 5	do.	.211
	——		——
——	for 145	do.	$6.108
	A's Cr. deducted,		5.58

In the same manner I find the shares of B, C, D, E and F.

		$
Due from A,	$0.528,	or 0.53 nearly.
—— B,	4.170,	4.17
—— C,	4.195,	4.20
—— D,	11.749,	11.75
—— E,	11.611,	11.61
	32.253,	32.26
Due to F,	1.506,	1.51
	30.747,	30.75
Amount of Cr. added,	49.750,	49.75
Proof,	$80.497	$80.50

The following questions are proposed to Algebraists.

If a solid globe of glass at the furnace, whose diameter is 8 inches, be blown into a hollow globe until the shell be but one-tenth of an inch in thickness, what will then be its diameter, and how much wine will it hold? Ans. Diameter of its cavity 29.1118+inches—content 55 gallons, 3qt. 1 pt.+

If two wheels, one 4 feet in diameter, and the other 5 feet, be connected together by an axletree one rod in length, and set a rolling on a plain until they describe two concentric circles, what will be the diameter of each of these circles. Ans. 8 rods, and 10 rods.

A ball, descending by the force of gravity from the top of a tower, was observed to fall half the way in the last second of time: Required the whole time of descent, and the height of the tower.

Ans. Time 3.414+seconds—height 186 feet, 6 in.+

ERRATA.

Page 27, first line below the Subtraction Table, for "may read thus," read "may be read thus."

" 27, line 3 from bottom, for "*downwards*," read "*upwards*."

" 57, line 4 from bottom, for "comes," read "will come."

" 58, Supplement to Division, 1st article of the Rule, in part of the edition, after the word "fraction," insert "to the product add the numerator."

" 161, at the beginning of the 2d line from the top, the number "64" is wanting, in a few copies.

" 162, *Ex.* 3, first line above the Ans., for "1qt." read "2qt."

" 162, line 2 from bottom, for "to," read "with."

In a few copies the answers to question 7, page 186, and question 1, page 194, are erroneous—the former answer should be "$2.3736," and the latter "$76.406+"

" 310, *Ex.* 3, in the question, for "$100," read "$200."

" 316, line 8 from top, for "*hexigon*," read "*hexagon*."

Page 374, Quest. 80th, for "two-fifths," read "one-twelfth."
" 377, Ans. to quest. 90th, for "25lb.," read "75lb."
" 387, Ex. 3d, for "18s. 4d." read "16s. 8d."
The two last errors are in but a few copies.
" 394, 3d line fr. bot. insert the article "a" after "into."

CONTENTS.

Lightning Source UK Ltd.
Milton Keynes UK
UKHW020027260219
337881UK00006B/504/P

9 780266 990857